# Plant Viruses and Virus Diseases

*by*

F.C. Bawden

2014

**BIOTECH BOOKS**

**DELHI - 110035**

ISBN 81-7622-064-7

*Published by* : **BIOTECH BOOKS**
1123/74, Tri Nagar,
Delhi - 110 035
Phone: 27382765
e-mail: biotechbooks@yahoo.co.in

*Showroom* : 4760-61/23, Ansari Road, Darya Ganj,
New Delhi - 110 002
Phone: 23245578, 23244987

*Printed at* : **Chawla Offset Printers**
New Delhi - 110 052

PRINTED IN INDIA

A NEW SERIES OF PLANT SCIENCE BOOKS

edited by Frans Verdoorn

Volume 23

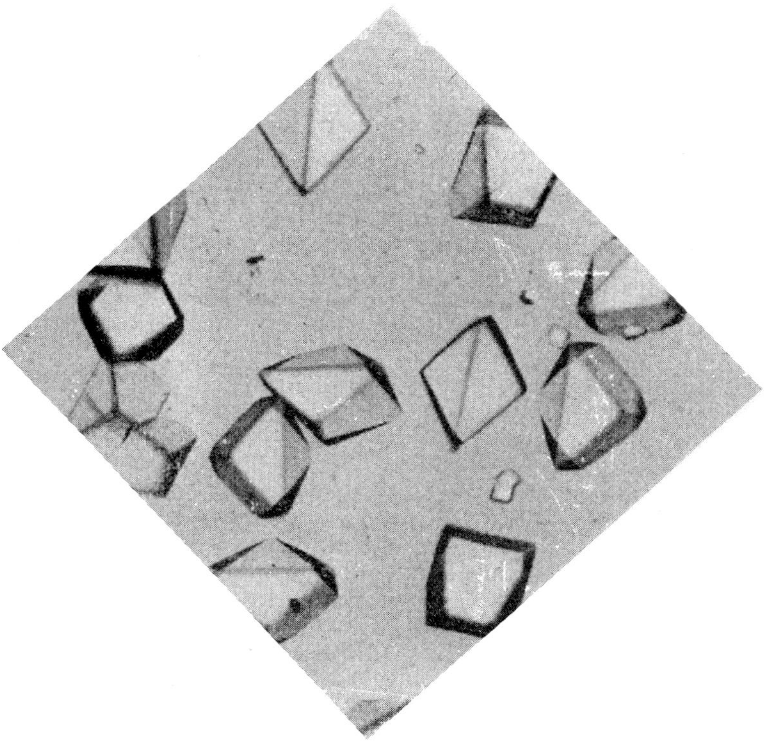

# PLANT VIRUSES

and

# VIRUS DISEASES

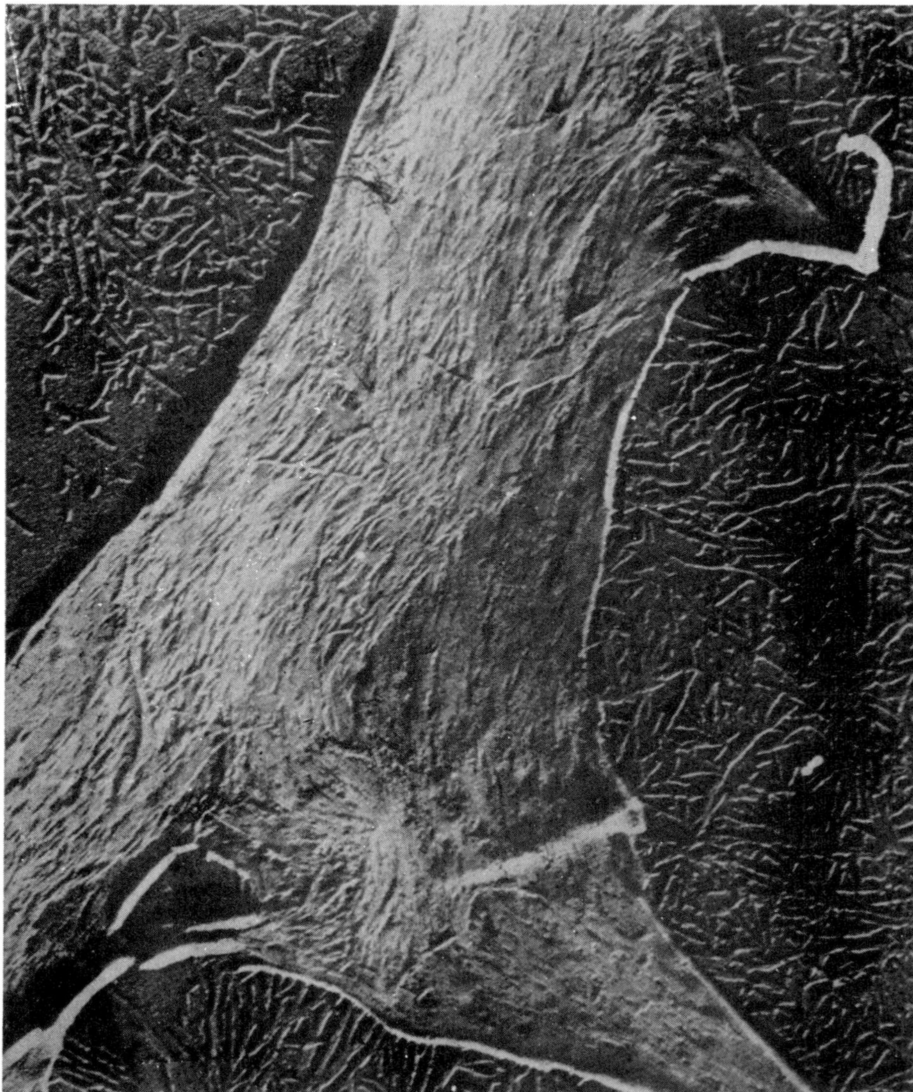

ELECTRON MICROGRAM OF A SHADOWED, FROZEN-DRIED, SOLUTION OF TOBACCO MOSAIC VIRUS, SHOWING AN EXTENDED SHEET OF ALIGNED PARTICLES OVERLYING SCATTERED PARTICLES OF VARIOUS LENGTHS. × 23,000. (WYCKOFF, R. W. G, 1947, Biochimica et Biophysica Acta, *2*, 139).

# PREFACE to FIRST EDITION

At the present time the study of plant viruses is in a transitional stage. Until recently, work on viruses has been left largely in the hands of pathologists, for, although the exact position of the subject has always been a little uncertain, viruses undoubtedly caused diseases and were therefore conveniently described with the pathogenic bacteria and fungi. In the past few years, however, the subject has received increasing attention from workers in other fields. Chemists, crystallographers, entomologists, geneticists, serologists, physicists, and others are becoming attracted by these intriguing pathogens. They are bringing fresh techniques to the subject and are greatly increasing our knowledge. One result of this is that publications on viruses are becoming even more widely scattered, in apparently unrelated journals, than before. Another is that the necessity for separating viruses from other recognised pathogens is becoming more obvious, and their adoption by protein chemists seems only a matter of time. This is not meant to be a text-book of virus diseases. Detailed descriptions of symptoms and host-ranges are not given, but all other aspects of the subject are treated. It is an attempt to describe and correlate the advances that have been made recently in the study of plant viruses. Many of the points described are still controversial, but I have tried to distinguish between facts and the deductions made from them.

*July 1939.*

# PREFACE to SECOND EDITION

With the invasion of the Netherlands the type of the first edition was lost, so that a second printing could not be made and the edition was out of print within a year of publication. Despite the war, work on viruses and virus diseases has made considerable progress in the last three years, and the need to set up new type has been made in the opportunity for bringing the text up to date and for adding new illustrations. An attempt has been made to include all new work, but the difficulties of getting literature in present circumstances almost inevitably means that some has been missed. More than half the chapters have been rewritten and all have been modified to some extent. The most extensive alterations have been needed in chapters dealing with the properties of viruses in vitro and with their relationships to their insect vectors, for in these aspects of the subject our knowledge has grown most rapidly. Some of the many controversies of 1939 have been resolved. In particular, few biologists now question that the specific proteins isolated from infected plants represent the viruses themselves and all workers are agreed that tobacco mosaic virus is a nucleoprotein, so that there has been less need to emphasise the evidence for these views in this edition. Unfortunately we still know little of the behaviour of viruses in their natural environment, that is, within the host plants: this aspect now calls for much increased attention and can be expected to provide the next major advances in the subject.

*Autumn 1942.*

# PREFACE to THIRD EDITION

*In preparing this edition, one additional chapter and many new illustrations have been added, and the text has been almost entirely rewritten. In other words, although the same general format has been maintained as in earlier editions, this is largely a new book rather than another edition. Such extensive revision has been needed, not because any single discovery has decisively altered the general direction of research, but because of steady advances in all the many kinds of work now being pursued in the study of viruses and virus diseases. During the last six years, many new virus diseases have been described, some of great economic importance and some with novel kinds of symptoms, much has been done on the factors that affect spread, and insect vectors have been identified in species of insects previously not incriminated. More viruses have been purified and their chemical constitution has been further elucidated, and advances in electron microscopy have provided much information on their morphology. Unfortunately, knowledge of the behaviour of viruses* in vivo *grows less rapidly than of their behaviour* in vitro, *but here, too, some progress has been made. Much more will be needed, however, before any definite conclusions can be drawn about such basic problems as what determines whether a plant is susceptible, or an insect is a vector, and the mechanism whereby viruses infect, multiply and vary.*

*The disciplines used in the study of viruses and virus diseases are so many and so varied, ranging from entomology to protein chemistry, epidemiology to crystallography and cytology to serology, and papers appear in such widely scattered journals, that the attempt to describe and correlate advances in all branches of the subject becomes increasingly difficult to achieve. Omissions and shortcomings are inevitable; that there are not more is largely because of the many discussions, suggestions and criticisms so generously provided by my colleagues.*

HARPENDEN, HERTS.

## CONTENTS

# ACKNOWLEDGMENTS

It is a pleasure to express my gratitude to my colleagues for their generous help and advice, especially to Mr. N. W. PIRIE, F. R. S., for his invaluable constructive criticisms. From Miss L. CUNOW and Dr. F. M. ROBERTS I have received much assistance with reading proofs and help in many other ways. I am, too, deeply indebted to Dr. F. VERDOORN for the great care he has taken in preparing the book for the press.

I also wish to thank authors and journals for permission to reproduce illustrations. The sources of these are acknowledged individually in the text. Acknowledgment also is made to "The Annals of Applied Biology" for the loan of a number of blocks. — The vignette on page xvi has been reproduced from HERMANN SCHACHT'S Bericht an das Königliche Landes-Oekonomie-Kollegium über die Kartoffelpflanzen und deren Krankheiten (Berlin, 1856, Taf. VI, 11) and is probably the first illustration of potato curl. The vignettes on pages 329 and 333 show the oldest known plant virus disease, tulip mosaic or break, after CLUSIUS' wood cuts of variegated early tulips (Rariorum aliquot stirpium per Pannoiam, etc., Antverpiae 1583; *cf.* MCKAY and WARNER, Nat. Hort. Mag., July 1933, p. 187). The other vignettes were made by Miss HILDE DE VRIES.

# LIST of TABLES

# LIST of ILLUSTRATIONS

*Chapter 1*

# INTRODUCTION

**Definition of Subject:** — Although the study of viruses and virus diseases has developed so rapidly in recent years that it now begins to approach the status of branches of science such as mycology and bacteriology, we still remain with no positive, generally-accepted, definition of the word virus. To describe a virus as an invisible pathogen is about the best that can be done, but this tells us nothing definite about its precise identity. Even so, this is far from its original meaning, and it is worth considering the changes the word has undergone, in an attempt to understand the current usage.

Pathologists seem rarely to have used the word virus for poison or toxin, which is apparently the precise translation of the original Latin. Instead, in the past they have applied it widely to all kinds of disease-producing agents, but especially to those about which little was known. Hence, until the germ theory of disease was established with certainty, even though the knowledge that some pathological states are contagious is age-old, virus was employed indiscriminately to cover both infective and non-infective agents. During the last 70 years or so, despite various fluctuations, its meaning has become increasingly restricted and has moved far from the original. With PASTEUR's unequivocal demonstration of the role of bacteria in causing disease, there seems to have been an abrupt change in usage and virus became restricted to infective entities and synonymous with microbe or germ. According to BULLOUGH (1938) in France around 1880 the term virus disease was currently used to describe such diseases as small pox which produce a lasting immunity in those who have once suffered, but it is improbable that this extreme restriction ever became general. Had it been widely adopted, it is difficult to see how plant diseases would have been included in the category, and it is clear that PASTEUR interpreted it more broadly. In 1889 he wrote "... virulent affections are caused by small microscopic beings which are called microbes. The anthrax of cattle is produced by a microbe.... The microbe of rabies has not been isolated as yet, but judging by analogy, we must believe in its existence. To resume: every virus is a microbe. Although these beings are of infinite smallness, the conditions of their life and propagation are subject to the same general laws which regulate the birth and multiplication of the higher animal and vegetable beings". In 1889, therefore, distinctions were not being made between viruses and bacteria, though it is perhaps significant that even then the word virus seems to have been applied more to those infectious diseases for which no causes had been isolated than to those known to be bacterial. Although for some time virus continued to be used as a generic term for all infectious diseases, it became increasingly restricted to those whose causes were undetermined. The first significant distinction between the pathogens we now call viruses and micro-organisms came with the demonstration by IWANOWSKI (1892) and LOEFFLER and FROSCH (1898) that the causes of tobacco mosaic and foot-and-mouth disease of cattle pass through bacteria-proof filters. Soon after these discoveries BEIJERINCK (1898) and BAUR (1904), in applying the term virus to the causes of tobacco mosaic and *Abutilon* variegation, were using it, not in the PASTEUR sense as synonymous with bacterium, but in contrast with micro-organism. To point this contrast it became customary to call such pathogens "filterable" viruses, but as the word virus also fell increasingly into disuse for microscopically visible organisms, the distinguishing epithet became redundant and the word virus alone came to stand for these obscure entities.

From this brief review we see that almost the only consistent usage for the word virus has been its association with the unknown; it has been a convenient cloak for ignorance, and remains largely so today. Put at its bluntest, virus diseases are simply those which remain after the removal of all those for which definite causes are known. We can take as a definition of disease in an organism any appreciable departure from the normal. Diseases have many different causes, but they divide sharply into two contrasting types, infectious and non-infectious, depending on whether or not the causes can be transmitted to healthy plants and the abnormal conditions reproduced. The infectious diseases in their turn we can also conveniently divide into two, using visibility of the causes as a criterion. First, there are those caused by obvious parasites, typified by insect, fungal and bacterial diseases, and, secondly, those for which no visible causes have been demonstrated. The second group comprises virus diseases as the term is used today, and the only significant feature common to the hundreds of different pathological states of animals, plants and bacteria, that are included under this name is that no causes resolvable by ordinary microscopy have been identified. In this respect virus diseases resemble nutritional disorders and the effects of toxins, but they are clearly distinguished from these by being highly infectious. Their causes, *i. e.*, the viruses, are obviously pathogens capable of multiplication in suitable hosts, for the diseases can be reproduced indefinitely in series by transmission from diseased to healthy organisms.

There is another feature common to the viruses. Although they obviously multiply readily in susceptible living cells, none has been cultivated on an artificial medium. The precise significance of this fact, however, is not easy to assess, for obligate parasitism is a negative feature that must always be postulated rather than proved. Actually no attempt has been made to cultivate the great majority of plant viruses, and it would be premature to assume that failure with the few that have been tried resulted solely from metabolic limitations of the viruses. Thus, although for various additional reasons it seems probable that obligate parasitism is a characteristic feature of viruses, the one feature certainly applicable to all virus diseases is that their causes are smaller than other recognised pathogens. It is, of course, this size difference that permits many viruses to pass through filters with pores small enough to stop all visible bacteria, and for a time passage through such filters was used as the decisive test for a virus. However, this criterion, which was always less applied in plant than in animal pathology, has now largely been abandoned, for it was found inapplicable and unreliable. Many factors other than the relative sizes of particles and pores can determine whether or not the particles will pass through the pores. An unsuitable suspension medium, a too small virus content, and adsorption on to either the filter or some constituent of the medium may all play important parts in preventing the passage of a virus through a filter. The fact that a pathogen is not filterable, therefore, cannot alone be taken as proof that it is not a virus. With many diseases of plants that are included in the virus group, filterability cannot even be tested. The only known method of transmitting these diseases is by grafting portions of diseased plants on to healthy ones, and there is no method at present of testing filtrates of sap from diseased plants for the presence of the viruses. Nor can passage through a filter alone be taken as proof that a pathogen is a virus. Filterable forms of some bacteria have been described, and some other visible organisms, such as the flexible spirochaetes, may also pass through filters usually regarded as impervious to bacteria.

The distinguishing features of viruses, therefore, can briefly be summarised as ability to cause pathological conditions, small size relative to other pathogens, and inability to multiply *in vitro* with ordinary bacteriological methods. Since filterability has been found to be unsuitable as a criterion of size, invisibility can be better and more accurately used in an attempt to define the characters of viruses with any precision. Visibility under the microscope depends on the refractive index of the particles being examined, the numerical aperture of the objective lens and the wavelength of the light used. Provided that their refractive index differs sufficiently from that of the mounting medium, spherical particles of 200 m$\mu$ diameter can be seen by means of their diffraction images with the best objective lenses and ordinary transmitted light. The shape of particles of this size cannot be determined, nor can the combined image of two such adjacent particles be resolved into its separate parts. For the particles to be clearly resolvable by customary microscopical methods, their dimensions must exceed 250 m$\mu$. With the knowledge that viruses are not resolvable by microscopes using visible light, therefore, we can proceed to define *a virus as an obligately parasitic pathogen with dimensions of less than 200m$\mu$.*

Such a definition clearly tells us nothing about the precise identity of viruses, for it could cover a multitude of different entities, ranging from complex bodies only a little smaller than some bacteria to relatively simple molecular substances. It serves to set some limitations on the use of the word, but it will almost certainly be modified to something more positive in the future. Eventually it is probable that viruses will be defined much more accurately on the basis either of their chemical constitution or biological activities, and the large number we now group under one heading may call for separate sub-divisions. A good deal is already known about the chemical constitution of a few plant viruses, but about most we know nothing of either their sizes or chemical natures, and the seeming uniformity of the few should not lead us to assume that all are necessarily similar. Those about which we know something definite may all eventually be found to be members of one of the sub-groups into which viruses may need to be divided.

The suggested definition has the great virtue of covering all the facts as they are known at present, but its limitations and artificiality are obvious. There is no apparent reason why the resolving power of light microscopes should have any particular biological significance. Also, although obligate parasitism may be a necessary consequence of small size, it is not solely a property of ultra-microscopic entities. There are many organisms that appear to be obligate parasites. Of plant pathogens, no rust or powdery mildew has yet been cultivated on an artificial medium; nor has *Plasmodiophora brassicae* or *Synchytrium endobioticum*. More fungi and bacteria are saprophytic than parasitic, and saprophytic entities resembling viruses may exist. BARNARD (1935), indeed, has already used the phrase "saprophytic viruses" to describe ultra-microscopic bodies which he found multiplying in horse- and rabbit-sera and which he photographed with the aid of an ultra-violet microscope. Nevertheless, until such bodies are shown to resemble the viruses in other characters than size, it would seem preferable to give them some other name. Furthermore, there is no obvious reason why all the entities resembling viruses should be actively pathogenic, but if non-pathogenic forms exist they would be indistinguishable from normal cell constituents and should be regarded as such. The possible relationships of such saprophytes and cell constituents to viruses is of great interest, and is a fertile field for speculation. However, to prevent the word virus from again becoming hopelessly indiscriminate, until there is good reason for considering any saprophytes or non-pathogenic cell components to be biologically related to viruses, we suggest restricting its use to agents that fulfil all three requirements of the proposed definition.

**Economic Importance:**—The existence of viruses was recognized only in 1892 and little attention was paid to them at first, but in the last thirty years they have been found to cause so many serious diseases that their economic importance can hardly be exaggerated. Despite the frequency with which "new" viruses are described and the fact that many workers consider virus diseases to be more prevalent now than in the past, there is no reason to regard them as afflictions peculiar to modern times. The names of but a few of the virus diseases from which man and animals suffer will be enough to indicate their importance in medical and veterinary science and to show that they include, not only some of the most troublesome complaints, but also some of the oldest known. Among the virus diseases of man are small pox, measles, mumps, herpes, yellow fever, influenza, poliomyelitis (infantile paralysis) and the common cold; those of animals include rabies, distemper, swine fever, louping ill, equine encephalomyelitis and foot-and-mouth; birds suffer from various kinds of pox, plague and tumour, as well as psittacosis, and insects from sacbrood and polyhedral diseases.

In the plant world the distribution of virus diseases seems to be more circumscribed than with animals, for several hundreds of diseases of higher plants (Angiosperms) have been described and also a large number of viruses (bacteriophages) that lyse bacteria, but none is known that attacks Gymnosperms or any cryptogam from the fungi to the Pteridophyta. Whether this restriction of the habitat of viruses is real and of any evolutionary significance, or only apparent because no adequate search of lower plants has been made, is uncertain. The number of viruses which attack angiosperms, that is, those with which this book is concerned, cannot be given even roughly, because it is unknown how many of the different diseases that have been described are in fact caused by different entities. However, in his *Handbook of Phytogenic*

FIG. 1. — Virus diseases in Majestic potatoes. Left-hand plant, severe mosaic, second-year infection with potato virus Y; centre plant, leaf roll; right-hand plant, healthy.

*Viruses,* HOLMES (1939) listed what he considered to be 82 distinct viruses and gave another 50 names about which there was insufficient evidence to decide whether or not they were also distinct, and since 1939 many more have been described. Some of these, of course, do little harm, but others are widespread and cause serious loss in all kinds of crops.

Although viruses regularly occur and often cause serious losses in annual crops raised from seed, they are most important in plantation crops or in those that are raised by vegetative methods of propagation. There is no recovery from most virus diseases. Once a plant becomes infected in nature it is usual for the virus to permeate the whole vegetative system and to remain active there indefinitely. Hence perennial plants (or plants propagated vegetatively) which are not killed by infection remain permanently affected and can go on accumulating viruses and becoming progressively weakened as long as they stay in cultivation. In such plants as fruit trees, hops and cacao, sugar cane, potato, strawberry, raspberry, or ornamentals like Dahlia and tulip, therefore, infection with viruses means not only a reduction in the current season's crop but also a loss of capital stock or a permanently reduced yield.

Precise figures of losses are difficult to obtain, and in many parts of the world there is little or no information on the prevalence of virus diseases, but a few illustrations will suffice to indicate the order of magnitude. Although most fruit trees are affected by virus diseases, the peach is possibly the worst sufferer. In north-east America there have been several great outbreaks of peach yellows, which have destroyed hundreds of thousands of trees, eliminating not only entire orchards but rendering peach growing impossible over large areas. Peach yellows is unknown in the southern U.S.A., but there phony disease is a major problem and in attempts to control it more than 1,000,000 trees have been removed in the State of Georgia alone. In addition to these, eighteen other virus diseases of peach trees have been described in north America, of which at least two, little peach and X or yellow-red disease, cause serious losses. (HILDEBRAND, BERKELEY and CATION, 1942). In West Africa a similar tale can be told of the cacao plantations, the existence of which is being threatened by the swollen shoot complex of viruses (POSNETTE, 1947). In the Gold Coast well over 1,000,000 trees have been destroyed and in the Eastern Province, which was the world's most intensive cacao-producing area, production has declined from 116,000 tons in 1936 to 64,000 tons in 1945. In strawberry and raspberry cultivation also, virus diseases have been responsible for the abandonment of many valuable varieties and for the loss of what was a thriving trade in some districts. Potato virus diseases are widespread and the use of infected stocks is responsible for the low yields obtained in many parts of the world. In the ware-growing districts of England losses from leaf roll and rugose mosaic (potato virus Y) (Fig. 1) are avoided only by the farmers annually purchasing some 500,000 tons of seed potatoes at a cost of more than £5,000,000. In addition to these two diseases, mild mosaic (potato virus X) is universally distributed through most widely-grown varieties, and must reduce the world's crop of tubers by at least 10%. In Britain alone, this virus, of whose existence most growers are unaware, probably causes losses of more than 200,000 tons a year.

In annual crops raised from seed the amount of loss depends greatly on the time at which infection occurs. Early attacks mean large losses, whereas if the plants are well developed before they become infected losses will be relatively small. There is also a good deal of seasonal variation in these diseases, which may be widespread over large areas in one year and infrequent or local in another. Tobacco mosaic virus occurs wherever tobacco and tomato are grown; in the U.S.A. it is estimated to cause an annual average loss of 40,000,000 pounds of tobacco, and in Britain it probably reduces the yield of tomatoes under glass by at least 10%. Curly top of sugar beet has on occasions rendered the crop valueless over large areas in the U.S.A., and in Europe the less severe yellows disease in some years reduces the yield of sugar by more than 20%. In other annual crops, such as lettuce, brassica of all kinds, legumes, cucurbits, and flowers as well as vegetables, virus diseases regularly take a toll and sometimes reach alarming proportions, but estimates of damage could be little more than wild guesses.

**Historical:**—Although many of the virus diseases of plants now recognised have undoubtedly existed for a long time, only a very few attracted attention before the twentieth century. The oldest of which we have any record is tulip mosaic in which the striking change in the appearance of the flowers (breaking) could hardly have been disregarded. Broken tulips feature prominently in many paintings of flowers by old Dutch masters and references to variegated flowers appear in botanical literature in 1576 (McKAY and WARNER 1933). According to VAN SLOGTEREN and OUBOTER (1941) some Dutch bulb-growers knew as early as 1637 that breaking could be produced in self-coloured tulips by grafting them with variegated ones and BLAGRAVE in 1675 gave precise details for grafting halves of bulbs. In spite of this early progress, however, breaking was widely thought to be a sign of maturity, which many considered could be hastened by planting bulbs in impoverished soil or some other unfavourable conditions, and it was not until 1928, when CAYLEY demonstrated transmission unequivocally, that it was generally accepted as an infectious disease.

Next to the tulip the potato is the oldest sufferer from virus diseases of which we have records. It was so severely attacked by 1775 that farmers in different parts of Europe had to abandon its cultivation, and prizes were offered to those who could find the cause of and remedy for the destructive ailment of such a useful plant. Of the numerous papers published around this time, the majority stress the degenerate state of affected plants, and the condition became widely known as "degeneration", "running-out" or "senility". It was early noticed that new varieties were, in general, more vigorous than old ones, which led to the belief that the condition was an inevitable consequence of continued propagation by "unnatural" asexual methods. Little or no experimental work was done, but ANDERSON (1792) in an essay written in 1778 suggested many things that modern work has confirmed. He noticed that tubers from affected parents always produced degenerate plants, and that stocks grown in the south of Britain suffered more than those in the north. He suggested that the condition was infectious, comparing it with small pox, and that it might be controlled by destroying all affected plants as soon as they could be identified. It seems likely that virus diseases of the potato were actually introduced into Europe at about this time, for most writers were agreed that the degeneration was unknown before 1770.

Peach yellows has almost as long a history as the potato degeneration, for it was described in 1791 and was probably doing damage as early as 1750 (SMITH 1888). Around 1870 the variegated *Abutilon* was popular as an ornamental plant in Europe, and it seems to have been widely appreciated that the condition could be reproduced in normal green plants by grafting scions from variegated ones. The first transmission of what were recognized as pathological conditions seem to have been achieved by SMITH (1891, 1894) who showed that healthy peach trees became diseased when grafted with scions from trees suffering from yellows and rosette. Shortly before this MAYER (1886) named tobacco mosaic, which was first described by SWIETEN in 1857, and showed that it could be transmitted by injecting sap from affected plants into healthy ones. These transmission studies can be considered as the starting point of modern work on virus diseases, although the distinction between viruses and other pathogens was not made until IWANOWSKI (1892) and BEIJERINCK (1898) demonstrated the filterability of tobacco mosaic virus.

During the last sixty years too much has been done and too many diseases listed as virus diseases for any comprehensive history to be attempted, but a few landmarks warrant mentioning. The early work was concerned chiefly with symptomatology, the identification of new diseases and methods of transmission. The importance of insects as vectors of virus diseases was first indicated in 1900 when a mosquito was found to transmit yellow fever, but only one year later TAKAMI showed that a leaf-hopper, *Nephotettix apicalis,* was the vector of the stunt disease of rice. It is rather surprising in view of these important early discoveries that their implications seem to have gone unappreciated and that progress for a time was so slow. It was not until 1913 that ORTON showed that the degeneration of the potato took more than one form. He then described three types under the names leaf roll, mosaic and streak, but leaf roll was not shown to be transmissible until 1916 (QUANJER), or its connection with aphids suspected until 1920 (OORTWIJN BOTJES), and it was 1923 before aphid vectors were identified by SCHULTZ and FOLSOM (1923) and MURPHY (1923). Similarly, tulip break, a condition known for more than three hundred years, was not shown to be aphid-transmitted until 1929 (HUGHES 1930).

The fact that plants might carry viruses, that is to say, be infected but show no abnormalities, was discovered in 1918, when NISHAMURA found that *Physalis alkekengi* developed no symptoms when inoculated with tobacco mosaic virus, although juice from these plants produced mosaic when inoculated to healthy tobacco plants. Naturally-infected carriers were first recognised in potatoes by SCHULTZ (1925) and JOHNSON (1925), who found one virus to occur with such regularity in American potatoes, whether apparently healthy or obviously diseased, that it became known as the "healthy potato virus". It is now well known that some distinct diseases are caused by simultaneous infection with two different viruses. The first example of this was described by VANTERPOOL (1926), who found that "winter blight" or "glasshouse streak" of tomatoes was caused by the joint action of tobacco mosaic virus and a potato virus, either of which alone caused only mosaic symptoms.

In the early work on plant virus diseases it was assumed that differences in symptoms were attributable to infection with distinct viruses, but as the properties of the viruses were studied more intensively it became clear that apparently different diseases could be caused by variants of one and the same virus. The first suggestions that plant viruses are not stable entities, but can undergo changes analogous to mutations and give rise to strains of different virulence, are found in the work of CARSNER (1925), JOHNSON (1926) and MCKINNEY (1926); since then most viruses that have been studied in any detail have been found to exist in a number of strains, differentiable by the symptoms they cause in the same hosts, by differences in host range, insect vectors, or some other property. Studies by THUNG (1931) and SALAMAN (1933) of the effects of mixed infections with strains of tobacco mosaic virus and potato virus X respectively, showed that the strains of one virus are antagonistic towards one another and led to the first examples in plants of acquired resistance to disease. The ability of one virus strain to protect plants against the effects of another has now been demonstrated with so many viruses that it seems probable that it is a general phenomenon, and it is widely used to determine relationships between viruses that cause different symptoms.

Before proceeding to consider the development of views on the nature of viruses, two other discoveries that have greatly affected the methods of studying plant virus diseases deserve mention. The first is the recognition of the importance of local lesions, *i.e.*, visible, discrete lesions produced at the sites of infection, by HOLMES (1928, 1929), who showed that the numbers were approximately proportional to virus concentration and so could be used in quantitative work. The second is the discovery that plant viruses are antigenic and produce, when injected into animals, antibodies with which they react specifically. Serological methods were first applied to virus infected plants by DVORAK (1927), but it was the pioneer work of PURDY BEALE (1928, 1929, 1931) that first showed their significance and their wide applications.

**Views on the Nature of Viruses:** — In the absence of any direct evidence about the nature of viruses, it is not surprising that speculation has been rife and that continued controversies have raged round the subject. The chief disagreement has been between those who considered viruses to be essentially similar to bacteria and those who considered them to differ from bacteria in properties more fundamental than size. The stage for this dispute was set by the first papers by IWANOWSKI (1892) and BEIJERINCK (1898). Although IWANOWSKI is usually credited with being the first to demonstrate the passage of a virus through a bacteria-proof filter, he was so convinced that tobacco mosaic was a bacterial disease that he argued that the symptoms were produced by bacteria penetrating the filter or by a toxin secreted by the causal bacteria. BEIJERINCK, on the other hand, fully appreciated the possible significance of filterability and was also impressed by the ability of tobacco mosaic virus to diffuse through agar gels. He considered that the virus was entirely different from a bacterium, was not even corpuscular but was what he called a "*contagium vivum fluidum*". Unless BEIJERINCK wished to convey the impression that the virus was small enough to form a stable solution, a view confirmed by modern work, it is difficult now to decide precisely what distinction he was making, for a non-corpuscular something rather baffles the imagination. Nevertheless, BEIJERINCK's statement had repercussions for many years, and until well after 1930 the particulate nature of viruses was seriously discussed. Unfortunately, no supporters of the non-particulate school have left us with a clear picture of their idea of viruses, or what they considered the fundamental qualitative difference between the corpuscular and fluid states of matter.

The discovery of viruses came from the break-down of an established bacteriological technique, the failure of a bacteria-proof filter to provide a sterile filtrate. Also, the invisibility of viruses and their apparent inability to multiply *in vitro* rendered customary bacteriological methods of little value in studying them. Yet the most generally accepted view of the nature of viruses was that they were ultra-microscopic organisms, for work on transmission and symptomatology brought nothing to conflict with it. However, other theories have been put forward from time to time, especially by those working on tobacco mosaic virus, but there was little positive evidence until the last ten years to suggest whether virus particles more nearly resembled bacteria or molecules. WOODS (1899, 1900) considered that tobacco mosaic was caused by an excessive accumulation of oxidising enzymes, and suggested that adverse growing conditions increase enzymic activity, so changing chlorophyll into xanthophyll and producing the characteristic variegation. HUNGER (1905) and FREIBERG (1917) again suggested that the virus was an enzyme or toxic principle, but ALLARD (1916) disproved WOODS' theory, and on reviewing the available evidence concluded that the virus was a microorganism.

Ideas on the nature of viruses have been influenced by the occurrence of peculiar bodies in the cells of affected organisms. Such bodies were first described as a feature of some animal diseases, such as molluscum contagiosum (HENDERSON, 1841), fowl pox (RIVOLTA, 1869) and vaccinia (GUARNIERI, 1894), long before the existence of viruses was recognised. The connection of such intracellular inclusion bodies with virus infections was first indicated by IWANOWSKI (1903) who described two kinds in plants suffering from tobacco mosaic. One was a vacuolate, amoeboid-like, body, often occurring in contact with the nucleus (Fig. 2), and the other a flat crystalline plate. Since then analogous bodies have been described accompanying many other virus diseases, but different workers have disagreed about their etiological significance. IWANOWSKI concluded that they could not be organisms responsible for the disease as they were too large to pass through the filters. He suggested that they were either abnormal products of nuclear division or reaction products of the cells to the disease. He was unable to isolate any bacteria, but he saw in stained sections short rods resembling bacteria and concluded that these were the cause of the disease. Although we now know they were wrong, most workers agreed with IWANOWSKI that the crystals were merely plant reaction products, and it was only on the amoeboid bodies that opinion was divided. SMITH (1924), HOGGAN (1927), HENDERSON SMITH (1930) and CLINCH (1931) considered these also to be plant products, whereas LYON (1910), KUNKEL (1921), MCWHORTER (1922), GOLDSTEIN (1924) and LIKHITE (1929) thought that they represented stages in the life cycle of some causal organism.

In addition to the intracellular inclusions, various other "organisms" have been described from time to time as the causes of plant virus diseases. NELSON (1923) first attributed mosaic of the bean to a protozoan and then (1932) to a small coccus. ECKERSON (1926) described small flagellates as the cause of tomato mosaic. v. BREHMER and BARNER (1930) pictured small oval bodies in potatoes and beet suffering from mosaic, which they named *Plasmodiophora solani* and considered responsible for the diseases. PAINE and BEWLEY (1919) stated that *Bacillus lathyri* produced "stripe" of tomato, but later BEWLEY (1931) suggested that a bacteriophage entering the plants with the bacteria was the direct cause and not the bacteria themselves. As all the organisms that have been described are obviously too large to be filterable, a life cycle has usually been postulated in which visible stages alternate with an invisible, filterable stage. None of these theories has been confirmed, and there is no reason to believe that an organism has been seen which is responsible for any of the recognised plant virus diseases.

With the failure of microscopic and cultural methods to reveal any cause for virus diseases, physical and chemical techniques used for investigating proteins have been increasingly applied. Most of the early work was done with tobacco mosaic virus, and the great stability of this virus was soon recognised. It was also found that it could be precipitated with various protein precipitants and resuspended without losing activity. This led to suggestions that the virus behaved more like molecules such as proteins (MULVANIA, 1926) than like organisms, and to attempts to isolate the virus by chemical methods. VINSON and PETRE (1929, 1931) obtained reasonably active, colourless preparations by the use of lead acetate, and from its behaviour concluded that tobacco mosaic virus was a nitrogen-containing chemical substance. BARTON-WRIGHT and MCBAIN (1933) claimed to have isolated this virus in the form of nitrogen-free crystals, but CALDWELL (1934) showed that their crystals were largely phosphate, contaminated with small amounts of virus.

The early attempts to purify tobacco mosaic virus failed to afford any positive evidence about its chemical nature largely because the workers did not concentrate it sufficiently to produce it in bulk. The failure to attempt this is hardly surprising when those who thought quantitatively could come to such depressing conclusions that more than 600,000,000 litres of sap might be needed to provide as little as 1 mg. of virus (ROBBINS 1934). On the data then available, such estimates were not altogether unreasonable, but CHESTER (1935) using the results of serological instead of infectivity tests as a basis for calculation, estimated the amount of tobacco mosaic virus more accurately at 0.1 to 1.0 mg. per ml. He seems to have been the first to suggest that inoculum might need to contain many millions of particles to produce a single infection, and his estimate of quantity was soon confirmed. STANLEY (1935, 1936), using a large volume of sap and precipitation methods similar to those by which enzymes were purified, obtained in substantial quantities what he claimed to be "a crystalline protein possessing the properties of tobacco mosaic virus." From his original description of this material, it is difficult to decide precisely what he had isolated, for many of the properties he ascribed to his product are certainly not those now recognised as typical of the virus. The needle-shaped bodies he described as crystals are characteristic forms of the precipitated virus, but his statements that his product contained 20 per cent. nitrogen, was free from both carbohydrates and phosphorus, and was a globulin that could be "partially digested with pepsin", strongly suggest that the bulk of it was something other than active virus. BAWDEN, PIRIE, BERNAL and FANKUCHEN

(1936) also showed that plants infected with tobacco mosaic virus contained considerable quantities of an unusual protein, but they identified their product as a liquid crystalline nucleoprotein, containing a nucleic acid of the ribose type. They showed that the particles were rod-shaped, substantiating an earlier suggestion by TAKAHASHI and RAWLINS (1932) who found that infective sap showed the phenomenon of anisotropy of flow, and that the particles in the purified preparations were at least ten times as long as they were wide. BAWDEN and PIRIE (1937*a*) found that plants infected with different strains of the virus contained similar but not identical liquid crystalline nucleoproteins and that nothing similar could be isolated from uninfected plants; they also suggested that the elongated particles characteristic of purified virus preparations might be artefacts, produced by the linear aggregation of smaller particles. From X-ray analysis of the purified preparations, BERNAL and FANKUCHEN (1937) obtained the first accurate measurement of the width of the particles and found that strains attacking solanaceous plants were wider than those attacking cucurbits. They also showed that the needle-shaped bodies produced by precipitating the preparations lack the three-dimensional regularity of true crystals and are more accurately described as fibres or para-crystals.

Since 1937 many workers have confirmed the presence of a specific protein in plants infected with tobacco mosaic virus and its identification as a liquid crystalline nucleoprotein. Specific proteins, either the viruses themselves or closely related to them, have also been isolated from plants infected with other viruses. Some of these resemble tobacco mosaic virus by occurring in rod-like particles that form liquid crystals, for example, cucumber viruses 3 and 4, potato viruses X and Y, and henbane mosaic virus (BAWDEN and PIRIE 1937*b*, 1938*a*, 1939). Others appear to have spherical particles and some of these form true crystals. Tomato bushy stunt virus was isolated by BAWDEN and PIRIE (1938*b*) as rhombic dodecahedra and later various tobacco necrosis viruses (PIRIE, SMITH, SPOONER and MACCLEMENT 1938; BAWDEN and PIRIE 1942, 1945), southern bean mosaic virus (PRICE 1946) and turnip yellow mosaic virus (MARKHAM and SMITH 1946) were obtained in characteristic crystalline forms. Two other viruses that have been purified, or partially so, without apparently forming either crystals or liquid crystals, are tobacco ringspot (STANLEY and WYCKOFF 1937; STANLEY 1939) and alfalfa mosaic (ROSS 1941). The one thing these various proteins have in common is that they are all nucleoproteins containing a ribose nucleic acid.

In deciding the nature of viruses the precise relationship of these proteins to the viruses themselves is obviously a vital problem. The absence of any conclusive test for purity, and the inability to cause infections with inoculum containing single particles, makes proof of their identity impossible, but there is a large mass of evidence suggesting that they are the viruses and little that conflicts with this idea. Similar proteins have not been found in uninfected plants, and the particular one obtained from an infected plant is determined solely by the infecting virus and not by the species of host or the symptoms it shows. From the same plant species infected with unrelated viruses, different proteins are obtained, whereas plants infected with strains of one virus give similar, though not identical, proteins. The different proteins denature under different conditions, and denaturation is always accompanied by loss of infectivity. The infectivity of some purified preparations is so high that, if it is not a property of the proteins but of some contaminant, this contaminant must be smaller and less complex than the specific nucleoproteins. Also, the homogeneity of some preparations is such that, when subjected to a variety of physical and chemical tests, they show no signs of containing more than one component, and where heterogeneity occurs, it can usually be attributed to variations affecting the protein particles themselves. For simplicity, in this book, the proteins are assumed to be the viruses and referred to as such. If this view is not adopted, there seem to be only two other possibilities. The first is that the viruses, though not identical with the proteins, nevertheless resemble them in all the properties yet studied. The second is that the viruses are minute contaminants in the purified preparations so firmly attached to the proteins that no methods now available can separate the two. Substances long regarded as pure have often later been demonstrably mixtures, and it may be so with these virus preparations. Adopting either of these views, however, unnecessarily complicates the problem. Viruses not only once again become mysterious entities but this time with the added complication that, in addition to multiplying themselves in susceptible cells, they also produce these specific and unusual nucleoproteins. Although this must be accepted as a possibility, until some supporting evidence is forthcoming, it is gratuitous to postulate two entities when one satisfies the data: applying the razor of OCCAM (*ca* 1330) *Entia non sunt multiplicanda praeter necessitatem.*

Although it would now be unreasonable to differentiate between the specific proteins and the viruses, it does remain uncertain to what extent results obtained by studying the purified preparations in the laboratory can be applied directly to the viruses as they are produced in infected cells. Also, in spite of the wide differences between the properties of the individual viruses so far purified, it would be rash to assume that they form a random and representative sample of the whole group. Because these are chemically similar, it can almost be taken for granted that some other viruses are also nucleoproteins, but to assume that all are would be decidedly premature. The methods of isolation so far used may be acting as a selective agency, succeeding only with those that are nucleoproteins. If this is so, and that it may be is perhaps suggested by the failures of attempts to isolate some viruses, the apparent chemical uniformity of plant viruses may be a temporary phase, reflecting merely the limited range of procedures yet applied to their study, and new methods may identify some that are chemically different. This needs stressing, but the identification of a plant virus as anything other than a nucleoprotein will be a major event, calling for much supporting evidence, whereas a claim to have isolated a new one as a nucleoprotein will be accepted as merely adding another to the growing list.

**Nomenclature: —** The absence of any systematic basis for the nomenclature of viruses has caused much confusion and is responsible for many of the apparent contradictions in published work. The same virus has often received different names from different workers, thereby leading to a multiplicity of synonyms, and sometimes the same name has been given to distinct viruses. This has come about largely as a result of naming viruses on the basis of the symptoms they cause, without appreciating that the same symptoms are not produced under all conditions by the same virus or that similar symptoms may be caused by different viruses.

At first, little attempt was made to distinguish between cause and effect, the virus and the disease usually being referred to as the same thing. Later a distinction was made by adding the word virus to the name of the disease to indicate the cause, for example, tobacco mosaic was stated to be caused by the tobacco mosaic virus. With increasing knowledge it became obvious that plants could be attacked by more than one virus, and that similar symptoms could be caused by different viruses. Also, it was found that the same virus might infect many plant species, inducing different symptoms in different plants. This showed that the system of naming on the basis of the host attacked and the symptoms produced was inadequate. Also, in plants such as the potato and tobacco, which are attacked by a number of viruses producing mottling or necrosis it became increasingly difficult to find sufficiently distinct descriptive names for all the viruses recognized. In 1927, therefore, JOHNSON suggested that descriptive names should be applied only to the diseases, the virus merely taking its name from the host in which it was first discovered, together with a number to indicate the specific virus. Thus the causes of tobacco mosaic and cucumber mosaic became Tobacco virus 1 and Cucumber virus 1 respectively. SMITH (1931), when he separated two viruses from a diseased potato, used letters instead of numbers and called them potato viruses X and Y respectively. This practice has been followed with other potato viruses several of which have since been named alphabetically.

At the International Botanical Conference at Cambridge in 1930 an International Committee on Virus Nomenclature was formed under the Chairmanship of Professor JAMES JOHNSON to consider the whole question of naming viruses. This committee reported to the Conference in Amsterdam in 1935 and proposed a scheme that was adopted in principle, but it has never been officially published. This was an elaboration of JOHNSON's original scheme, designed to cover names for strains as well as for separate viruses. In addition to the common name of the host followed by the term virus and a number, strains were indicated by a capital letter and sub-strains by a small letter. Thus the common cucumber virus became cucumber virus I, the strain producing a yellow mottle IB and the variants isolated from this by PRICE (1934) 1Ba, 1Bb and so-on. Working with potato virus X, SALAMAN (1933) used a different method of indicating strains; a further letter was allotted to each strain and written as an index, the severe strain causing ringspot symptoms in tobacco becoming $X^s$ and the mild one causing a faint mottle $X^g$.

Since 1935 new systems of nomenclature have been put forward with such frequency by individual workers that a state approaching chaos has been reached, for even well-established names have been replaced by complicated unknowns, and the name used by one virus worker may convey little or nothing to another.

SMITH (1937) introduced a modification of the scheme tentatively proposed by the International Committee, in which the Latin generic name of the host is used instead of the common English name. Thus tobacco mosaic and aucuba mosaic viruses become Nicotiana virus I and

IC, respectively, and potato virus X Solanum virus I. SMITH added to the confusion by keeping some of the numbers for individual viruses that had been proposed in JOHNSON's scheme and altering others.

HOLMES (1939) introduced a binomial system of nomenclature, similar to that in use for plants and animals, with a trinomial for indicating strains. In this scheme ordinary tobacco mosaic virus becomes *Marmor Tabaci,* var. *vulgare,* aucuba mosaic virus *Marmor Tabaci,* var. *Aucuba,* cucumber mosaic virus *Marmor cucumeris* var. *vulgare,* and the strain of potato virus X causing ringspot in tobacco *Marmor dubium* var. *annulus.*

Both SMITH and HOLMES claim that their systems of naming are based on a classification of viruses, but for reasons discussed in Chapter 14 it is clear that the systems of grouping adopted hardly warrant the term classification.

The limited use of numbers or letters worked reasonably well, but the extension of such a system to all viruses renders it unwieldy. Also, as a number bears no relationship to any property of the virus to which it is given, the difficulties in remembering what are the specific properties of viruses with such names as tobacco virus 18 or Solanum virus 17 are considerable. BENNETT (1939) has pointed out the many difficulties of a nomenclature based on numbers and suggested that names might replace numbers with considerable advantages. For example, *Nicotiana virus altathermus,* obviously tells more than *Nicotiana virus* 1. Such a name could readily be turned into a true Latinized binomial, if it is ultimately decided that viruses should be classified into genera and species. As an illustration, BENNETT suggests that tobacco mosaic virus might then become *Paracrystalis altathermus,* a name that really carries some information about the virus.

On the other hand, should it be decided that a nomenclature more in keeping with those in use by chemists, rather than those in use by biologists, should be applied to viruses, then the specific name could be given an appropriate suffix. The designation of a particular class of substances by a common suffix, for example the use of "ase" to denote enzymes, is an established practice in chemistry, and BENNETT suggested that the suffix "vir" might be adopted to denote virus, so that tobacco mosaic virus could become "altathermovir".

FAWCETT (1940) adopted the suffix principle and prepared a scheme that he claimed combined the best features of JOHNSON's, SMITH's and HOLMES' schemes. The rules are that the stem "vir" for virus is added to the Latin generic name of the genus in which the virus was first discovered and the specific name is taken from the list devised by HOLMES. As examples FAWCETT gives *Rubivir orientale* for raspberry streak virus and *Solanivir vastans* for potato yellow dwarf virus, but it is not clear what tobacco mosaic virus would become; presumably it would be *Nicotianivir Tabaci,* which, to say the least, would seem to emphasize the host plant unnecessarily.

VALLEAU (1940) prepared a key for the identification of 8 viruses commonly found attacking tobacco crops, and used the Holmes-type of name, except that he omitted the trinomial for strains. Also some of HOLMES' names were modified because VALLEAU rightly considered that HOLMES' "generic" names indicated false relationships; in this scheme tobacco mosaic virus, for example, becomes *Musivum Tabaci* and cucumber mosaic virus *Murialba cucumeris*.

Each of these schemes has some adherents, and it appears that Latin binomials are becoming increasingly favoured by pathologists, but none has yet gained international acceptance. In previous editions of this book a conservative attitude was adopted and viruses were described under the common name first given to them or under those used by the workers who had studied them most. As an interim measure, until there is international agreement about the most suitable type of nomenclature, it was suggested that a list of approved names would be generally useful and help to check the growing number of synonyms. The Imperial Mycological Institute (1946), after consulting organisations in various countries, has issued a list of names under which viruses will be indexed in the Revue of Applied Mycology. With one or two exceptions, the names used are the common names of the diseases with the word virus added, and the list gives synonyms and viruses believed to be related strains. In this book, as far as is possible, the names used will be taken from that list.

### *References:-*

ALLARD, H. A. (1916) : J. Agr. Res. *7*, 481.
ANDERSON, J. (1792) : Bath Soc. Papers *4*, 92.
BARNARD, J. E. (1935) : Brit. J. Exp. Path. *16*, 129.
BARTON-WRIGHT, E. and MCBAIN, A. M. (1933) : Nature *132*, 1003.
BAUR, E. (1904) : Ber. der Deutsch. Bot. Gesell. *22*, 453.
BAWDEN, F. C., PIRIE, N. W., BERNAL, J. D., and FANKUCHEN, I. (1936) : Nature *138*, 1051.
BAWDEN, F. C. and PIRIE, N. W. (1937*a*) : Proc. Roy. Soc. *123*, 274.
—— and —— (1937*b*) : Brit. J. Exp. Path. *18*, 275.
—— and —— (1938*a*) : *ibid. 19*, 66.
—— and —— (1938*b*) : *ibid. 19*, 251.
—— and —— (1939) : *ibid. 20*, 322.
—— and —— (1942) : *ibid. 23*, 214.
—— and —— (1945) : *ibid. 26*, 277.
BEALE, H. PURDY (1928) : Proc. Soc. Exp. Biol. and Med. *25*, 702.
—— (1929) : J. Exp. Med. *49*, 919.
—— (1931) : *ibid. 54*, 463.
BEIJERINCK, M. W. (1898) : Verh. Akad. Wet. Amst. *6*, 1.
BENNETT, C. W. (1939) : Phytopath. *29*, 422.
BERNAL, J. D. and FANKUCHEN, I. (1937) : Nature *139*, 923.
BEWLEY, W. F. (1931) : Nature *127*, 442.
BOTJES, O. J. G. (1920) : Inaugural Dissertation, Wageningen (In Smith, K. M. (1931) Biological Rev. *6*, 302).
v. BREHMER, W. and BARNER, J. (1930) : Arb. Biol. Reichsanstalt *18*, 1.
BULLOUGH, W. (1938) : The History of Bacteriology, Oxford University Press, London.
CALDWELL, J. (1934) : Nature *133*, 177.
CARSNER, E. (1925) : Phytopath. *15*, 745.
CAYLEY, D. M. (1928) : Ann. Appl. Biol. *15*, 429.
CHESTER, K. S. (1935) : Science *82*, 17.
CLINCH, P. (1931) : Sci. Proc. Roy. Dublin Soc. *20*, 143.
DVORAK, M. (1927) : J. Inf. Dis. *41*, 215.
ECKERSON, S. (1926) : Bot. Gaz. *81*, 204.
FAWCETT, H. S. (1940) : Science *92*, 559.
FREIBERG, G. W. (1917) : Ann. Missouri Bot. Gard. *4*, 175.
GOLDSTEIN, B. (1924) : Bull. Torrey Bot. Club *51*, 261.
HENDERSON SMITH, J. (1930) : Biol. Rev. *5*, 159.
HILDEBRAND, E. M., BERKELEY, G. H., and CATION, D. (1942) : Miscellaneous Publication, Michigan Agr. Exp. Station.
HOGGAN, I. A. (1927) : J. Agric. Res. *35*, 7.
HOLMES, F. O. (1928) : Bot. Gaz. *86*, 66.
—— (1929) : *ibid. 87*, 56.
—— (1939) : Handbook of Phytopathogenic Viruses. Burgess, Minneapolis.

HUGHES, A. W. MCKENNY. (1930) : Ann. Appl. Biol. *17*, 36.
HUNGER, F. W. T. (1905) : Ber. Deutschen Bot. Ges. *23*, 415.
Imperial Mycological Institute. (1946) : Rev. Applied Myc. *24*, 575.
IWANOWSKI, D. (1892) : Bull. Acad. Imp. Sci. Petersburg *35*, 67.
—— (1903) : Zeit. Pflanzenkrht. *13*, 1.
JOHNSON, J. (1925) : Wis. Agric. Exp. Stat. Res. Bull. *63*.
—— (1926) : Science *64*, 210.
—— (1927) : Wis. Agric. Expt. Sta. Res. Bull. *87*.
—— (1935) : Rep. to Inter. Bot. Conf. Amsterdam.
KUNKEL, L. O. (1921) : Bull. Exp. Stat. Hawaiian Sugar Planters' Ass. Bot. Ser. *3*, 108.
LIKHITE, V. (1929) : Med. Landb. hoogesch. Wageningen *33*, 1.
LOEFFLER, F. and FROSCH, P. (1898) : Zbl. Bakter. *23*, 371.
LYON, H. L. (1910) : Hawaiian Planters' Record *3*, 200.
MCKAY, M. B. and WARNER, M. F. (1933) : Nat. Hort. Magazine.
MCKINNEY, H. H. (1926) : Phytopath. *16*, 893.
MCWHORTER, F. P. (1922) : Philippine Agr. *11*, 103.
MARKHAM, R and SMITH, K. M. (1946) : Nature *157*, 300.
MAYER, A. E. (1886) : Landw. Versuchsstation *32*, 450.
MULVANIA, M. (1926) : Phytopath. *16*, 853.
MURPHY, P. A. (1923) : Sci. Proc. Roy. Dublin Soc. *18*, 169.
NELSON, R. (1923) : Mich. Agr. Exp. Sta. Tech. Bull. *58*.
—— (1932) : *ibid. 118*.
NISHAMURA, M. (1918) : Bull. Torrey Bot. Club *45*, 219.
ORTON, W. A. (1913) : Phytopath, *3*, 69.
PAINE, S. G. and BEWLEY, W. F. (1919) : Ann. Appl. Biol. *6*, 183.
PASTEUR, L. (1889) : New Reviews *505*.
PIRIE, N. W., SMITH, K. M., SPOONER, E. T. C. and MACCLEMENT, W. D. (1938) : Parasitology *30*, 543.
POSNETTE, A. F. (1947) : Ann. Appl. Biol. *34*, 388.
PRICE, W. C. (1934) : Phytopath. *24*, 243.
—— (1946) : Amer. J. Bot. *33*, 45.
QUANJER, H. M., LEK, A. H. and OORTWIJN BOTJES, J. (1916) : Meded. R. Hoog. Land- Tuin-, en Boschbouwsch. Wageningen *10*, 1.
ROBBINS, W. J. (1934) : Science *80*, 275.
ROSS, A. F. (1941) : Phytopath. *31*, 394.
SALAMAN, R. N. (1933) : Nature *131*, 468.
SCHULTZ, E. S. (1925) : Science *52*, 571.
SCHULTZ, E. S. and FOLSOM, D. (1923) : J. Agr. Res. *25*, 2.
VAN SLOGTEREN, E. and OUBOTER, M. P. (1941) : Medel. der Landbouwhoogeschool, Wageningen, Deel 45.
SMITH, E. F. (1891) : U. S. Dept. Agr. Dis. Veg. Path. Bull. *1*.
—— (1894) : U. S. Dept. Agr. Farmers Bull. *17*.
SMITH, K. M. (1924) : Ann. Bot. *38*, 385.
—— (1931) : Proc. Roy. Soc. B. *109*, 251.
—— (1937) : Textbook of Plant Virus Diseases. Churchill. London.
STANLEY, W. M. (1935) : Science *81*, 644.
—— (1936) : Phytopath. *26*, 395.
—— (1939) : J. Biol. Chem. *129*, 405.
STANLEY, W. M. and WYCKOFF, R. W. G. (1937) : Science *85*, 181.
TAKAHASHI, W. N. and RAWLINS, T. E. (1932) : Proc. Soc. Exp. Biol. Med. *30*, 155.
THUNG, T. H. (1931) : Hand. 6th med. ind. Naturw. Cong. p. 450.
VALLEAU, W. D. (1940) : Phytopath. *30*, 820.
VANTERPOOL, T. C. (1926) : Phytopath. *16*, 311.
VINSON, C. G. and PETRE, A. W. (1929) : Bot. Gaz. *87*, 14.
—— and —— (1931) : Boyce Thompson Inst. *3*, 131.
WOODS, A. F. (1899) : Centralbl. Bakt. *5*, 745.
—— (1900) : Science *11*, 17.

*Chapter 2*

# EXTERNAL SYMPTOMS OF INFECTED PLANTS

The importance of symptomatology in the study of virus diseases can hardly be over-stressed. Before the existence of a virus can be suspected, let alone established, there must be some change in the appearance of a plant; also, though unreliable for the purpose, symptoms will for long provide the main basis for identification and for distinguishing between different viruses. Viruses cause many different symptoms and from symptomatology alone a pathological condition cannot conclusively be identified as a virus disease. For this, the condition must be demonstrably transmissible in the absence of any visible pathogen. Nevertheless, there are some symptoms that are common to many virus diseases, a fact that limits the utility of symptoms in identifying specific viruses but is helpful in indicating the etiology of newly-discovered diseases. It is the purpose of this chapter to outline the range of symptoms produced by viruses and to indicate the many factors that influence the symptoms shown by infected plants. In doing this some diseases will be described, but no attempt is made to give a systematic account of the many recognised virus diseases. For this, Dr. KENNETH M. SMITH's ***Textbook of Plant Virus Diseases***, which contains many excellent illustrations, should be consulted.

**Range of Host Plant Reactions:** — The reactions of plants to infection with viruses range from no perceptible change, through diseases of different degrees of severity, to rapidly fatal conditions. Viruses are obligate parasites, and if they kill their hosts they also eliminate themselves. Hence in the field, viruses occur most commonly in plants that to some extent tolerate their presence, and it is customary for them to cause chronic rather than lethal diseases. When a virus is widespread in a host which dies soon after infection, it either has an unusually efficient mode of transmission or, more probably, is being maintained in an alternative, tolerant host. The acme of tolerance is the ability to be infected and suffer not at all, to act as a symptomless carrier, a condition that is far from uncommon, for many plants can carry one or more viruses. As we depend on symptoms for recognising the existence of viruses, their occurrence in carriers is usually demonstrable only by transmissions to intolerant hosts.

In susceptible hosts other than carriers, the most universal symptom is a reduction in size and cropping power, but before this becomes obvious changes usually occur in the appearance of some individual plant parts. Roots, stems, leaves, flowers or fruits may become discoloured or deformed, but an alteration in leaf colouring is most often the first indication of virus infection. Instead of being uniformly green, the leaves develop spots, blotches, streaks or rings of light-green, yellow, white, brown or black; or, in default of any definite pattern, the leaves may become uniformly chlorotic. The confusing repetition of the words mosaic, streak, ringspot and yellows in virus nomenclature is one unfortunate result of the frequent occurrence of such symptoms. Deformation of the leaves varies from a mere modification of the outline, so that the leaves are waved or crinkled, to changes that alter the leaf type beyond recognition. Leaves normally entire may become dentate, or the laminae may become so reduced that what should be broad leaves consist of little more than the main veins and resemble tendrils.

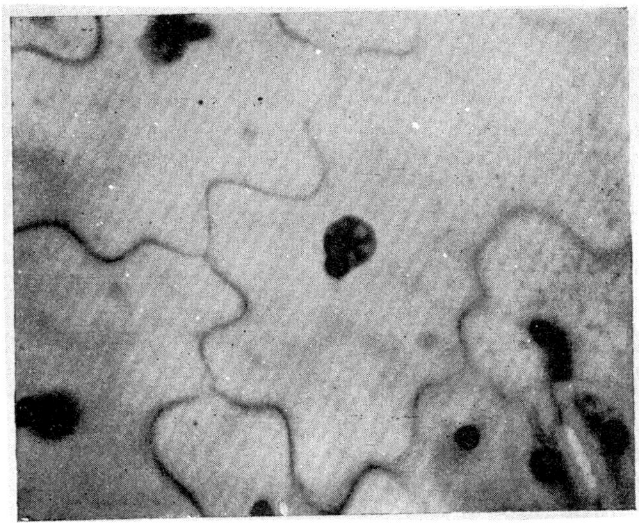

FIG. 2. — Epidermal strip from *Solanum nodiflorum* infected with tomato aucuba mosaic virus, a strain of tobacco mosaic virus. The vacuolate inclusion is in contact with the nucleus, which has stained deeply with Feulgen's reagent. × 480. (J. HENDERSON SMITH, Ann. appl. Biol. *17*, 213, 1930). — *Cf.* p. 10.

Viruses usually cause systemic infections, that is to say, from single entry points they spread internally through the host plants to invade all the susceptible tissues, most often the whole of the vegetative system, in which they persist as long as the tissues remain alive. There are, however, exceptions to this, for some host plants can localise infection with one or more viruses. Symptoms in such plants are nearly always necrotic, the reaction of the first invaded cells being so severe that they die before the viruses have time to move from them. The phenomenon is comparable with the hypersensitiveness of some wheat varieties to *Puccinia graminis*; viruses and rust fungi, being obligate parasites, are unable to multiply in dead tissues, and remain isolated in the necrotic local lesions. Hypersensitiveness is almost as valuable a defence mechanism as immunity, and in the field plants are rarely found infected with viruses to which they are hypersensitive, so that systemic infection is the normal condition of naturally-infected hosts. The only exceptions seem to be the tobacco necrosis viruses, for which no hosts are known that become systemically invaded,* and it is uncertain how these are perpetuated. As with other viruses, localisation of these causing tobacco necrosis is in most hosts associated with the production of necrotic lesions, but in *Primula obconica,* which is a symptomless carrier, no true systemic infection occurs and virus multiplication is largely confined to cells immediately adjacent to entry points (BAWDEN and KASSANIS, 1946).

Because viruses most often produce systemic infections, symptoms tend to occur more or less uniformly over the whole plant. Consequently the clinical picture presented by leaves of virus-infected plants is more often likely to be confused with pathological conditions produced by toxins, deficiencies of mineral nutrients or damage at or below ground level, than with infections by other kinds of pathogens, the effects of which are usually localised. For example, the symptoms of poisoning by copper (CALDWELL, 1935) and chlorate (OWEN, 1937) in some plants closely resemble mottling caused by virus infections. In tomato plants an overdose of the growth hormones now widely used for setting fruit, can produce deformity and reduction of the leaves indistinguishable from the "fern-leaf" caused by cucumber mosaic or tobacco mosaic virus. Fumigation with tetrachlorethane, or damage at the base of the stem, causes the leaves of potato plants to roll in much the same manner as when they are infected with leaf roll virus. The necroses of potato leaves when potassium is deficient can also be confused with leaf-drop streak caused by potato virus Y, and in sugar beet the necrosis and chlorosis caused by an inadequate supply of potash, manganese and magnesium are often difficult to distinguish in the field from symptoms caused by beet yellows and mosaic viruses (HALE, WATSON and HULL, 1946). A further confusing similarity between such physiological diseases and virus infections is the absence of a visible cause, but the two are sharply separated by the transmissibility of virus diseases. Transmissibility also distinguishes virus diseases from genetical abnormalities, many of which resemble effects produced by viruses.

It has already been said that, because of the systemic nature of virus diseases, symptoms usually occur over a whole plant, but it cannot be assumed that a virus is not concerned when parts only of a plant are affected. It takes time for viruses to invade the whole of a plant, and in large trees symptoms may appear on some branches long before others. Also, leaves that are mature at the time that a plant becomes infected rarely show any symptoms, and current season infection of a well-established plant may show only at the apex or on shoots that develop subsequent to infection. Regardless of where infection occurs, the first symptoms are most usually seen on the young leaves at growing points, for viruses soon move into them and they are in a favourable condition to react. The viruses invade these leaves from the vascular system and the most frequent first symptom of systemic infection is a pallor of the parenchymatous tissue immediately adjacent to the veins of the young leaves. This clearing of the veins is a transitory symptom that is

---

*Since this was written tobacco necrosis viruses have been found causing systemic necrotic diseases in tulips (KASSANIS, B. (1949): Ann. Appl. Biol. *36,* 14) and French bean (BAWDEN, F. C. and VAN DER WANT, J. P. H. (1949): Tijdschr. Plantenziekten *55,* 142).

**Fig. 3.** — *Rumex acetosa* with wound tumour disease. Root system of old diseased plant showing many tumours of various sizes. (L. M. Black, Sixth Growth Symposium, 79, 1947).

soon replaced by more generalised reactions when the whole leaf tissue becomes invaded.

Reduction in size of plants is the general result of virus infections, but this is often accompanied by local hyperplasia. On the leaves the most common form is the production of ear-shaped or frilly outgrowths, so-called enations, which sometimes develop to such an extent from the veins that they may provide the bulk of the leaf tissue. Proliferation of lateral buds may also occur to produce the bushy type of growth known as witches' broom. In cacao trees suffering from swollen shoot there is often excessive local development of xylem tissue in stems. This may occur at any position, causing club-like protuberances at the apices or spindle-shaped swellings at any point along the stems. Local over-production of wood, mainly in the roots, also characterises wound tumour disease (BLACK, 1945). Above ground the main symptoms of this are leaf distortion, irregular enlargements of the under-surfaces of the veins and the production of enations, but the roots carry many spherical woody tumours, varying in size on different hosts from minute pustules up to bodies 1 cm. in diameter. (Fig. 3). The production of tumours is stimulated by wounding and uninjured roots form few or none (BLACK, 1946). Similar swellings on the roots are also the main symptoms of tobacco club root (VALLEAU, 1947), but few other viruses have been found to cause any clearly defined external symptoms on the roots. This applies even to some viruses that may be confined to the root system. Peach trees suffering from phony disease, for example, are dwarfed, have unusually deep green leaves and smooth bark, and produce a light crop of small fruit, but the roots seem to be the only parts of the trees to contain virus and these appear normal (HUTCHINS, 1933). Similarly, tobacco necrosis viruses often occur in apparently normal roots of plants whose leaves may or may not have shown any symptoms (SMITH and BALD, 1935; SMITH, 1937), though these viruses do produce local discolourations when inoculated artificially into roots of *Phaseolus vulgaris* (BAWDEN and KASSANIS, 1946). Black root rot of snap beans, however, is an exception, for pathological changes originating in the vascular system extend to the surface and show as black streaks (JENKINS, 1940, 1941). In some tree diseases also, the lateral roots die, and this seems to be a primary symptom of elm phloem necrosis (MCCLEAN, 1944), but in others it is a secondary symptom, the roots being starved either because the leaves are photosynthesising inadequately or normal food transport is interrupted. An interesting example of the last effect is given by the decline or tristeza disease of citrus (FAWCETT and WALLACE, 1946), which presents several unusual features. The disease occurs only when sweet orange scions are grafted on to sour orange stocks and the trees apparently die from a root rot, without showing any of the external symptoms customary with virus infections. However, the rotting of the roots is only a secondary effect, following on the failure of food to pass from the scion to stock, because of the death of the phloem cells at the scion-stock junction. It seems that the sweet orange is a carrier of a virus to which the sour orange is hypersensitive. When the scion becomes infected it shows no symptoms, but the passage of the virus into the stock causes local necrotic reactions, which interfere with the passage of food so that the whole tree rapidly dies.

Symptoms on flowers are frequent; the commonest are reduction in size and deformation, but the most striking are the colour changes leading to variegation or breaking (Fig. 4). This occurs in many different ornamental species, but is most well known in tulips, where infection often greatly enhances the beauty of the blooms. Seedling tulips have self-coloured flowers, but virus infection alters the normal distribution of anthocyanin in the petals, which become splashed or striped with different colours. According to McWhorter (1938) two antagonistic viruses are concerned in breaking, one of which prevents normal antho-

Fig. 4. — Breaking in wall-flower *(Cheiranthus Cheiri)*: Left-hand plant showing variegated petals caused by infection with cabbage blackring virus; right-hand plant healthy.

cyanin production and also causes a leaf mottle whereas the other adds flower colour and has no effect on the leaves, but van Slogteren and Ouboter (1941) consider that variations in host reaction and in one virus are enough to account for the many types of breaking known. Once established the condition is permanent, propagation by bulbs continuing indefinitely to produce broken blooms. The character of the flower is so altered that bulbs from infected plants were often sold as new varieties (Rembrandt, for example, is a broken Princess Elizabeth) and some have been in cultivation for more than three hundred years. Infected plants are less vigorous than their healthy counterparts, but infection clearly has no cumulative weakening effect. Only in tulip has the breaking of self-coloured flowers been exploited commercially, but it occurs in many different plant species and provides a much more specific indication of virus infection than do most leaf symptoms.

In recent years many abnormal conditions of fruits have been attributed to viruses, as increasing numbers of fruit-tree disorders have been identified as virus diseases. In stone and pome fruits symptoms range from simply a reduction in size, or early maturation, with no other appreciable change, to considerable alterations in the colour, shape, texture and flavour. Swollen shoot virus of cacao mottles the pods, reduces their size and makes them more nearly spherical than usual. The fruits of annual crops also often show symptoms. For example, infection with cucumber mosaic virus or tobacco ringspot viruses causes cucumber fruits to become mottled and develop wart-like outgrowths; a number of viruses produce chlorotic rings, spots or streaks on tomato fruits, and tomato bushy stunt virus causes ulcer-like lesions which sometimes extend far enough into the flesh to render the seeds visible (Fig. 5).

**Type Diseases:**—There are commonly three distinct phases during the course of a virus disease. There is first the development of a local lesion at the site of infection. This is followed by the development of systemic symptoms, which usually appear first on leaves surrounding growing points, and the early systemic symptoms often differ strikingly from those shown later. However, no universal rule can be laid down, for variability is the normal with viruses and the same virus may do different things in different plants. The degree of variation in type and sequence of symptoms caused by viruses in different hosts is probably best indicated by describing a few selected examples.

First, tobacco mosaic virus; in tobacco this usually causes no clearly visible local lesions, though there are sometimes faint chlorotic spots. Systemic symptoms start as a clearing of the veins on the young leaves and later change to an uneven pattern of various shades of green or yellow, often accompanied by leaf deformation. In tomato, there are no local lesions and systemic symptoms consist of a light-green or yellow mottling and some distortion of leaf shape. In *Nicotiana glutinosa*, by contrast, only local lesions, black necrotic spots up to 0.5 cm. in diameter, are produced, and uninoculated parts of the plant remain uninfected and symptomless.

The production of necrotic local lesions is not always a safeguard against systemic infection. For example, potato virus Y, the cause of leaf-drop streak and rugose mosaic in Majestic, President and many other potato varieties, produces black, necrotic spots or streaks on inoculated leaves, but these are only preliminary to the development of two kinds of systemic reaction. The uppermost leaves of the plants develop a blotchy veinal mottle and become crinkled, while black necrotic stripes appear on the veins of the lower leaves. The necroses spread along the undersides of the veins and also involve the leaf parenchyma; they cause the petiole to collapse, so that the leaves fall and shrivel but usually remain suspended from the main stem. From being restricted at first to the lower leaves, the necrosis and leaf-fall affects successively higher leaves until the whole plant may be involved and killed. More often, however, the youngest leaves do not become necrotic, but show only mosaic, waving and ruffling. This is also the condition in subsequent years of infection, that is, when new plants are raised from infected tubers. There are now few or no black necroses and little leaf drop. Instead, plants are dwarfed, tend to sprawl untidily and mature early, while the leaves are small, crinkled and twisted. In other potato varieties, for example, Arran Victory and Arran Pilot, virus Y causes quite different effects. Inoculated leaves show no lesions, and systemic symptoms, in both the first and subsequent years of infection, are those of a mild crinkle, with some dwarfing of the plant while the leaves are mottled and waved. Different symptoms and different sequences of events are

FIG. 5. — Fruit from plant infected with tomato bushy stunt virus, showing ulcerous lesion.

produced by infecting species other than the potato. In *Lycium barbarum* (DENNIS, 1938) virus Y causes only necrotic local lesions. In tobacco, on the other hand, it usually causes no local lesions and systemic symptoms pass through two phases; about a week after inoculation the young leaves show clearing of the veins, but this is later replaced by an interveinal mottle and the formation of green bands of tissue along the veins.

There are similar differences between the diseases caused by potato virus X in different hosts. This is the commonest of all potato viruses and is widespread in commercial stocks of many common British varieties and in all American ones. In such varieties, for example, Majestic, Up-to-Date and Green Mountain, it may be carried or produce a faint mottle, but other varieties, such as King Edward, Epicure and Arran Crest, are hypersensitive and rarely occur naturally infected. When leaves of hypersensitive varieties are inoculated, they develop necrotic local lesions, which is often the only result, but sometimes infection is not localised, presumably because the cells do not die rapidly enough to prevent the virus moving from the initially infected cells. When this happens, the plants are killed by a generalised necrotic disease, top-necrosis. A fortnight or so after infection, many minute black necroses appear on the leaves surrounding the growing point; these increase in size and involve the stem so that the plants die back from the top and may be killed within a month of infection. Many of the tubers set by such plants show deep fissures or ulcers. The eyes become necrotic and most fail to sprout; those that do, give rise either to healthy plants or to ones that are smothered with necrotic spots and soon die. In tobacco plants virus X produces strikingly different effects. Inoculated leaves develop local lesions consisting of a central white necrotic spot surrounded by a variable number of white concentric rings. Systemic symptoms also first consist mainly of necrotic rings and spots, but the severe reaction is usually a transitory phase to be replaced by mosaic symptoms, which themselves become less bright as the plant ages.

A reduction in severity of symptoms with increasing length of infection is common to many virus diseases. It seems that healthy cells are in a condition in which they react violently to the virus whereas tissues that have been long infected are better able to tolerate its presence. This recovery is probably shown most strikingly by species of *Nicotiana* when infected with tobacco ringspot virus (PRICE, 1932). Large necrotic spots and rings develop on the inoculated leaves within a fews days, and a little later the young leaves also become severely necrotic. The first reactions to systemic infection are so violent that affected plants look as though they could not survive. However, young leaves put out subsequently show progressively fewer lesions, so that after a few weeks new leaves are being produced that would pass as healthy, though they are infected. Cucumber plants show much the same type of recovery; soon after infection with tobacco ringspot virus, they become severely mottled and very necrotic, but new leaves formed after infection show few or no such lesions. When viewed from above the leaves of plants that have been infected for some weeks may appear symptomless, though the under-surfaces often carry numbers of enations.

These few examples are probably enough to indicate the great variability of symptoms that can be produced by one virus and the consequent uncertainty in diagnosing from the appearance of infected plants. The same virus may cause a mosaic in one plant, ringspot in a second, a lethal disease in a third, and be carried by a fourth; and in one and the same plant the condition may vary during the course of infection from acute ringspot almost to the symptomless state of a carrier. Also, different viruses may cause similar symptoms in the same plant. There are some viruses, mainly those with narrow host ranges, which produce more consistent effects. Potato leaf roll, for example, takes much the same course in all potato varieties, variations being of degree rather than of kind. In the first year of infection there may be no symptoms, or the young leaves may become chlorotic and cupped, while the under-surfaces become unusually coloured. Plants raised from infected tubers are dwarfed and the lower leaves roll upward and inwards; the upper leaves are pale and more erect than normal, but do not usually roll. Similarly, the known hosts of sugar beet yellows virus all react in much the same manner. Yellow or orange-yellow patches at the infection sites are the first symptoms, which are followed by vein clearing of the young leaves and this in turn is replaced by a general yellowing of the leaves. Because of their narrow host ranges and the relative constancy of the symptoms caused, the names of these two viruses have some significance, but the names of many others are merely historical accidents, which serve as convenient tags but as little else. Such viruses as tomato spotted wilt and cucumber mosaic have host ranges extending to hundreds of plants in many different families, including both monocotyledons and dicotyledons, and in some host or other these two viruses cause most of the different types of symptoms yet recognised.

**Virus Strains:** — In the preceding section, for the sake of simplicity when describing diseases, several assumptions were made, of which perhaps the least justified was to treat named viruses as though they are single, invariable entities, which always cause the same clinical picture in any one host. The description of a single disease as typical of infection by, say, either tobacco mosaic virus or potato virus X is an over-simplification that is liable to be more confusing than helpful in diagnosis. Such descriptions apply only to the effects caused by a specific strain of the named virus, and each is made up of many strains which individually may cause widely differing symptoms in the same host. For instance, there are many strains of tobacco mosaic that do cause the kinds of symptoms previously described for infected tobacco and tomato, but there are many others that do not. Some strains produce only local lesions in tobacco, others cause systemic necrosis, brilliant yellow mottle, or acute deformity with little mottling. Similarly in tomato, some strains of tobacco mosaic virus cause predominantly necrotic symptoms (streak), others produce mosaic patterns varying greatly in brilliance, while the main effect of others is to cause fern-leaf or to produce enations. There is not necessarily any agreement between the symptoms caused by individual strains in the two hosts; one producing a yellow mosaic in tomato may cause only local lesions in tobacco, and one causing a mild mosaic in tobacco may produce enations in tomato.

Similarly with potato virus X, the symptoms described in the preceding section do not apply to most of the known strains. In tobacco plants, the symptoms produced by different strains vary from the severe ringspot type, through mosaic diseases of varying degrees of severity, to barely perceptible effects on the colour of the leaves or vigour of the plants (SALAMAN, 1938). In one potato variety such as Majestic, some strains are carried, some cause mild mosaic, others bright mottlings, sometimes accompanied by severe necrotic spotting and crinkling of the leaves, and still other strains cause the lethal disease top-necrosis. To indicate such differences in behaviour it is usual to say that the strains differ from one another in virulence, the words virulent and avirulent often being used as though they were fixed intrinsic properties of individual strains. But this is rarely justified. Virulence is a reflection of host-parasite interaction and cannot strictly be applied to a strain without simultaneous reference to a specific host plant. One strain may cause a serious disease in one host, but not in another, and whether a strain is regarded as unusually virulent or a host as unusually susceptible is often a matter of convenience or of terminology rather than a true distinction. The behaviour for different strains of potato virus X illustrate this point vividly. By their reactions in tobacco plants the strains can be graded in order of increasing virulence from those that cause faint mottles to those that cause crippling ringspot diseases, but this order will not necessarily still apply when the strains are transferred to potato varieties. Of two strains that differ in the reactions produced in tobacco, the more virulent may also cause a more severe disease in potato, but it is almost equally likely that it will not. Some strains that cause faint mottling in tobacco cause top-necrosis in potato, and some that cause ringspot in tobacco may cause only a mosaic in potato. However, the symptoms resulting from infection depend so much on individual reactions between host and parasite that to state that a strain of virus X causes a certain type of symptom in potato is largely meaningless unless the variety of potato is also specified. The same strain of virus X may affect different varieties differently, and a strain that is virulent towards one variety may be avirulent towards another. For example, there are strains that produce only a faint inter-veinal mottle in Majestic and kill President with top-necrosis, and there are others that reverse this procedure (BAWDEN and SHEFFIELD, 1944).

Much the same range of variation also occurs with strains of potato virus Y (BAWDEN and KASSANIS, 1947), for its description in the preceding section as a virus that causes leaf-drop streak in Majestic and mosaic in Arran Victory is far from the whole story. The diseases caused by different strains in the variety Majestic alone vary from mild mosaic with no necrosis, through leaf-drop streak reactions of various degrees of severity, to a lethal top-necrosis. In the variety Arran Victory, on the other hand, all strains of virus Y so far tested produce mosaic symptoms only, the reactions to different strains differing in degree rather than kind. Similarly, all the strains common in Britain, whether they cause mosaic, leaf-drop streak or top-necrosis in Majestic, produce the same type of vein-banding symptoms in tobacco plants, although an additional strain found in South America (SMITH and DENNIS, 1940; NOBREGA and SILBERSCHMIDT, 1944) causes a necrotic disease of tobacco and only slight symptoms on the potato varieties yet infected with it.

Thus it is obvious that two strains may be indistinguishable by the reactions they cause in one host, but produce strikingly different symptoms in another. Also, the diseases caused by unrelated viruses in the same host may resemble one another more closely than those caused by strains of one virus. For example, the range of symptoms caused by strains of virus Y in potato plants overlaps that caused by strains of virus X; in the variety Majestic, only the characteristic veinal necrosis and leaf-drop streak symptoms can confidently be diagnosed as infection with virus Y. Top-necrosis or mosaic symptoms may be caused by other strains of virus Y, by virus X, or by other viruses unrelated to either. Before symptoms can be relied upon to identify the causes of such diseases, transmissions must be made to a range of suitable indicator plants that act as differential hosts.

The behaviour of potato viruses X and Y has been described in some detail because it not only illustrates the degree and kind of variation in symptoms that can be caused by strains of one virus, but also shows that unrelated viruses may cause diseases in the same host that resemble one another more closely than do diseases caused by related strains. Although all viruses may not occur in such a bewildering range of clinically distinct strains as these two potato viruses, there is no reason to consider them as exceptional. An essentially similar story could have been told of tobacco mosaic and cucumber mosaic viruses, each of which occurs in strains that produce widely different symptoms in tobacco; and the appearance of plants infected with a strain of cucumber mosaic virus may much more nearly resemble that caused by some strain of tobacco mosaic virus than that caused by other strains of cucumber mosaic virus. With only a few exceptions, for example, tomato bushy stunt, all those viruses that have been studied in any detail have been found to exist in strains differing in virulence towards certain hosts. The idea of a virus as something that produces constant effects largely derives from experimental work in which single isolates have been intensively studied and propagated continuously in the same host grown in relatively constant environmental conditions. In such conditions constancy and reproducibility of results are to be expected, but this expectation should not be extended to other isolates propagated in other hosts and in different environments. Virus literature contains many detailed descriptions of symptoms, usually first accounts of "new" viruses, in which differences from clinical pictures previously recorded are stressed as evidence of newness. How valueless such comparisons can be is evident from the number of times potato viruses X and Y, cucumber mosaic and tobacco mosaic viruses, have been identified as new viruses and given new names. Nevertheless, virus workers seem reluctant to appreciate the fact that variability is the normal, and they continue to describe with a wealth of picturesque detail symptoms that may never again be exactly reproduced.

**Effects of Environment on Symptoms:** — Even when the strain of a virus and the variety of the host plant are both specified, it is still impossible to make categorical statements about symptoms unless the environmental conditions are also given, for changes in these may greatly modify the appearance of infected plants. Variations in conditions under which plants are grown can affect their susceptibility to infection, the time taken for symptoms to develop and the severity of symptoms shown. The environment influences some virus-host interactions so profoundly that it determines whether infection is localised or systemic, and whether symptoms are produced or the plant acts as a carrier. No safe generalisations can be made, however, for different virus-host combinations are affected differently by changing conditions. Perhaps the statement applying most widely is that time for appearance of symptoms is shortened by increases in temperature and light intensity.

The early literature on virus diseases contains many statements about the effect of temperature and light intensity on virus diseases, but their significance is often uncertain. Most of these statements derive from field observations and not from experiments; often the identity of the virus concerned is far from sure, or it is uncertain whether the prevalence of the disease or the reactions of the infected plants influenced observers in deciding whether to call it severe or not. Hence many of the statements about the effect of environment on virus diseases reflect conditions that affect the activities of insect vectors rather than the susceptibility of plants to infection or their reactions to infection.

In latitudes where there are considerable differences in climate between the seasons, ample evidenc of the great effects of environment on virus diseases can be obtained merely by maintaining different viruses throughout the year. In the glasshouses at Rothamsted, for example, some viruses produce more severe symptoms during summer and others during winter. The first seem to be largely those that cause the accumulation of carbohydrates in the leaves, such as potato leaf roll and sugar beet yellows; during summer infected plants show clear and unmistakable symptoms, but under the reduced light intensity of winter infected plants may be indistinguishable from uninfected controls. On the other hand, most of the viruses that cause mosaic symptoms are more virulent during the winter. Strains of potato virus X that produce only a faint mosaic in tobacco seedlings inoculated during summer may cause brilliant mottles or ringspot during the winter; similarly, strains of cucumber mosaic, henbane mosaic, and tobacco etch viruses, which cause moderately severe mosaic symptoms during summer, may kill tobacco seedlings inoculated in the winter. Symptoms take longer to appear in plants inoculated during winter, but transmission is easier and the host reactions are more acute.

An extreme example of seasonal effect on reaction is shown by *Phaseolus vulgaris*, var. Bountiful, and cucumber mosaic virus. In the glasshouses at Rothamsted from November to February, the bean leaves are readily infected and react by producing discrete, black, necrotic, local lesions, whereas plants inoculated during summer produce no symptoms and seem to be immune from infection. Seasonal variations in local reactions of tobacco plants also occur with cucumber mosaic and various other viruses, such as potato Y, henbane mosaic and tobacco etch. Strains of these viruses that during most of the year produce no local lesions sometimes do so during spring and autumn. Tobacco etch virus, for example, usually produces local reactions that become visible only when leaves are decolourised and stained with iodine, but sometimes it causes diffuse chlorotic spots or necrotic rings. It is unknown whether temperature, light intensity or length of day, is primarily concerned in influencing the local reaction. Only one host reaction seems to have been attributed definitely to an effect of photoperiodism. McKINNEY (1937) finds that wheat infected with mosaic virus shows rosette symptoms only in short days; a temperature around 16°C. and a daylight period of 8 hours are optimal for the rosette phase of the disease, longer day length or higher temperature suppressing this symptom and leaving only mosaic.

Two other virus diseases that are greatly affected by environment and favoured by winter conditions are tobacco necrosis and tomato bushy stunt. In naturally-infected plants symptoms of tobacco necrosis are restricted to the old leaves in contact with the soil, and consist of necrotic lesions that spread until the whole leaf may be involved. If infected early the loss of the first formed leaves is sufficient to kill many tobacco seedlings. During the months November to February the disease is common under glass, but plants are rarely attacked during summer (SMITH and BALD, 1935). Tobacco leaves inoculated artificially with the virus develop white necrotic rings and spots and do not become systemically infected. During the summer relatively few, scattered lesions are produced by the same inoculum that during winter will produce so many lesions that they coalesce and cause the whole leaf to collapse. Similarly, in tomato plants during summer tomato bushy stunt virus causes a relatively mild disease, the symptoms consisting of isolated necrotic and chlorotic blotches with some twisting of the petioles, whereas in the winter it will rapidly kill plants of the same age. The main factor determining the seasonal reactions of plants to infection with these two viruses is light intensity, and BAWDEN and ROBERTS (1947) have shown that the typical winter reactions can be produced during summer if plants are raised under appropriate shade.

With other viruses the effect of light intensity on susceptibility to infection has yet to be tested, but there are a number of observations on the effects of shading on the severity of symptoms shown by infected plants. Various workers have stated that variegated *Abutilon* plants produce uniformly green leaves if grown in the shade, and the severity of certain other diseases has also been found to be reduced by shading. Tomato plants in the field usually die when infected with sugar beet curly top virus, but SHAPOVALOV and LESLEY (1931) found that shaded plants tolerate infection sufficiently well to produce some crop. During summer tomato aucuba mosaic virus produces a bright yellow mottle in tomato, but during winter the mottling is indistinct and the main symptom may be leaf deformity, so that infected plants show the fern-leaf condition usually produced by cucumber mosaic virus (AINSWORTH, 1935). Similarly, during summer cucumber viruses 3 and 4, two strains of tobacco mosaic virus that infect cucumber, are readily differentiable because the first produces a mild green mottling and the second a bright yellow mottle, but during winter the two cause identical symptoms. KNIGHT (1941) has found that cucumber virus 4 and tomato aucuba mosaic virus both produce their bright mottling during winter if plants are given extra light; infected plants showing only faint mottles develop yellow lesions when illuminated for a few days by 500 or 1000 watt bulbs.

BEST (1936) found that sunlight inhibits the production of local lesions in tobacco by tomato spotted wilt virus, continuous illumination reducing the number formed and delaying the onset of symptoms. He found no such effect with tobacco mosaic virus in *Nicotiana glutinosa*, and suggested that the more unstable tomato spotted wilt virus is inactivated by some photo-chemical reaction before it develops sufficiently in the host to cause visible lesions. BAWDEN and ROBERTS (1947, 1948) also found that the light intensity to which plants are exposed after inoculation has little effect on the numbers of local lesions produced by tobacco mosaic and tomato bushy stunt viruses on *N. glutinosa* or by a tobacco necrosis virus on tobacco or *Phaseolus vulgaris*. The illumination of the plants before inoculation, on the other hand, has large effects, the number of lesions produced by all three viruses being greatly increased by placing the plants for periods of 24 to 72 hours in the dark, or by shading for longer periods (Fig. 6). The treatment does not alter the type of lesions produced but only the number, the increased susceptibility presumably occurring because some products of photosynthesis are removed which normally accumulate in the leaf cells and interfere with the establishment of infection.

In most conditions light intensity and temperature fluctuate together, and from observations in the field it is often impossible to decide which variant, if either, is responsible for the changes in appearance of infected plants. There are, however, ample experimental data to show the importance of temperature in determining the kind of reactions of some plants to viruses. In general, increasing the temperature at which plants are grown reduces the severity of the symptoms shown, the so-called "heat-masking", but the optimum temperature for expression of symptoms, and the temperature range over which obvious symptoms are produced, varies considerably with different diseases. Potato viruses A and X produce crinkle in many potato varieties at temperatures around 16°C., but as the temperature is increased the symptoms fade, and above 20°C. there may be little obviously wrong with the plants; similarly, the mosaic symptoms caused by virus X alone may be obvious below 20°C. but undetectable at higher temperatures. By contrast, there is no diminution of the symptoms caused by virus Y in potatoes when the temperature is raised from 16° to 20°C.

FIG. 6. — Leaves of *Nicotiana glutinosa* showing effect of shade in increasing susceptibility to tomato bushy stunt virus. *Left:* leaf from shaded plant. *Right:* leaf from unshaded plant. Left-hand half-leaves inoculated with virus at 10 mg./l. and right-hand halves with virus at 1 mg./l.

Tobacco plants with mosaic also show increasingly milder symptoms as the temperature is raised; if the temperature is not too high and the exposure too long, the plants regain their usual symptoms when returned to lower temperatures. When symptoms are masked by exposure to 35°C. for long periods, however, the plants continue to show only mild symptoms when grown at the normal temperature of 15° to 20°C. Not only do the treated plants continue in this way, but plants infected from them and kept continually at the lower temperatures also show only mild symptoms, the treatment having selected a less virulent strain of the virus (HOLMES, 1934). Tobacco mosaic is also masked by low temperatures, and below 10°C. infected plants show few or, no symptoms (GRAINGER, 1936). The effects of tobacco mosaic virus on *Nicotiana glutinosa* also depend greatly on temperature; at 20°C., 30 per cent. more lesions are produced than at 15°C., and the lesions also appear more quickly (BEST, 1936). Raising the temperature also alters the type of symptom; at low temperatures, the local lesions are small and wholly necrotic, but with increasing temperature the lesions increase in size. Above 28°C., although still necrotic the lesions are less localised and tend to run together instead of being discrete, and at 35°C. or higher there is no necrosis, the inoculated leaves showing only chlorotic blotches, and systemic infection occurs (SAMUEL, 1931).

Some of the most detailed work on the effect of temperature on symptoms has been done with viruses that infect cruciferous plants, and the results suggest that variations in the conditions under which different workers have grown their test plants may account for many differences that have been attributed hitherto to infection with separate viruses. POUND and WALKER (1945, 1945*a*), working with cabbage black ring and cauliflower mosaic viruses, found that the symptoms caused by the former in cabbage increase with increasing air temperature and are most severe at 28°C. and least at 16°C., whereas increasing temperature over this range decreases the symptoms caused by cauliflower mosaic virus, with which masking occurs above 24°C. Different strains of cabbage black ring virus differ in the temperature at which they produce maximum effects, one strain being more virulent towards cabbage than another at 28°C. but less at 20°C. The effects of varying temperature differ with different hosts; in contrast to the behaviour of infected cabbage, *Nicotiana glutinosa* reacts more severely towards the cabbage black ring virus at 16-20°C., and strains of the virus that can be readily differentiated by the symptoms they cause in *N. glutinosa* at higher temperatures produce identical symptoms at lower temperatures. The symptoms produced by cabbage black ring and cauliflower mosaic viruses in cabbage are correlated with the virus concentration reached in infected plants, the multiplication of black ring virus being increased by increasing temperatures from 16° to 28°C. and that of cauliflower mosaic virus being decreased.

When peach trees suffering from yellows, rosette, little peach, and red suture are grown at 35°C. they recover, not because of masking of symptoms, but because the viruses are inactivated. Leaves already showing symptoms apparently continue to do so, although free from virus, but new growth produced after treatment is normal (KUNKEL, 1936). The length of time required to cure trees depends on the temperature to which they are exposed and varies with the size of the trees. Much longer treatment is needed to free the roots of potted plants from virus than to free the tops. KUNKEL suggests that the geographical distribution of peach yellows in the U.S.A. may result from this susceptibility to heat. The disease is prevalent in the north, but is not found in the Southern States, where temperatures of around 35°C. prevail for long periods, except in the mountainous regions of Virginia where summer temperatures are relatively low. The occurrence of yellows virus in the roots but not the tops of infected peach trees that have been exposed to heat for short periods is similar to the distribution of virus in peach trees suffering from phony disease (HUTCHINS, 1933). Phony disease is common only in the Southern States and HUTCHINS worked in Georgia where the summer temperature is high. KUNKEL makes the interesting suggestion that the apparent restriction of virus to the roots of phony trees may not necessarily mean that the virus cannot infect aerial parts, but that it is destroyed by the high temperature and can persist only in the cooler roots.

Potato yellow dwarf is a striking exception to the generalisation that high temperatures decrease the severity of symptoms (WALKER and LARSON, 1939). Below 16°C. infected plants may be symptomless, but raising the temperature leads to the rapid development of severe symptoms, and high soil temperature may prevent the sprouting of infected tubers which grow normally at lower temperatures. The effects of soil conditions on other virus diseases have been little studied, but the seasonal occurrence of symptoms in strawberry plants with yellow edge depends on soil moisture and temperature (KING and HARRIS, 1942). Symptoms are obvious only when there is adequate moisture and the air temperature is above 16°C., so that infected plants are usually most easily detected in the field during spring and autumn. Excessive watering or a high water-table increase both the prevalence and severity of mosaic in tomato plants, whereas symptoms are suppressed by drought, nitrogen deficiency, or excessive dressings of soluble nitrogen (SELMAN, 1947).

The growth of plants is so profoundly influenced by soil conditions that it can safely be assumed that these also exert a powerful effect on virus diseases, and there are many statements in the literature relating the severity of certain diseases with certain soil types or nutrition of the host plant. The precise significance of many observations, however, is uncertain, for there has been little experimental work and it is often difficult to decide whether the particular soil conditions have been affecting symptom expression or the prevalence of the disease. The nutrition of the host may affect reaction to infection, susceptibility to infection and, in the field, the chances of infection, by creating conditions more or less favourable for the activity of vectors. Disentangling the various factors is often impossible and may be difficult even in controlled experiments. For example, BRIERLEY and STUART (1946) find that increasing the nitrogen level of manuring for onions increases both their susceptibility to infection with yellow dwarf virus and the severity of symptoms produced. A direct comparison of the numbers of plants showing symptoms at high and low levels of nitrogen suggests that nitrogen has a greater effect than it really has in increasing susceptibility to infection, for at low nitrogen levels some infected plants will show no symptoms and so be passed as uninfected. In general it can be said that the more favourable the soil conditions are for growth of the host, the more definite will be the reactions of infected plants. On the other hand, susceptibility to infection may not increase as conditions for growth improve; nor may conditions become increasingly favourable for vectors, though, other things being equal, larger plants are more likely to become infected than smaller ones, and, once infected, also supply a more abundant source of virus to infect other plants. The interactions of these factors call for much more study than they have so far received and promise to be a fertile field for future research.

Before leaving the subject of variables that affect the reaction of plants to viruses, a few words should be said on the condition of the host plant at the time of infection. Almost invariably, the more vigorously a plant is growing at the time of infection, the more pronounced will be the symptoms and the more generally will they occur over systemically infected plants. Leaves that have ceased to grow rarely show symptoms unless they are actually inoculated, and localisation of infection often becomes increasingly common as plants age. Strains of potato virus X that readily infect young tobacco seedlings and cause a systemic ringspot, produce only local lesions in older, more slowly growing plants. Similarly, even in the summer, tomato bushy stunt virus becomes systemic in young tomato seedlings and causes a severe disease, whereas slightly older plants react less severely, systemic symptoms consisting of scattered yellow or necrotic blotches, and plants a foot or more high often produce only local lesions. In *Nicotiana glutinosa* both susceptibility to infection with bushy stunt virus and the type of local lesion are profoundly affected by age of the leaf. The young leaves produce many local lesions, necrotic circles up to 0.5 cm. diameter with brown centres surrounded by black rings, whereas the oldest leaves produce very few lesions in the form of small, diffuse chlorotic spots; the lesions on the middle leaves are intermediate in number and type, consisting of a small necrotic centre with a chlorotic halo. The physiological age rather than the actual age is the determining factor, and any condition that simulates old age in a plant may lead to an increased resistance to systemic infection. An older plant that is still green and actively growing is more likely to become systemically infected with potato virus X than a younger plant that is maturing because it is pot-bound or nitrogen-deficient. The greater susceptibility of plants to tomato bushy stunt virus in winter, or when raised under shade in summer, is also correlated with the physiological immaturity of the tissues, for exposure to brighter light causes plants to become mature, tough, and nitrogen-deficient much more rapidly.

At first sight the statement that young plants are, in general, more susceptible to virus diseases than old ones, conflicts with the often repeated idea that virus diseases are in some way connected with senility, but there is no necessary contradiction. The connection with senility has been noticed mainly with vegetatively propagated plants or in plantation and orchard crops, and with these it merely means that the older a variety is, or the longer a crop has stood, the more chances it has had of becoming infected and of accumulating an increasing number of viruses. The apparent connection of virus diseases with increasing age in such plants is no evidence that they have become increasingly susceptible; indeed, when comparative tests have been made it has been found that young seedlings are more susceptible than older plants, even towards viruses like swollen shoot of cacao, which, under natural conditions of spread, occur at all commonly only in long-established crops (POSNETTE, 1947).

A further condition of the host that may be all-important in determining its reactions to a virus is whether or not it is virus-free, for the presence of one virus can greatly modify the symptoms produced by another. If the two viruses are related, then inoculation with the second will usually produce no effect; for example, tomato plants infected with a strain of tobacco mosaic virus that causes only a faint mottle, will not develop any characteristic yellow mottling when re-inoculated with aucuba mosaic virus. On the other hand, if the two viruses are unrelated, the inoculation with the second may produce a disease quite different from that caused by either alone. Strains of virus X that alone cause only mild mosaic in the potato variety President, will cause crinkle if the host plants are infected with virus A, which alone the plants carry without showing symptoms (MURPHY and MCKAY, 1932). Here these factors merely need mention, for they are dealt with in more detail in later chapters.

*References:-*

AINSWORTH, G. C. (1935) : Ann. Appl. Biol. *22*, 55.
BAWDEN, F. C. and KASSANIS, B. (1946) : Ann. Appl. Biol. *33*, 46.
—— and —— (1947) : *ibid. 34*, 172.
—— and ROBERTS, F. M. (1947) : Ann. Appl. Biol. *34*, 286.
—— and —— (1948) : *ibid. 35*, 418.
—— and SHEFFIELD, F. M. L. (1944) : Ann. Appl. Biol. *31*, 33.
BEST, R. J. (1936) : Austral. J. Exp. Biol. Med. *14*, 223.
BLACK, L. M. (1945) : Amer. J. Bot. *32*, 408.
—— (1946) : Nature *158*, 56.
BRIERLEY, P. and STUART, N. W. (1946) : Phytopath. *36*, 297.
CALDWFLL, J. (1935) : Ann. Appl. Biol. *22*, 465.
DENNIS, R. G. W. (1938) : Nature *142*, 154.
FAWCETT, H. S. and WALLACE, S. M. (1946) : Calif. Citrogr. *32*, 50.
GRAINGER, J. (1936) : Nature *137*, 31.
HALE, J. B., WATSON, M. A., and HULL, R. (1946) : Ann Appl. Biol. *33*, 13.
HOLMES, F. O. (1934) : Phytopath. *24*, 845.
HUTCHINS, L. M. (1933) : Office State Entom. Georgia Bull. *78*, 1.
JENKINS, W. A. (1940) : J. Agr. Res. *60*, 279.
—— (1941) : *ibid. 62*, 683.
KING, M. E. and HARRIS, R. V. (1942) : J. Pomol. *19*, 212.
KNIGHT, C. A. (1941) : Arch. ges. Virusforsch. *2*, 26.
KUNKEL, L. O. (1936) : Phytopath. *26*, 809.
MCCLEAN, D. M. (1944) : Phytopath. *34*, 818.
MCKINNEY, H. H. (1937) : U. S. Dept. Agr. Cir. 442.
MCWHORTER, F. P. (1938) : Ann. Appl. Biol. *25*, 254.
MURPHY, P. A. and MCKAY, R. (1932) : Sci. Proc. Roy. Dublin Soc. *5*, 227.
NOBREGA, N. R. and SILBERSCHMIDT, K. (1944) : Arq. Inst. Biol. S. Paulo *15*, 307.
OWEN, O. (1937) : J. Min. Agric. *44*, 866.
POSNETTE, A. F. (1947) : Nature *159*, 500.
POUND, G. S. and WALKER, J. C. (1945) : J. Agr. Res. *71*, 255.
—— and —— (1945*a*) : *ibid. 71*, 471.
PRICE, W. C. (1932) : Contrib. Boyce Thompson Inst. *4*, 359.
SALAMAN, R. N. (1938) : Phil. Trans. Roy. Soc., *B*, *229*, 137.
SAMUEL, G. (1931) : Ann. Appl. Biol. *18*, 494.
SELMAN, I. W. (1947) : J. Pomol. *23*, 50.
SHAPOVALOW, M. and LESLEY, J. W. (1931) : Phytopath. *21*, 83.
VAN SLOGTEREN, E. and OUBOTER, M. P. (1941) : Meded. der Landbouwhoogeschool, Wageningen, Deel 45.
SMITH, K. M. (1937) : Parasitology *29*, 86.
SMITH, K. M. and BALD, J. G. (1935) : Parasitology *27*, 231.
—— and DENNIS, R. G. W. (1940) : Ann. Appl. Biol. *27*, 65.
VALLEAU, W. D. (1947) : Phytopath. *37*, 580.
WALKER, J. C. and LARSON, R. H. (1939) : J. Agr. Res. *59*, 259.

*Chapter 3*

# INTERNAL SYMPTOMS OF INFECTED PLANTS

Internal changes of two kinds are common in plants suffering from virus diseases. The first kind is the destruction or modification of normal tissues or cell contents, and the second is the production of peculiar inclusion bodies not found at all in healthy cells. The second is the more characteristic, for similar bodies have not been found accompanying any infectious diseases except those caused by viruses. They have been seen in both animals and plants suffering from a number of different virus diseases, and their etiological significance has been a subject of continual speculation.

It is not doubted that they are a direct result of virus infection, but some workers have regarded them as stages in the life history of causative organisms and others merely as masses of coagulated plant protoplasm. Workers on animal viruses have adopted a rather different view; using ultra-violet light microscopes of high resolving power they have shown that some inclusions are aggregates of elementary bodies of approximately the same size as that estimated for the virus particles themselves by filtration experiments. The inclusions in infected animals are therefore regarded as visible aggregates of virus particles, although there is as yet no adequate explanation for the aggregation in the infected cells of particles which form stable suspensions *in vitro*.

Although the intracellular inclusions appear to be specific to virus infections, occurring in some with sufficient frequency to be of considerable diagnostic value, they have not been found accompanying all virus diseases. Their production depends more upon the infecting virus than upon the host plant infected; for example, they have been found in a large number of species infected with tobacco mosaic virus but in none of the same species infected with cucumber mosaic virus. The host plant, however, plays some part, the formation of inclusions being correlated with external symptoms. A virus may produce many inclusions in all species in which it causes mosaic symptoms, but it is unusual for it to produce them in hosts in which it is carried or causes only necrosis. The presence of inclusion bodies in diseased plants is, therefore, strong evidence that the cause is a virus, but their absence cannot be taken as evidence to the contrary. More especially is this so, as inclusions often persist only for a part of the time during which affected plants show symptoms.

**Intracellular Inclusions Formed by Tobacco Mosaic Viruses:** — Most work on intracellular inclusions has been done with strains of tobacco mosaic virus, which was the first found to cause their production. In 1903 IWANOWSKI described two kinds of inclusions in the cells of plants suffering from tobacco mosaic, one, amoeboid-like, consisting of amorphous material, and the other flat, crystalline plates. These observations have been repeatedly confirmed, but interest has centred chiefly on the amorphous type of inclusion, for at first sight this differs more definitely from the constituents of normal cells. Crystalline material is common in both healthy and infected plants, and the plate-like inclusions, although recognised as peculiar to the infected tissues, have been unanimously regarded in the past as merely reaction products of the cell to the disease, with no etiological significance. GOLDSTEIN (1924, 1926) described both inclusions in great detail, and aptly summarised the conflicting views on the nature of the amorphous inclusions by naming them X-bodies.

The inclusions are most abundant in the epidermal and hair cells and are conveniently examined in living preparations made by stripping the epidermis from the leaf. The X-bodies resemble compact masses of cytoplasm. Typically they are rounded or oval, but they may be quite irregular, for as they are carried round the cell by the streaming cytoplasm they readily change their shape on coming in contact with the cell wall or nucleus. They vary in size considerably; in large cells the diameter may be as much as $30\mu$, in small ones as little as $5\mu$. They are partially translucent and granular, and usually contain one or more clearly visible vacuoles. GOLDSTEIN claims to have seen external membranes, but HENDERSON SMITH (1930) and SHEFFIELD (1934) were unable to confirm this.

The X-bodies formed by the different strains of tobacco mosaic virus differ slightly in appearance. For example, those formed by aucuba mosaic virus in *Solanum nodiflorum* are granular and more sharply differentiated from the cytoplasm than the bodies formed by tobacco mosaic virus. By making a cinematograph of cells of *S. nodiflorum* infected with aucuba mosaic virus, SHEFFIELD (1931) has determined the sequence of events leading to the formation of X-bodies. Soon after the cells become infected the rate of cytoplasmic streaming is increased and minute granules appear in the cytoplasm. These are carried passively around the cell, and when two particles meet, they fuse to form larger bodies. The process of fusion continues until all the particles are contained in one, or occasionally two, granular masses. These masses then become rounded and vacuolate, and appear less granular. They frequently contain mitochondria and oil globules, which are acquired during the circulation around the cell. The fully formed X-body and the nucleus are often found in contact (Fig. 2), but the association is accidental. It is a consequence of the tendency of circulating particles to accumulate at the junction of cytoplasmic strands, several of which usually converge on the nucleus.

SHEFFIELD's results clearly showed that the X-body is neither a breakdown product of the nucleus nor an organism of the amoeboid type. The latter claim was often made by previous workers who stated that the body possessed the power of autonomous movement, grew in size, put out pseudopodium-like projections, or divided by fission. The bodies move about no more than any other of the cell contents such as nuclei, and the other effects claimed are satisfactorily explained by the mode of formation described. A small aggregate of particles joining up with a larger may present an appearance simulating pseudopodia. Also, two large masses may come together but fail to fuse completely, and the two may later separate again. Both the fusion and the separation will give appearances superficially resembling multiplication by division.

Although SHEFFIELD was unable to see any signs of an external membrane to the bodies, she has obtained results suggesting that there may be one. She removed the intact bodies from infected cells and subjected them to various osmotic pressures. In salt solutions of 0.1 M the body persists indefinitely, at 0.07 M it disintegrates slowly, while below 0.05 M disintegration occurs rapidly. The bodies also disintegrate when mechanical pressure is applied to the cells containing them or when they are pricked with a micro-needle. With these treatments they break down into particles or granules, morphologically indistinguishable from those that fuse together to form the body, which are soon dispersed throughout the cell (Fig. 7).

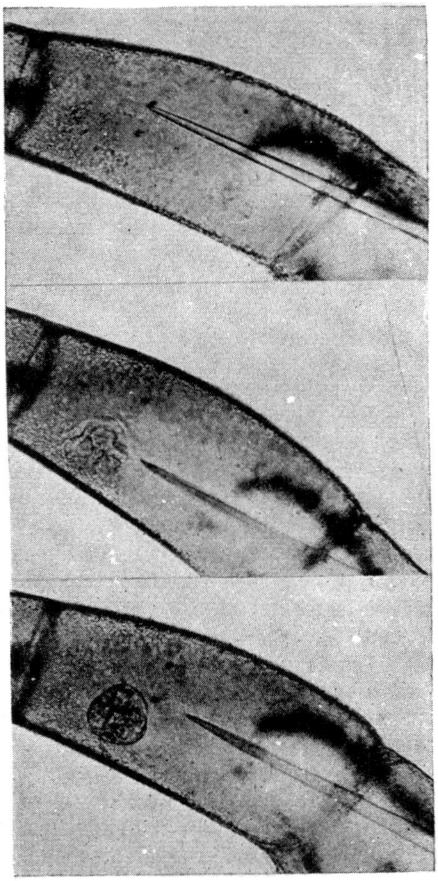

FIG. 7. — Cell from hair of *Solanum nodiflorum* infected with aucuba mosaic virus. *Below*, a needle has been inserted through the transverse septum into the cell, leaving the inclusion body unaffected. *Centre*, this photograph was taken 15 seconds after the body was punctured by the needle. The body is rapidly changing its consistency, forming a mass of clear bubbles in which a few granules are visible. *Above*, photographed 10 minutes later. The inclusion has almost disappeared, only a few scattered granules being visible. × 300. (SHEFFIELD, F. M. L., 1939, Proc. Roy. Soc. B, *126*, 529).

The X-body does not remain unchanged for long; two to four weeks after it is formed, crystals appear inside it and gradually it loses its amorphous character and becomes wholly crystalline. The crystals produced in this way do not seem to differ in any essential manner from the crystalline inclusions described below, which are formed without the intervention of the X-body (Fig. 8).

Although the isolated X-bodies are susceptible to changes in osmotic pressure, they resist *p*H changes. They are unaffected by the addition of acid to *p*H 2 and are preserved by the ordinary fixation methods. Microchemical tests have been made on them, and they give all the usual protein colour reactions. Apart from occasional oil globules enclosed, they do not give any positive test for fats. They are sharply differentiated from the nucleus when stained by Feulgen's reagent, for the nuclei take on a deep red colour, while the bodies are unaffected (Fig. 2).

The X-bodies produced by other strains of tobacco mosaic virus look slightly different, but give identical colour reactions. They also break down when pricked with a micro-needle (LIVINGSTONE and DUGGAR, 1934) and are unaffected by weak acids. There is no reason to think that they differ significantly from those produced by aucuba mosaic virus, but SHEFFIELD (1934) was unable to follow their method of formation and disappearance as she did with the latter.

The crystalline inclusions differ in many ways from the X-bodies and similar types are formed by many different strains of tobacco mosaic virus. They may occur either alone or in the same cell as the X-bodies, and in plants infected with aucuba mosaic virus they may be produced either directly without the previous formation of X-bodies or indirectly from the degenerating X-bodies. They are best examined in fresh living cells, for fixation either destroys them or alters their appearance.

The crystalline plates have various shapes. Some are irregular, others are perfect hexagons (Fig. 9), but intermediate types with one or two angles of 120° are the most common. They are colourless and transparent, and have a refractive index higher than the cell sap. They are undoubtedly true crystals with a three-dimensional regularity, for as they turn over in the sap they show both side and end faces. When seen edgeways they are rectangular, often showing faint striations, especially if the untreated crystals are examined in polarised light. IWANOWSKI (1903) showed that the plates develop striations and seem to be composed of needle-like crystals when acidified, a feature responsible for the name "striate-material" that has been applied to this type of inclusion.

When viewed between crossed Nicol prisms the appearance of the plates depends on whether they are lying flat or edgeways. If edgeways they are highly birefringent with straight extinction, but if flat they are not birefringent (*See* Chapter 11). This difference is shown in Fig. 10. As many have angles of 120°, it is highly probable that they are true hexagonal crystals, for these are birefringent only when viewed along the transverse axes. The presence of the birefringent cell wall, coupled with the small size of the crystals, makes a critical examination in polarised light difficult, and proof of the crystal class by means of an interference figure has not yet been obtained.

The crystals also give all the usual colour reactions for proteins, but their instability makes them difficult to examine in detail. Any damage to the cell containing them, the addition of acid or pricking with a micro-needle, causes them to break down into small needles or to dissolve (Fig. 11). Soon after infection several crystals can often be found in a single cell. Each then may be definitely hexagonal, but later they coalesce to form a single, rather shapeless mass.

Intracellular inclusions have been found in the roots, stems, leaves and flowers of infected plants, but they are most common in the leaves, especially in the epidermal cells. They cannot be found at all times or necessarily in all infected cells. Some parts of the leaf may have inclusions in nearly every cell and a nearby part have none. In good growing conditions the first appearance of incipient X-body or crystal formation can be seen a week or so after infection, at the same time as external symptoms appear on the young leaves. The inclusions rapidly increase in number, reaching a maximum about a month after infection. After a further month the X-bodies begin to disappear, those of aucuba mosaic being replaced by crystals. The crystalline inclusions persist for longer than the amorphous bodies, but after a few months of infection both types have usually disappeared.

The exact association of the inclusions with symptoms is not clear. Some workers state that they have found them only in chlorotic areas, and others that they occur also in the intervening green areas. In tobacco infected with a masked strain of tobacco mosaic virus, BEALE (1937) states that the crystalline inclusions are definitely restricted to the occasional scattered areas that show symptoms. Tobacco mosaic virus produces inclusions in all hosts in which it causes systemic infection and symptoms, but SHEFFIELD (1936) found none in the necrotic local lesions on *N. glutinosa*.

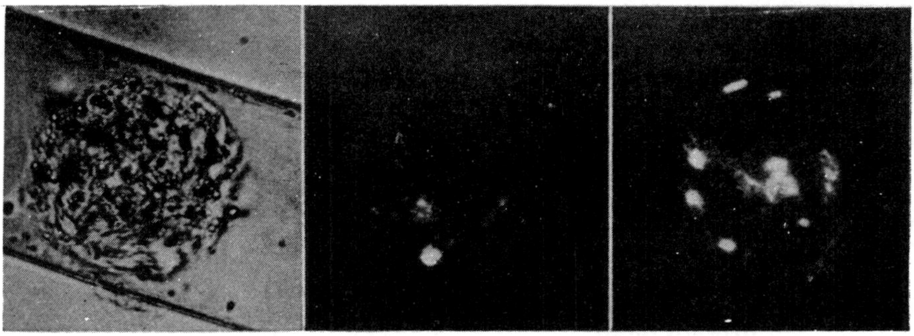

FIG 8. — A large X-body in hair cell of tomato infected with aucuba mosaic virus, showing the development of crystalline material. *Left:* photographed in transmitted light shows a few crystals embedded in the body. *Centre:* photographed in polarised light showing the presence of a little birefringent material. *Right:* photographed 30 hours later showing the presence of more birefringent material. (B. KASSANIS and F. M. L. SHEFFIELD, Ann. appl. Biol. 28, 360, 1941).

Although the flat plates are the best known type of crystalline inclusion produced by strains of tobacco mosaic virus, many other crystalline or fibrous forms also occur. At Rothamsted the types, and relative proportions of different types, of inclusions have varied considerably. In 1930 and 1931 HENDERSON SMITH and SHEFFIELD found that aucuba mosaic virus produced large numbers of X-bodies in *Solanum nodiflorum* and that the crystalline plates were formed only from disintegrating X-bodies. Crystal-like spikes, often as long as the cell, were also frequently seen. Between 1932 and 1940 no such spikes were found and relatively few amorphous bodies; instead the infected cells contained large numbers of crystalline plates, produced as such and not indirectly from disintegrating X-bodies. In 1940 the spike-like body was found again with great frequency, as well as a number of other, previously undescribed, fibrous forms (KASSANIS and SHEFFIELD, 1941). These consisted of birefringent, spindle-shaped bodies, masses of short needle-like fibres, and long fibres often curved into figures-of-eight (Fig. 12). The reasons for these variations are unknown, but they do not appear to be related to any change in the virus strain, for a number of different strains all produced similar forms of fibrous inclusions simultaneously. KASSANIS and SHEFFIELD suggest that varying environmental factors, which are known to affect external symptoms, may also determine the type of intracellular inclusion.

Until recently the varied types of inclusion body found in plants infected with tobacco mosaic virus had at least one feature in common, they all occurred in the cytoplasm. Now, however, WOODS and ECK (1948) have described one strain of the virus that produces fibrous and crystalline inclusions in the nuclei of infected cells.

**Possible Relationship Between Tobacco Mosaic Viruses and the Intracellular Inclusions:** — SHEFFIELD's results (1931, 1934) suggested that the X-bodies of tobacco and aucuba mosaic consisted of material coagulated and precipitated from the normal plant cytoplasm, and the view became generally adopted that both kinds of inclusion were plant reaction products. Recent work has necessitated a considerable modification of this view. Many of the properties of tobacco mosaic virus have been investigated in considerable detail and a knowledge of these makes it possible to put forward a possible explanation for the formation of both kinds of inclusion. In doing this it will be necessary to refer briefly to properties of the virus which are not described in detail until later chapters.

Neutral solutions of purified tobacco mosaic virus are faintly opalescent and show a slight satin-like sheen. On the addition of acid both the opalescence and the sheen increase. Between about *p*H 4 and *p*H 3 the virus is insoluble and precipitates in the form of microscopically visible, birefringent, needle-shaped bodies. Below *p*H 3 the preparations clear as the protein again dissolves, and below about *p*H 1 the virus is destroyed. BEALE (1937) has pointed out that the needles of the precipitated purified virus closely resemble in appearance the needles or "striations" produced by IWANOWSKI (1903) and GOLDSTEIN (1924) by adding acid to cells containing crystalline intracellular inclusions. She has also further shown that the *p*H stability and solubility ranges of the needles produced in the cells by the breakdown of the inclusions, are approximately the same as those of the purified virus *in vitro*. There is little doubt, therefore, that crystalline inclusions contain virus and it is highly probable that the virus is a major, though perhaps not the only, constituent.

Granted that the crystalline and fibrous inclusions are largely virus, an explanation is still needed for its deposition in solid or quasi-solid forms. The behaviour of purified virus in different conditions provides phenomena that could adequately account both for the variable forms of inclusions and for their separation and disappearance. First, the inclusions may consist solely of virus that has been precipitated because of the action of electrolytes in the cell sap. Neutral solutions of the virus do not give any solid phase when concentrated, they merely become increasingly viscous and turn to gels. In the presence of salts, however, solid phases of various types are produced from solutions containing no more virus than infective sap, *i.e., circa* 0.2%. Precipitation occurs rapidly with strong salt solutions and produces small paracrystalline needles, but with concentrations around 0.5 N it occurs slowly and produces long flexible fibres resembling the inclusions sometimes seen in infected cells. The character of the fibres varies somewhat with the nature of the salt, as does the concentration needed to produce them

(BEST, 1940, 1947). Tobacco mosaic virus is also precipitated in the form of fibres from neutral solutions as a result of combining with substances as varied as nicotine, arginine, protamines and histones, trypsin, ribonuclease, and certain other proteins (Fig. 13). The speed of precipitation varies with the different substances, occurring immediately with some but taking some hours with nicotine; the fibres produced by nicotine, like those produced by slow precipitation with dilute salts, can be dispersed by shaking. Also, the complexes are soluble in salt solutions, though the concentration needed to dissolve them varies with the different substances and with the *p*H value. With some substances, for example the virus-inhibiting glycoprotein present in *Phytolacca* sp. (KASSANIS and KLECZKOWSKI, 1948), the fibres are produced only when the virus and other substance are present in certain definite

**FIG. 9.** — Hair cell of *Nicotiana Tabacum* infected with tobacco mosaic virus. Two plate-like crystals are seen, one being almost a perfect hexagon. × 450. (BAWDEN, F. C. and SHEFFIELD, F. M. L., 1939, Ann. appl. Biol. *26*, 102).

ratios, and variations in the concentration of one or other component will lead to the appearance and disappearance of fibres. The virus is the major constituent of most of the fibrous complexes; the one formed with clupein is more than 95% virus, and the fibres dissolve in salt solutions more concentrated than 0.1 N, but more salt is needed to dissolve them at *p*H 5.5 than at *p*H 6.5. So many different materials can combine with and precipitate the virus that it must be rare for cells not to contain substances capable of doing so, and the fluctuating conditions of living cells could provide ample changes for the separation and resolution of the inclusions. Thus changes in the composition of the cell sap during metabolism, such as variations in the type and concentration of electrolytes or of many metabolites, or pH changes, could all determine whether the virus remains in solution, or separates in some solid form, which might in turn become soluble if conditions change.

The precipitates that have so far been produced artificially, whether of virus alone or of a complex with another substance, differ from the inclusions formed in infected cells in one significant property. All the precipitates produced *in vitro* are fibrous and not true crystals (BERNAL and FANKUCHEN, 1937), whereas some of the inclusions are true crystals with a perfect three-dimensional regularity of structure; this difference can probably be interpreted as an indication that the virus particles in purified preparations are in a different physical state from that in which they are formed in the host cells. There is evidence that the processes of extracting and purifying the virus cause the particles to aggregate linearly and to occur in widely different sizes. It is possible that the particles are more uniform as they occur inside cells and in this state can fit into a three-dimensional crystalline lattice, whereas this is impossible with the longer and more variable particles obtained *in vitro.*

At first sight the amoeboid-like X-body seems to have little in common with the virus and it is more difficult to relate the two than with the crystalline inclusions. However, the formation of crystals, indistinguishable in appearance and behaviour from the others, within the ageing X-bodies of plants infected with aucuba mosaic virus, points to a relationship and suggests that the two types of inclusion have a common constituent (Fig. 8). There is also direct evidence that the X-bodies contain considerable amounts of virus. LIVINGSTONE and DUGGAR (1934) found that the protoplasmic contents of the cell contained more virus than the sap, while SHEFFIELD (1939) has isolated

TABLE 1. — *Aucuba mosaic virus. Infectivity of isolated inclusions compared with purified virus:—*

| *Inoculum* | *Concentration* | *No. of leaves inoc.* | *Total no. of lesions* |
|---|---|---|---|
| Water | — | 66 | 0 |
| X-bodies | 10 in 4 ml | 66 | 16 |
| Purified virus | $10^{-8}$ gms per ml | 66 | 20 |
| Water | — | 6 | 0 |
| X-bodies | 10 in 0.8 ml | 6 | 7 |
| Purified virus | $2 \times 10^{-7}$ gms per ml | 6 | 4 |

individual X-bodies and shown them to be infective. She removed the bodies from infected cells by means of fine micro-pipettes, washed them thoroughly in buffer solutions, and then suspended them in water before inoculating them to *N. glutinosa*. From a consideration of the amount of purified virus necessary to give infections it is apparent that only few lesions could be expected from a single body. Depending on the host plant and method of inoculation, from $10^{-8}$ to $10^{-10}$ gms of purified aucuba mosaic virus are required to give infection. If the bodies are assumed to be spherical with an average radius of $10\mu$, then they will have a volume of $4 \times 10^{-9}$ ml. Within the accuracy required for the comparison the bodies can be taken as solid with a specific gravity of 1. Then the weight of a single body will be approximately $4 \times 10^{-9}$ gms, which is of the same order as the weight of purified virus necessary to give infection. SHEFFIELD suspended lots of 10 isolated X-bodies in water and compared their infectivity with solutions containing a known weight of purified aucuba mosaic virus. The results obtained are summarised in TABLE 1.

In addition to demonstrating the presence of virus by infectivity tests, SHEFFIELD (1946) had identified virus particles in X-bodies and crystalline inclusions by means of the electron microscope. When inclusions of both types are isolated from cells, dispersed in water and dried to form suitable mounts for electron microscopy, they contain, among other unidentifiable material, many rod-like particles of varying lengths but the same width as those characteristic of the virus. Although it can be accepted that the X-bodies contain virus, this fact does not prove that the virus is an essential constituent; the bodies frequently contain chondriosomes and oil globules, which are presumably accumulated as the growing body moves around the cell, and the virus particles might be acquired in the same manner. However, as the washed bodies are, weight for weight, approximately as infective as purified virus, and their content of rod-shaped particles is greater than other parts of infected cells, it seems likely that there is too much virus present to be accounted for by accidental adsorption. Hence they are probably composed largely of an insoluble complex produced by the union of the virus particles with a constituent or constituents of the host cells. The stability of the X-bodies towards $p$H changes and their isotropic character suggest a different type of complex from those already mentioned in discussing the crystalline inclusions, and probably one containing a smaller ratio of virus to other material. Analogies for such complexes can be supplied from work with purified virus.

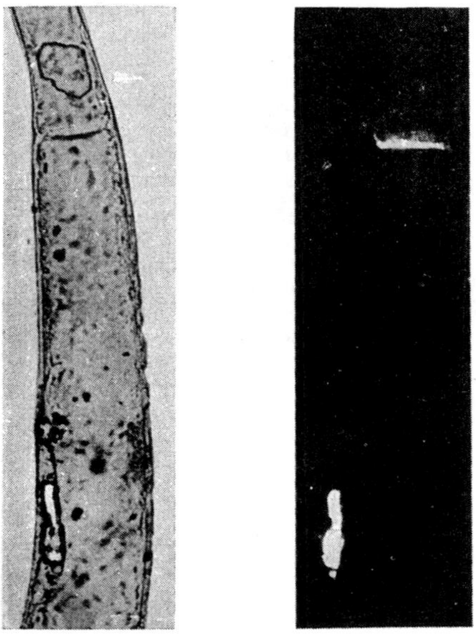

FIG. 10. — Hair of *Lycopersicum esculentum* infected with tomato enation mosaic virus. In the upper cell an irregularly shaped, crystalline inclusion is seen lying flat; in the lower cell a similar inclusion is seen lying edgeways. *Left:* Photographed by transmitted light. *Right:* Photographed between crossed Nicol prisms. The cross cell wall and the inclusion lying edgeways are birefringent, but the inclusion lying flat is not. × 450. (BAWDEN, F. C. and SHEFFIELD, F. M. L., 1939, Ann. appl. Biol. *26*, 102).

KLECZKOWSKI (1946) has shown that tobacco mosaic virus combines with and is precipitated by many proteins and other materials that carry an opposite electric charge. The virus is by far the greater component of most complexes which usually form fibrous precipitates, but it combines with a quarter of its weight of globin to give an amorphous precipitate. Similarly, amorphous precipitates, which are unaffected by moderate $p$H changes or salt solutions, are formed when tobacco mosaic virus combines with its antibodies. Like the X-bodies, these precipitates are dispersed by distilled water. In virus-antiserum mixtures kept circulating by convection currents, there is a sequence of events superficially resembling that which occurs in cells during the formation of X-bodies. Small particles first appear distributed uniformly throughout the fluid. On coming together these fuse to form large aggregates, which finally settle at the bottom of the tube as one amorphous clump. We have seen that the X-bodies tend to be formed relatively early during infection when virus content is low and to disappear later with increasing virus content. If the bodies are composed of a virus-host complex with solubility relationships similar to those of the virus-antibody precipitates, this behaviour can also be explained. For these precipitates are formed only when virus and antibody are mixed in certain proportions, their formation is prevented by the presence of too much virus and they dissolve in the presence of a great excess of virus.

The similarities between the X-bodies and the virus-antibody complexes have been described here merely to show that phenomena in the plant can be simulated at least in part *in vitro*. It is not intended to

TABLE 2

| *Viruses known to cause the formation of intracellular inclusions:—* | *Viruses which apparently do not cause the formation of intracellular inclusions:—* |
|---|---|
| Dahlia mosaic | Cucumber mosaic |
| Dwarf (Stunt) of rice | Cucumber 3 and 4 |
| Henbane mosaic | Potato A |
| Hippeastrum mosaic | Potato leaf roll* |
| Iris stripe | Potato paracrinkle |
| Maize mosaic | Tomato big-bud |
| Oat mosaic | Tomato spotted wilt* |
| Onion yellow dwarf | |
| Pea 2 | |
| Phaseolus 2 | |
| Potato X (many strains) | |
| Potato Y | |
| Sandal-Spike | |
| Sugar beet curly top | |
| Sugar-cane chlorotic streak | |
| Sugar-cane Fiji disease | |
| Sugar-cane mosaic | |
| Tobacco mosaic (many strains) | |
| Tobacco ringspot | |
| Tobacco etch | |
| Tomato bushy stunt | |
| Wheat mosaic | |

*Since the table was compiled, J. G. BALD (Amer. J. Bot. (1949) *36*, 335) has described new techniques for staining inclusion bodies. He considers that these also stain viruses in the cytoplasm and by their use he claims to have demonstrated inclusion bodies in plants infected with potato leaf roll and tomato spotted wilt viruses.

suggest that the inclusions are actually complexes formed by the union of virus and antibody, for there is no valid evidence that plants contain or produce antibodies. It is probable, however, that they do contain proteins capable of uniting with the virus and rendering it insoluble. Any such combination might also be expected to act in part as a protective mechanism, for it is likely that the biological activity of the virus would be reduced by being rendered insoluble.

**Intracellular Inclusions Caused by Viruses Other Than Tobacco Mosaic:** — It is impossible to give a complete list of the viruses which do and which do not cause the production of intracellular inclusions in infected plants. In many published accounts of symptoms no mention is made of their occurrence but there is no indication as to whether they have been sought. Also, the contents of uninfected cells are extremely varied. In addition to the nucleus, cytoplasm and cell sap, the living cell contains plastids, chondriosomes, oil globules, crystals of calcium oxalate, and numerous particles of unknown composition, many of which are birefringent. Unless adequate comparisons are made between healthy and infected plants, therefore, it is possible for normal constituents of cells to be mistaken for virus inclusion bodies. This is especially so as infection often increases both the rate of cytoplasmic streaming and the numbers of crystals and particles of the types that occur in healthy plants. Most normal cells contain only one nucleus, but it is not uncommon for those of virus-infected plants to contain two (SHEFFIELD, 1936; SALAMAN, 1938). Hence it is necessary that a clear distinction should be drawn between the nucleus and any suspected inclusion bodies. This is conveniently done with Feulgen's reagent, which stains the

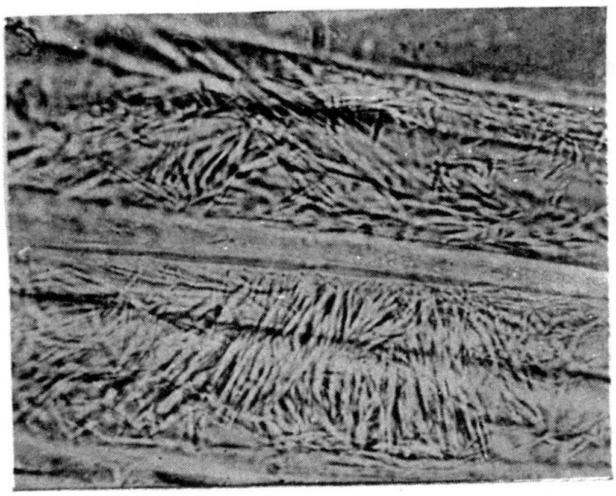

FIG. 11. — Epidermal cells from Turkish tobacco plant infected with tobacco mosaic virus after treatment with dilute HCl. The crystalline plates previously present have been replaced with numbers of paracrystalline needles. × 660. (BEALE, H. P., 1937, Contrib. Boyce Thompson Institute *8*, 413).

nuclei red but has no effect on any of the inclusion bodies yet tested. This would be expected from our chemical knowledge of the plant viruses, for all those so far isolated have contained nucleic acid of the ribose type and this is unaffected by Feulgen's reagent, which reacts only with the desoxy-pentose nucleic acid characteristic of nuclei.

In TABLE 2 are listed the viruses which have been found to cause undoubted inclusions and those causing diseases which have been carefully examined for inclusions and none found.

The viruses listed are far from all that have been recognised, but many have not been studied in enough detail to warrant including. Both lists contain viruses possessing widely different characters *in vitro*, and it is impossible at present to indicate with any certainty what features determine the production of inclusions. The absence of inclusions from some diseases may simply mean that the virus content of infected cells is insufficient to produce aggregates large enough to be microscopically visible. Alternatively, the isoelectric points of some viruses may be too close to those of possible complex-forming substances for these to combine with and precipitate the viruses, or some property of the host cell may inhibit their formation. The last may explain the absence of inclusions in plants infected with cucumber viruses 3 and 4. Both these occur in sap in high concentrations and in their general properties and the conditions in which they precipitate, they closely resemble tobacco mosaic virus (BAWDEN and PIRIE, 1937). However, only members of the cucurbitaceae have been infected with them, and all these are immune from tobacco mosaic virus, so that comparisons in the same host are impossible. Expressed sap of cucumber plants is more alkaline than tobacco sap, *p*H about 7.8 compared with 5.8, and this difference could alone account for the absence of inclusions. Increasing *p*H increases the solubility of the viruses and of many complexes of the type discussed; it would also prevent the formation of complexes between

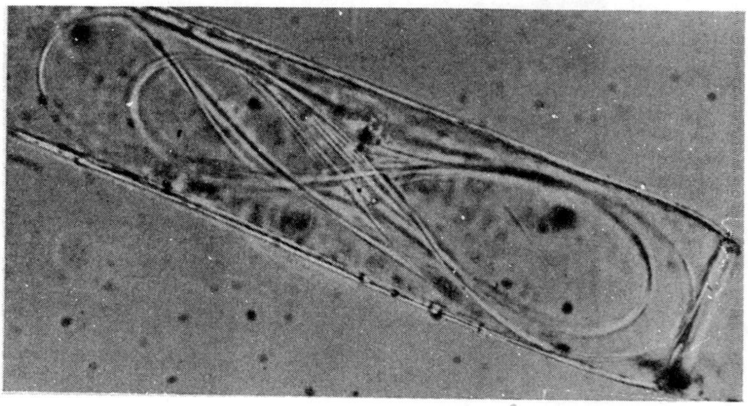

FIG. 12. — A hair cell from tomato plant infected with aucuba mosaic virus containing long, fibrous inclusion bodies curved to form figures of 8. × 500. (KASSANIS, B. and SHEFFIELD, F. M. L., 1941, Ann. appl. Biol. *28*, 360).

the viruses and materials isoelectric between 5.8 and 7.8, for in this range the viruses will combine with substances on the acid side of their isoelectric points, because they are oppositely charged, but they will not combine on the alkaline side, because they carry a similar charge to the virus particles themselves.

That cucumber is perhaps an unsuitable host for the formation of inclusions is suggested by work on tobacco ringspot virus. In tobacco this produces many inclusions, but in cucumber it produces only occasional and rather indefinite bodies (WOODS 1933; BAWDEN and SHEFFIELD, 1939). Tobacco ringspot virus resembles tobacco mosaic virus in that it produces both amorphous and crystalline inclusions. The former are amoeboid and vacuolate, closely resembling those in tobacco mosaic plants. The crystalline inclusions have been seen only in infected solanaceous plants. They occur less frequently than in plants with tobacco mosaic and have a different crystalline form, most being rectangular blocks. They are biaxial crystals, for when seen between crossed Nicol prisms they are birefringent along all axes (Fig. 14). Crystals not seen in healthy plants of the same species have also been described in oats, millet, barley and maize suffering from mosaic (SUKHOV and VOOK 1938). The commonest are small needles, but the epidermal cells contain large protein crystals, which sometimes break up into small aciculate crystals.

Although intranuclear inclusions have been described in about thirty different virus diseases of animals, there was no convincing demonstration of their occurrence in virus-infected plants until KASSANIS (1939) found that the nuclei of solanaceous plants infected with tobacco severe etch virus invariably contain abnormal bodies (Fig. 15). These give the usual protein reactions and with most stains behave like the nucleoli. They are thin flat crystalline plates, of various sizes and birefringent when viewed edgeways. The crystals are first seen about a week after

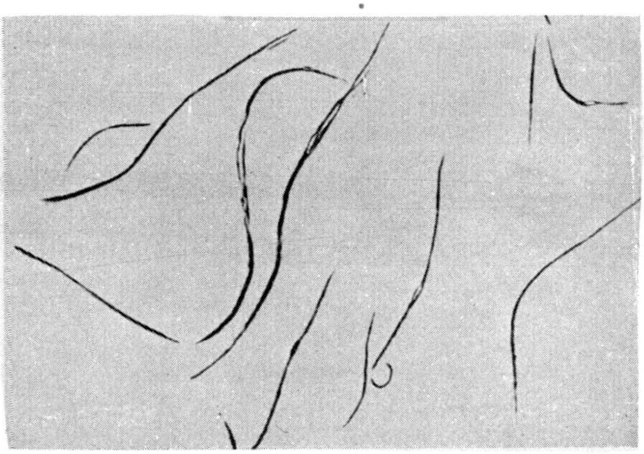

FIG. 13. — Insoluble mesomorphic fibres produced by the addition of a neutral solution of clupein sulphate to a purified preparation of tomato aucuba mosaic virus. × 450. (BAWDEN, F. C. and SHEFFIELD, F. M. L., 1939, Ann. appl. Biol. *26*, 102).

external symptoms become obvious, and they occur in all tissues except the growing points; as many as thirty may be produced in a single nucleus.

Similar inclusions are also produced in the nuclei of plants infected with mild etch virus (BAWDEN and KASSANIS, 1941), but these are larger and fewer than in plants infected with severe etch virus. Not all the intranuclear inclusions formed by mild etch virus seem to be flat plates, for some show a characteristic cross suggesting that they are probably eight-sided bi-pyramids (Fig. 16). The intranuclear inclusions are all remarkably stable; they can be extracted from the nuclei, when their structure is unaffected by *p*H changes between 2 and 10 or by the addition of alcohol. Except for the presence of these crystals, the nuclei appear normal and they still contain nucleoli. SHEFFIELD (1941) found that infection with severe etch virus sometimes stimulated the nuclei of fully differentiated cells to divide, so that binucleate cells were produced. Any intranuclear crystals were then extruded into the cytoplasm, but as soon as the daughter nuclei were reconstituted further crystals appeared in them.

McWhorter (1940) has found isometric crystals in nuclei and cytoplasm of leguminous plants infected with pea virus 2 and Phaseolus virus 2. Similar crystals were also seen inside the nucleoli. Usually five or more occurred in a nucleolus, but sometimes one large crystal filled it. They vary from 0.3 μ to 4 μ in diameter, and give the usual protein reactions. Like those in plants suffering from etch, these inclusions can be removed from the nucleus and remain stable in saline.

In infections with other viruses only the X-body type of cytoplasmic inclusion has been reported. The differences in morphology that have been described between inclusions formed by different viruses are about as great as those between the bodies formed by tobacco mosaic and aucuba mosaic viruses. Wherever microchemical tests have been made, positive tests for protein have been obtained and the inclusions have failed to

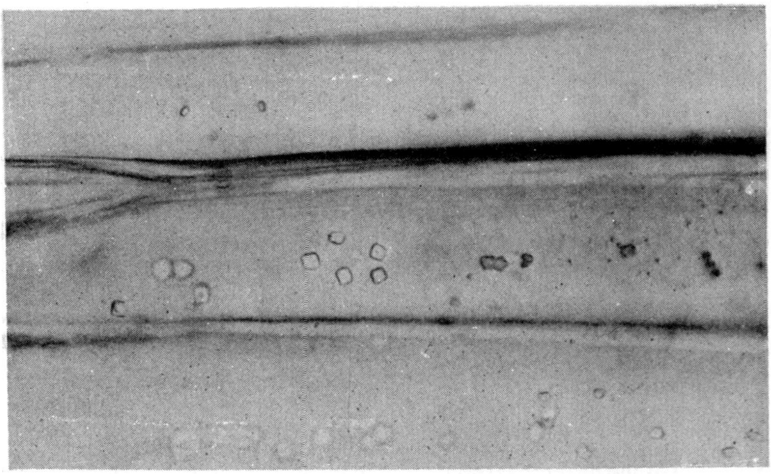

Fig. 14. — *Nicotiana Tabacum* infected with tobacco ringspot virus. Epidermal cells from beneath a vein containing many small crystalline blocks. × 450. (Bawden, F. C. and Sheffield, F. M. L., 1939, Ann. appl. Biol. *26*, 102).

stain with Feulgen's reagent. Sheffield (1934, 1941) has shown that the X-bodies formed by henbane mosaic virus and severe etch virus result from the aggregation of small particles circulating in the cytoplasm; Clinch (1932) and Salaman and Hurst (1932) suggest that those produced by potato virus X are also aggregates.

Where the history of the inclusion body has been followed in detail, it has been found to persist for only a limited period, usually changing in appearance before disintegrating and disappearing. Disintegrating bodies in plants infected with henbane mosaic virus often contain long, thin, crystal-like needles, resembling those formed by precipitates of tobacco mosaic virus. They show no extinctions in the polarising microscope, but whether this is because of their small size or because they are really isotropic is unknown. Old cytoplasmic inclusions in severe etch plants may also crystallise, especially in leaves that become deformed. The usual form is again needle-like, varying from 1 μ to 10 μ long (Fig. 17), but these are birefringent; birefringent particles of other shapes are also formed. Whether these are true crystals or paracrystalline fibres is unknown. They can sometimes be seen in malformed leaves within a month of inoculating young seedlings, but they usually need longer than this to develop in quantity. All the bodies in a malformed leaf may not contain crystals, and some parts of a single large body may have them while other parts remain granular and isotropic. These needles are less stable than the intranuclear inclusions; they dissolve and the X-body regains its earlier appearance when treated with weak acid. SHEFFIELD (1941) has extracted these cytoplasmic inclusions and shown that they are infective. They differ from the inclusions of aucuba mosaic in that they can be pricked and cut into pieces without disintegrating. They are chemically complex, for they contain fats and oil globules as well as proteins. The formation of microscopically visible crystals in

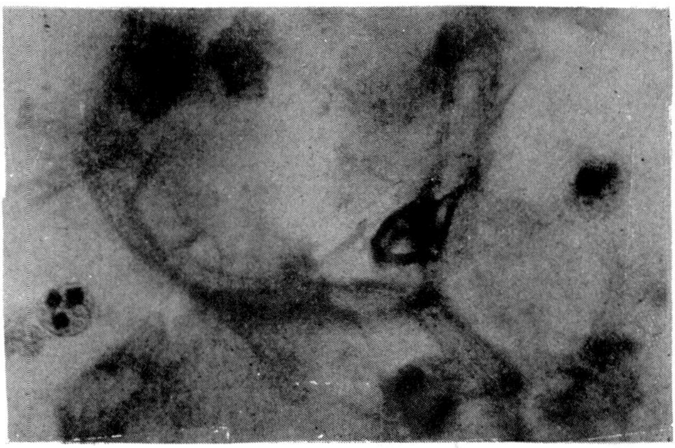

FIG. 15. — Pith cells from *Nicotiana Tabacum* infected with severe etch virus, stained with haematoxylin. All the nuclei contain darkly stained, plate-like inclusions. × 500. (*Photograph by* B. KASSANIS).

the old inclusions is the end-stage of a process, for electron micrograms of isolated inclusions show the presence of sub-microscopic crystals of various forms (SHEFFIELD, 1946).

No appreciable differences have been found between the properties of severe etch virus, henbane mosaic virus and potato virus Y *in vitro,* yet they differ widely in their behaviour in the plant. Severe etch virus produces large numbers of inclusions in both nuclei and cytoplasm and henbane mosaic virus produces many cytoplasmic inclusions but none in the nuclei. For long no inclusions were found in plants infected with virus Y, and now only a few, small ones have been seen. If the inclusion bodies are insoluble complexes of viruses and host constituents, this suggests that the substances with which the viruses unite may not be normal host constituents but products specific to the particular disease, for these three viruses infect the same plants and precipitate in such similar conditions *in vitro* that any normal host constituent precipitating with one would also be expected to precipitate with the others.

Although X-bodies have been found in all tissues except the apical meristem, different viruses produce them in greater abundance in different tissues. In tobacco mosaic plants they are most common in the epidermal layers of the leaf, with potato virus X the palisade cells contain the greatest number, and in sugar beet suffering from curly top they are restricted to the phloem and adjacent tissues. By contrast, henbane mosaic virus appears to produce them equally in all tissues. Most workers state that the inclusions are associated with external symptoms, and BEALE (1937) found that tobacco plants infected with a "masked" strain of tobacco mosaic virus contained inclusions only in the occasional areas showing symptoms. But this association would not seem to be essential, for bodies have frequently been pictured in the roots that show no external symptoms. Also, SALAMAN (1938) states that tobacco plants infected with a masked strain of potato X contain more inclusion bodies than those infected with virulent strains that produce severe symptoms. BAWDEN and SHEFFIELD (1944) also found inclusions in otherwise symptomless potato plants infected with virus X; in general they were produced most abundantly in plants that reacted with a yellow mottle with little or no necrosis, but the numbers and size of inclusions varied with different virus strains and host plants. In tobacco the inclusions are about the same size as the nuclei, but in potato they may occupy half the volume of cells and be many times larger than the nucleus. The inclusions in the epidermis occur equally in the green and yellow areas, but in the palisade cells and spongy parenchyma they are usually restricted to yellow areas.

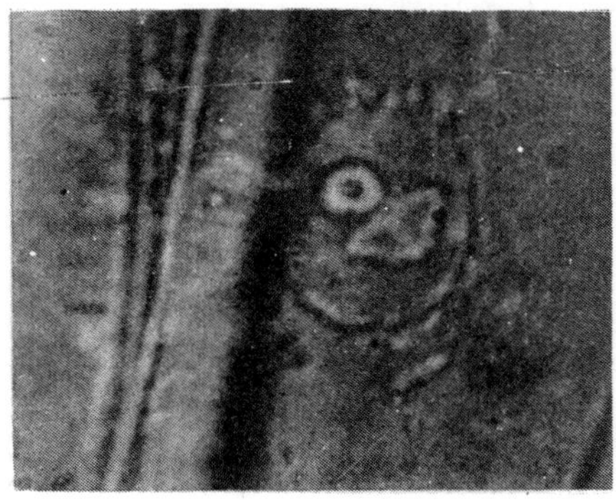

FIG. 16. — An epidermal cell from a tobacco plant infected with mild etch virus. The intracellular inclusion shows a definite cross, suggesting that the crystal is a bipyramid. × 900.

There are slight morphological differences between the inclusions formed by unrelated viruses, but these are no greater than those between the inclusions formed by different strains of one virus, or by the same strain in different hosts. It is reasonable to suppose, therefore, that the various bodies described will have a similar mode of formation. HARTZELL (1937) has described inclusions in plants suffering from peach and aster yellows which differ from other intracellular bodies. They are much smaller than those found in other virus diseases and, according to HARTZELL, are actively motile. They were also seen in infective insect vectors. However, these bodies are so ill-defined that until they are more clearly differentiated from the contents of uninfected cells, and their movement from Brownian movement, it would seem premature to regard them as virus inclusion bodies.

**Internal Changes Other Than Intracellular Inclusions:** — The formation of intracellular inclusions depends more upon the infecting virus than upon the type of external symptoms produced. Two viruses, for example tobacco mosaic and cucumber mosaic, may cause very similar external symptoms, but inclusions will be formed only in plants infected with the former. Other types of internal changes in virus-diseased plants much more accurately reflect the external symptoms, and in general viruses causing similar external symptoms also cause similar histological changes.

The leaves of plants showing mosaic have been studied by many workers. Usually both the size of the cells and their contents are altered. The chlorotic areas are thinner than the green areas, usually because of a reduction in the length of the palisade cells, and the intercellular spaces are smaller. The following measurements made by CLINCH (1932) on the potato variety President suffering from crinkle show this reduction:-

| | *Thickness of leaf* | *Length of palisade cells* |
|---|---|---|
| Yellow area of crinkled leaf | 113 μ | 45 μ |
| Dark green area of crinkled leaf | 137 μ | 57 μ |
| | 148 μ | 72 μ |
| Healthy leaf | 140 μ | 52 μ |
| | 147 μ | 56 μ |

The chloroplasts are smaller than usual, more granular, and less deeply coloured, and often have irregular outlines. The whole effect is one of reducing the assimilatory tissues. How this reduction is brought about is uncertain. The view has frequently been put forward (COOK, 1930; SHEFFIELD, 1933) that the viruses do not destroy chlorophyll or plastids, but act only by inhibiting plastid formation. This is largely based on the fact that systemic symptoms are always more definite on leaves which are developing at the time of infection, mature leaves often being quite unaffected. It seems that some viruses at least, especially tomato stripe in tobacco (SMITH, 1935), must affect mature chloroplasts, for if fully mature leaves are rubbed with a concentrated virus preparation they may become chlorotic. Mature leaves which are not inoculated rarely develop a high concentration of virus. Thus it is probable that the absence of symptoms in leaves mature when plants are infected can be explained on the basis of their low virus content rather than on the inability of the virus to affect mature plastids.

Although the assimilatory tissues are reduced, there is often an abnormal accumulation of starch in cells of chlorotic areas. This condition is found in many diseases of the yellows type and is especially definite in potato leaf roll, where the chloroplasts are frequently so full of starch that they may burst (CLINCH, 1932). In this disease and in many others, such as sereh disease of sugar-cane, curly top of sugar beet, raspberry curl and aster yellows, this excessive accumulation of carbohydrates is accompanied by abnormalities in the phloem. In the potato the sieve tubes and companion cells of affected plants became necrotic (QUANJER, 1913). QUANJER (1931) considered that the death of the phloem elements is responsible for the accumulation of carbohydrate in the leaf, but MURPHY (1923) and BAWDEN (1932) could often find no necrotic phloem in plants recently infected though their upper leaves were rolled and contained much starch. SHEFFIELD's (1943) results, however, support QUANJER's view, for she invariably found necrotic phloem elements somewhere in infected plants before there was any rolling of the leaves. Staining sections of the lower part of the stem with phloroglucinol and HCl is a useful diagnostic procedure, for the necrotic primary phloem elements turn yellowish red, an effect not found in plants whose leaves roll for other causes. Leaf roll virus produces a necrosis in the tubers of some varieties, but this is a variable symptom; it may be associated with the simultaneous presence of another virus, such as potato virus A, and it seems to occur only in tubers set by plants infected during the current season. FOLSOM (1946) finds that the tuber net-necrosis also depends on temperature, the incidence being greatest when tubers are stored around 7° to 10°C. The tuber tissue adjacent to necrotic phloem elements fluoresces in ultra-violet light, whereas normal tissue and the necrotic phloem elements themselves do not (SANFORD and GRIMBLE, 1944).

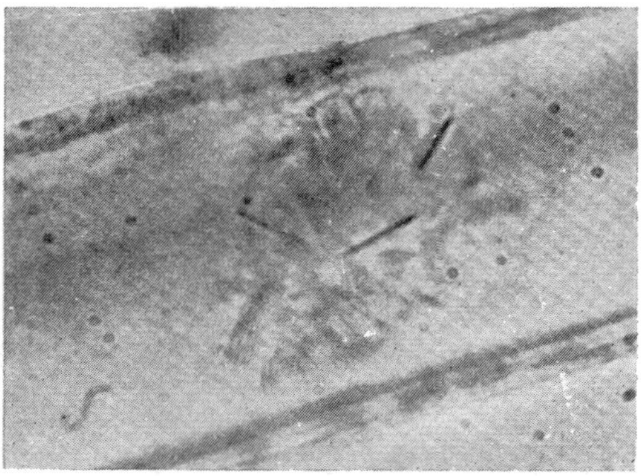

FIG. 17. — An epidermal cell from a malformed leaf of a tobacco plant which was infected with severe etch virus when a young seedling. The cytoplasmic inclusion body has crystallised and contains needles; these are birefringent when viewed in polarised light. × 900.

A necrosis of the phloem, which may extend into the pericycle, also occurs in sugar beet plants suffering from curly top. Near the degenerate areas there is often hypertrophy and hyperplasia, leading to a thickening or distortion of the veins and to a formation of protuberances (ESAU, 1933). According to ARTSCHWAGER and STARRETT (1936) the nuclei of sugar beet cambium are also affected. Early in infection there is an increase in nuclear and chromatin material, the nuclei enlarging and becoming mis-shapen. Later the hypertrophied nuclei break down liberating fragments which remain unaltered in the cytoplasm. In elm trees suffering from phloem necrosis (MCCLEAN, 1944), the first changes occur in the fibrous roots, where hypertrophy and hyperplasia of the parenchymatous cells crush the sieve tubes and companion cells. The affected phloem becomes first yellow and then brown, and the degeneration progresses to the larger roots and stem, when the trees die. Associated with the visible changes is a faint odour of wintergreen.

Potatoes suffering from top-necrosis also show a degeneration of the phloem elements in leaves, stems and tubers. The necroses arise in the phloem, but once originated spread to other tissues. Large intercellular spaces may be formed which are filled with gum-like deposits. The cell walls become thickened, because of the deposition of lignin or suberin, while the cell contents either disappear or are changed into darkly-staining substances rich in pectin. Phellogens are frequently formed around necrotic areas, especially in the tubers, and these form regular layers of cork cells which sometimes localise the destructive processes. Around affected areas in the tubers are starch grains in all stages of dissolution. This disease can be caused by a number of different viruses, but the symptoms, both external and internal, are the same regardless of the virus. The frequency with which degenerative changes arise in the phloem has often been advanced as an argument in favour of the view that viruses move through the plant in the phloem. But in leaf-drop streak of the potato changes occur which, although chemically resembling those in top-necrosis, do not affect the vascular tissues at all. The necroses appear to start in the mesophyll of the leaves and spread from thence down the petioles into the main stem. The collenchyma is chiefly affected, but the whole of the cortex may be destroyed while the wood and phloem remain apparently normal. (QUANJER, 1931; BAWDEN, 1932).

In Fiji disease of the sugar-cane a rather different effect has its origin in the phloem, where elongated galls are formed by the proliferation of the sieve tubes. Adjacent tissues may also be affected, the hyperplasia being sufficient to show externally as long swellings on the under sides of the leaves. The cytoplasm of the affected cells is also altered, for it appears denser and it stains more deeply (KUNKEL, 1924). SAMUEL, BALD and EARDLEY (1933) have described a somewhat similar condition in tomatoes suffering from bigbud. There is an unusually large amount of internal phloem tissue, but it is unlike normal phloem, for the bulk of it consists of small cells with prominent nuclei; sieve tubes are rare and isolated tracheids occur. Perhaps the most extensive development of new tissues is found in tobacco plants suffering from kroepoek (KERLING, 1933). Not only does the primary phloem increase, but so does the pericycle, and within the enlarged pericycle cambium arises which forms new vascular tissues. Also, the type of leaf is altered, for the spongy parenchyma is no longer produced and palisade tissue is formed on both the upper and lower sides of the leaf. Palisade tissue is formed in the lobed veins and this proliferates to form new leaflets or "enations".

*References:-*


ARTSCHWAGER, E. and STARRETT, R. C. (1936) : J. Agr. Res. *53*, 637.
BAWDEN, F. C. (1932).: Proc. Roy. Soc. B. *111*, 74.
—— and KASSANIS, B. (1941) : Ann. Appl. Biol. *28*, 107.
—— and PIRIE, N. W. (1937) : Brit. J. Exp. Path. *18*, 275.
—— and SHEFFIELD, F. M. L. (1939) : Ann. Appl. Biol. *26*, 102.
—— and —— (1944) : *ibid. 31*, 33.
BEALE, H. P. (1937) : Contrib. Boyce Thompson Inst. *8*, 413.
BERNAL, J. D. and FANKUCHEN, I. (1937) : Nature *139*, 923.
BEST, R. J. (1940) : Austral. J. exp. Biol. *18*, 307.
—— (1947) : *ibid. 25*, 283.
CLINCH, P. (1932) : Sci. Proc. Roy. Dublin Soc. *20*, 143.
COOK, M. T. (1930) : J. Dept. Agric. Porto Rico *14*, 69.
ESAU, K. (1933) : Phytopath. *23*, 679.
FOLSOM, D. (1946) : Phytopath. *36*, 1016.
GOLDSTEIN, B. (1924) : Bull. Torrey Bot. Club *51*, 261.
—— (1926) : *ibid. 53*, 499.
—— (1927) : *ibid. 54*, 285.
HARTZELL, A. (1937) : Contrib. Boyce Thompson Inst. *8*, 378.
HENDERSON SMITH, J. (1930) : Ann. Appl. Biol. *17*, 213.
—— and SHEFFIELD, F. M. L. (1930) : Nature *125*, 200.
IWANOWSKI, D. (1903) : Zeitschr. Pflanzenkr. *13*, 1.
KASSANIS, B. (1939) : Ann. Appl. Biol. *26*, 705.
—— and KLECZKOWSKI, A. (1948) : J. Gen. Micro. Biol. *2*, 143.
—— and SHEFFIELD, F. M. L. (1941) : Ann. Appl. Biol. *28*, 360.
KERLING, L. C. P. (1933) : Phytopath. *23*, 175.

KLECZKOWSKI, A. (1946): Biochem. J. *40*, 677.
KUNKEL, L. O. (1924): Bull. Exp. Sta. Hawaiian Sugar Plant. Assoc. *3*, 99.
LIVINGSTONE, L. G. and DUGGAR, B. M. (1934): Biol. Bull. *67*, 504.
McCLEAN, D. M. (1944): Phytopath. *34*, 818.
McWHORTER, F. P. (1940): Phytopath. *31*, 760.
MURPHY, P. A. (1923): Sci. Proc. Roy. Dublin Soc. *17*, 163.
QUANJER, H. M. (1913): Meded. R. Hoog. Land-, Tuin- en Boschbouwsch. Wageningen *6*, 41.
—— (1931): Phytopath. *21*, 577.
SALAMAN, R. N. (1938): Phil. Trans. Roy. Soc. B. No. 559, *229*, 137.
—— and HURST, C. C. (1932): J. Micr. Soc. *52*, 237.
SAMUEL, G., BALD, J. G. and EARDLEY, C. M. (1933): Phytopath. *23*, 641.
SANFORD, G. B. and GRIMBLE, J. G. (1944): Can. J. Res. *22*, 162.
SHEFFIELD, F. M. L. (1931): Ann. Appl. Biol. *18*, 471.
—— (1933): *ibid.* *20*, 57.
—— (1934): *ibid.* *21*, 430.
—— (1936): *ibid.* *23*, 498.
—— (1939): Proc. Roy. Soc. B. *126*, 529.
—— (1941): J. Roy. Micr. Soc. *61*, 30.
—— (1943): Ann. Appl. Biol. *30*, 131.
—— (1946): J. Roy. Micr. Soc. *66*, 69.
SMITH, K. M. (1935): Parasitology *27*, 450.
SUKHOV, K. S. and VOOK, A. M. (1938): C. R. Acad. Sci. U. S. S. R. *20*, 745.
WOODS, M. W. (1933): Contrib. Boyce Thompson Inst. *5*, 419.
—— and ECK, R. V. (1948): Phytopath. *38*, 852.

*Chapter 4*

# METHODS OF TRANSMISSION

The demonstration that an abnormal condition is a virus disease needs proof of transmissibility in the absence of a visible pathogen. The three main methods used for transmission are, first, by intergrafting healthy and diseased plants, second, by inoculating healthy plants with sap from a diseased one, and third by transferring insects to healthy plants after they have fed on a diseased one. All three have been successful with some viruses, for example, potato Y, cucumber mosaic and spotted wilt. but with others only one or two of the methods have yet succeeded. Potato virus X and tomato bushy stunt virus, for example, have been transmitted by grafting and mechanical inoculation of sap, but not by insects; potato leaf roll, cacao swollen shoot and strawberry yellow edge have been transmitted by grafting and by insects, but not by inoculation, whereas apple mosaic, peach rosette, hop nettle-head, tomato big-bud, and many others, have so far been transmitted only by grafting. The methods whereby a virus has been transmitted play a prominent part in descriptions of virus diseases and are often used as diagnostic characters, but they are probably less valuable for this purpose than has hitherto been recognised. There is no doubt that many viruses which as yet have been transmitted by grafting only, or by grafting and inoculation, do have insect vectors awaiting discovery. *Abutilon* variegation, for example, has been known to be graft-transmissible for more than 70 years, but its transmission by an insect, *Bemisia tabaci*, was claimed for the first time in 1946 by ORLANDO and SILBERSCHMIDT. Also, it is known that some viruses are readily transmitted by inoculation between some hosts, but not between others, so that it remains uncertain to what extent mechanical transmission fails because of some intrinsic property of the viruses themselves or because of properties of the host plants in which they occur.

**Grafting and the Use of Dodder:** — From the few examples quoted in the preceding paragraph it will be seen that grafting is the only method of transmission shared by all the viruses mentioned. It is, indeed, so characteristic that it has largely become the criterion of a virus disease. It can be taken as axiomatic that viruses will be transmitted by grafting from infected to healthy individuals of plants in which they cause systemic infections, for once union occurs between scion and stock transmission is an inevitable sequel to the ability of the virus to move through vegetative tissues. Grafting may also produce systemic infections in hosts that give only local lesions when they are inoculated with sap. For example, potato varieties that produce only black necrotic local lesions when inoculated with certain strains of potato viruses X and Y will die from top-necrosis if the strains are introduced by grafting. Similarly, if *Nicotiana glutinosa* is grafted with a scion from a tobacco or tomato plant infected with tobacco mosaic virus, it will die with a systemic disease resembling top-necrosis, whereas inoculation produces only local lesions. Such effects can be expected only when a hypersensitive host is grafted with one that is more tolerant; even then systemic infection of the hypersensitive host does not always occur, for the reaction at the scion-stock junction may be so violent that the tissues die and prevent the virus from moving. Of the viruses yet studied, only those causing tobacco necrosis would fail to be graft-transmissible between any known hosts, for no plants have yet been found in which they spread systemically. With diseases like phony peach, in which the virus occurs only in the roots, transmission, as might be expected, is not obtained by using scions from the aerial portions of diseased trees, but only by root grafting.

The normal methods of grafting used in horticulture are often applicable to virus work, but modifications are sometimes needed to deal with plants not normally intergrafted. For herbaceous plants, the cleft graft is the most suitable. The top of a healthy plant is cut off, preferably at or immediately above a node, and the stem is then slit down the centre for an inch or so. An apical shoot is cut from the diseased plant, trimmed to a few small leaves, and the stem cut to a wedge shape. The scion is inserted in the slit stock, so that the two cambial layers are in contact, the junction is lightly bound with bast or elastic tape, and the plant is placed in damp, warm conditions until a firm union is established. In these conditions the stock produces rapidly-growing, succulent side shoots, ideal material for reacting to infection.

For woody plants or trees, budding is often preferable to grafting. A bud is cut from a diseased plant so that it carries cortical tissues and phloem; two slits at right angles to one another are made in the stem of the healthy stock reaching to the cambium, and the bud is inserted in the opening so that the cambial layers of bud and stock are in contact. The presence of a bud is not essential for the transmission of all viruses. Spike-disease of sandal, which has not been transmitted by sap inoculation, can be transmitted by inserting pieces of stem tissue from diseased trees between the wood and bark of healthy trees, callus formation between the two apparently sufficing for the passage of virus from the infected to healthy tissue (SREENIVASAYA, 1930). Similarly, peach mosaic (COCHRAN and RUE, 1944) and psorosis of citrus (WALLACE, 1947) have been transmitted by placing pieces of infected leaves under the bark of healthy trees. Only cut pieces of leaf are successful, for the cut edges develop callus which unites with cambial callus from the host, whereas whole leaves form no callus.

In working with strawberry, HARRIS (1932) has used a method of inarching runners. Equal areas of epidermis and cortex are removed from runners of healthy and diseased plants and oblique cuts are made into them. The pieces are interlocked so that the cut surfaces of the two runners are in contact, and the junction is bound. The virtue of this method is that the scion is nourished by the parent plant until union with the stock is firm. A variation of inarching that is useful with some plants, such as raspberries, is "bottle-grafting". The lower end of the scion is kept in water or a nutrient solution, which keeps it alive, until cut surfaces of the scion and stock unite firmly. Viruses can also be transmitted by grafting infected and healthy tubers and bulbs. With potatoes, core grafting is a convenient method, which can be done during the dormant season, and symptoms show soon after the shoots emerge. A piece of a healthy tuber is removed by means of a cork borer, and into the hole so made is inserted a piece, carrying a bud, that has been taken from an infected tuber by means of a slightly larger cork borer (MURPHY and MCKAY, 1926). For most transmissions, infected scions have been taken and grafted to healthy stocks, but the reverse procedure sometimes has advantages. For instance, when a tree is suspected of being virus-infected, more rapid and reliable results may accrue from grafting it with several healthy scions, than by taking one or two scions from it and grafting to healthy trees. Systemic infection is not always complete and small scions may be virus-free, or delay in passage of virus from scion to stock may be considerable, whereas with many scions grafted onto an infected stock passage of virus into some is likely to occur relatively soon.

An ingenious method of transmission, which can be regarded as a variant of grafting and promises to prove of considerable value in increasing our knowledge of some viruses, is the use of the plant parasite, dodder (*Cuscuta* sp.) as a vector. The writer saw this method attempted by Dr. D. J. MACLEOD in 1937 when he was working at the Potato Virus Research Station, Cambridge, but the first published accounts of its use are by BENNETT (1940) and JOHNSON (1941). The method consists of first establishing the parasite on an infected plant and then allowing the dodder stems to attach themselves to healthy plants. BENNETT found that *Cuscuta californica* transmitted cucumber mosaic virus and *C. subinclusa* sugar beet curly top virus. BENNETT (1944) later observed that symptomless plants of *C. californica* found parasitising some desert shrubs often contained a previouly undescribed virus, which he named dodder latent mosaic, and which passed to and caused serious diseases in sugar beet, potato, cantaloupe and many other plants when parasitised by the dodder. Using *C. campestris*, JOHNSON transmitted aster yellows, bushy stunt, cucumber mosaic, beet curly top and tobacco mosaic viruses, but not tobacco ringspot or pea wilt viruses. Dodder is most successful in transmitting viruses to which it is susceptible, for then the systemic invasion of the parasite by the multiplying virus will ensure its passage from plant to plant, but it can also transmit viruses, like tobacco mosaic virus, which do not multiply in it, and for which it presumably acts merely as a conducting channel. COCHRAN (1946) has shown that tobacco mosaic virus readily passes from infected to healthy tomato plants through connecting dodder if the dodder is heavily pruned and if the healthy plants are shaded while the diseased ones are exposed to bright light, whereas few transmissions occur if the dodder is allowed to proliferate unchecked and the healthy plants are allowed to photosynthesise normally.

KUNKEL (1943) has eloquently argued the value of this method of transmission in providing some viruses with host plants that are more suitable for experimental work than those in which they occur naturally. Viruses that can be transmitted only by grafting, or by insects with a restricted host range, have been in the past susceptible of study only in plants that can be inter-grafted or are hosts for the specific vectors. By using *C. campestris* as a vector, KUNKEL (1945) has transmitted cranberry false blossom virus from cranberry, its only previously known host and an extremely difficult plant to work with experimentally, to 28 different species, belonging to 10 different families, and including such convenient experimental plants as tomato and tobacco. The viruses causing potato witches'-broom and peach mosaic, both previously thought to have very narrow host ranges, were also transmitted by dodder to such varied plants as *Vinca rosea,* tomato and sugar beet. There would seem to be great potentialities in the use of dodder as a vector, for there are many species of *Cuscuta* with different host ranges, and there are probably few species of plants that cannot be simultaneously parasitised by one species or another. The single species *C. campestris* is known to be capable of parasitising more than 100 different plants.

There is little doubt that the use of dodder as a vector to provide new hosts for viruses whose host ranges were previously restricted will greatly facilitate the study of these viruses, and will make possible comparisons between them and other viruses which have up to now been impossible. However, the method needs to be used with circumspection, for it has some potential dangers. By providing new hosts for these viruses, the research worker may also be providing them with new methods of spread, for their new hosts may have insect pests that are capable of acting as vectors but so far have not functioned as such because they do not feed on plants found naturally infected. For instance, some of the many insects that infest the new hosts to which cranberry false blossom, potato witches'-broom, or peach mosaic viruses have been transmitted, may be vectors of one or other virus; if such an insect should chance to feed on experimentally infected plants, it might start to spread what would be in effect a new disease. This would, no doubt, add to the virus worker's knowledge, but the cost might be high.

Attempts to use other plant parasites as vectors of viruses have as yet failed. JOHNSON and JONES (1943) failed to transmit pea mosaic and pea enation mosaic viruses by nodule-forming bacteria (*Rhizobium leguminosarum*) and by six different fungi, namely, *Ascochyta pisi, Cladosporium pisicola, Erysiphe Polygoni, Rhizoctonia solani, Fusarium* sp. and *Pythium* sp. Similarly, HANSEN (1947) failed to transmit sugar beet yellows virus with *Ramularia betae, Peronospora Schachtii,* or *Uromyces betae,* and at Rothamsted we have consistently failed to transmit potato virus X and the tobacco necrosis viruses by soil-inhabiting fungi like *Rhizoctonia solani* and *Thielaviopsis basicola.*

**Inoculation With Plant Extracts:** — One of the chief values of finding new hosts for plant viruses that can now be transmitted only by grafting or by insects, is the possibility of discovering hosts between which they may be transmitted mechanically by rubbing extracts of diseased plants on to healthy leaves, for only when a virus is sap-inoculable can its properties *in vitro* be investigated conveniently. Whether or not a virus has been transmitted by inoculation of sap is often taken as an intrinsic property of the virus itself. It may be true that properties of some viruses prevent mechanical transmission, but there is now abundant evidence that properties of the host plant often determine transmissibility. Failure of sap-transmission may simply mean that the virus content of the sap is too small, *i.e.*, that the concentration is below the dilution end-point; the concentration reached by the same virus in different hosts may vary widely, and different hosts also vary in their susceptibility to infection when rubbed with the same virus-containing fluid. Dandelion yellow mosaic virus, for example, has not been transmitted from infected to healthy dandelion, but it has occasionally been transmitted from dandelion to lettuce and from lettuce to dandelion; it is readily transmitted from infected to healthy lettuce for this host is more easily infected than dandelion and also supports a higher concentration of virus (KASSANIS, 1947). Again the host can influence inoculability because its sap may contain substances that either inhibit infection or inactivate the virus. Cucumber mosaic virus is readily inoculable between most hosts, but it is difficult to transmit from *Phytolacca* sp. (DOOLITTLE and WALKER, 1925) or sugar beet (HOGGAN, 1933) to tobacco because extracts of these plants contain substances that interact with the virus and render it largely non-infective. No virus that infects rosaceous plants has yet been transmitted by sap-inoculation. This may be a natural property of these viruses, but the sap from strawberry leaves contains so much tannin that it inactivates tobacco mosaic virus so that the failure of sap inoculation with strawberry viruses ceases to have any relevance in considering the properties of the viruses themselves (BAWDEN and KLECZKOWSKI, 1945).

In early work on the inoculation of viruses it was customary to inject sap into the plants or to place drops of sap on the surface of leaves and scratch them with a needle. These methods were reasonably successful with some viruses, but with others gave inconsistent results and are now rarely used. Better results are obtained by a gentle rubbing that produces no obvious damage to the leaves. A convenient method for transferring virus from one plant to another is to macerate infected leaves with a pestle and mortar and then rub the extracted sap gently with the pestle, forefinger or a piece of cheesecloth, over the upper surface of the leaves of the test plant. Precisely how infection occurs is uncertain, but presumably virus particles enter cells through injuries produced by rubbing, but, viruses being obligate parasites, injuries severe enough to kill cells would be expected to prevent infection. The breaking of hair cells is often thought to be the most important type of wounding, but some results of BOYLE and MCKINNEY (1938) throw doubt on this. They find that the number of infections obtained with tobacco mosaic virus does not depend on the number of hair cells and in pepper leaves they obtained as many infections on parts of leaves without as with trichomes. They obtained few infections with inoculation methods that affected only the trichomes and they suggest that bruising the epidermal cells is more important than injuring the trichomes.

Some viruses that are difficult to transmit by rubbing leaves with infective sap are readily transmitted if a suitable abrasive is incorporated in the inoculum. Sand was the first to be used (FAJARDO, 1930; SAMUEL and BALD, 1933), but this is much less effective than carborundum, kieselguhr, animal charcoal and some other substances since discovered. The best way to use these abrasives is to dust them over the leaf surfaces and then lightly rub with the inoculum. The value of carborundum was first demonstrated by RAWLINS and TOMPKINS (1934, 1936) who found that sap from lettuce infected with tomato spotted wilt virus, and from beans, celery and cabbage infected with various viruses, rarely gave infections with ordinary inoculation methods but readily did so when carborundum was present. The small carborundum crystals pierce the epidermal cells apparently without killing the parenchymatous cells beneath. The particle size of the carborundum is important, 400-500 mesh being more effective than either smaller or larger particles (COSTA, 1944). How the abrasives facilitate transmission is uncertain, but their effects are not restricted to viruses normally difficult to transmit. From TABLE 3 (KALMUS and KASSANIS, 1945) it will be seen that the use of carborundum increases the number of local lesions produced by tobacco mosaic virus on *Nicotiana glutinosa*, the increase being approximately the same as if the virus concentration were increased a hundred-fold.

These results do not explain the mechanism of the abrasive action, but they offer an explanation for the usefulness of abrasives in giving positive results with viruses normally difficult to transmit. It is probable that sap from plants infected with such viruses has a virus content at or below the normal dilution end-point, and it is significant that all such viruses have on occasion been transmitted without abrasives. A comparison of the behaviour of potato virus A and Y in transmission tests to tobacco and President potato plants will illustrate the factors involved. Without abrasives sap from tobacco plants infected with virus Y will usually give infections in tobacco when diluted to 1/10,000, whereas potato ceases to be infected when sap is diluted more than 1 in 500, that is, inoculum needs to be about twenty times as concentrated to produce infection in potato as in tobacco. The dilution end-point is increased about fifty-fold for inoculations to both hosts if an abrasive is used. Potato virus A is sap-inoculable to tobacco, giving a dilution end-point of less than 1 in 50, but not to potato, which is infected only if an abrasive is used (BAWDEN, 1936). Clearly if potato and tobacco plants have the same relative susceptibilities to virus A as to virus Y, the failure of virus A to infect potato without abrasives is simply explained, for the virus content of sap will be just below the dilution end-point for this host, whereas with abrasives it will be slightly above.

TABLE 3. — *Effect of carborundum 400-mesh on the dilution end-point of tobacco mosaic virus:—*

| | *Average number of lesions per leaf at different virus concentrations (g./ml.)* | | | | |
|---|---|---|---|---|---|
| *Inoculum:* | $10^{-5}$ | $10^{-6}$ | $10^{-7}$ | $10^{-8}$ | $10^{-9}$ |
| With carborundum | 407 | 166 | 26 | 5 | 1 |
| No carborundum | 24 | 4 | 2 | 0 | 0 |

As abrasives obviously cannot increase the virus content of an inoculum, they must act by reducing the resistance of the host. They may do this merely by increasing the number of cells injured by rubbing and so increase the likelihood of virus particles finding entry points. The effect of abrasives in increasing the permeability of leaves has been demonstrated by KALMUS and KASSANIS (1945) who treated leaves with ammoniacal silver hydroxide, which turns black when it reacts with reducing substances under exposure to light. When unwounded leaves are placed in the reagent, there is no blackening, and when leaves rubbed with the forefinger are treated they become stippled with black spots, in hairy leaves the spots being mainly at the bases of trichomes. After rubbing with abrasives, on the other hand, treated leaves become uniformly blackened (Fig. 18). The abrasives remove the cuticle and the trichomes and temporarily make leaves highly susceptible to infection when they are simply sprayed with inoculum, an effect not obtained by ordinary rubbing. Within three hours of rubbing with abrasives, the leaves regain their resistance to infection by sprayed virus solutions, though they remain permeable to salts, presumably because the cuticle is not repaired. It seems possible that, in addition

FIG. 18. — Effect of abrasives in facilitating transmission and in increasing permeability of leaves. Left-hand half-leaves rubbed with water; right-hand half-leaves rubbed with water and celite. Leaf *at left* sprayed with suspension of tobacco mosaic virus and leaf *at right* treated with ammonical silver hydroxide (argentaffin test). Note many more lesions and much more uniform staining of half-leaves rubbed with celite. (H. KALMUS and B. KASSANIS, Ann. appl. Biol. *32*, 230, 1945).

to increasing the number of wounds, abrasives may facilitate infection by allowing the inoculum to reach cells which are unaffected by ordinary rubbing and which are more susceptible than the trichomes or cuticle.

There is evidence from many different sources that cells differ in their susceptibility to infection in the sense that inoculum that will cause infection in one may not do so in another. Also, the conditions that favour infection with one virus do not necessarily favour infection with another. For example, if a *Nicotiana glutinosa* plant in the 8 to 10 leaf stage has opposite halves of its leaves rubbed with tobacco mosaic and tomato bushy stunt viruses, the former will cause lesions on all the leaves, though most will be produced on the middle and lower ones. By contrast, tomato bushy stunt virus will produce no lesions on the lowest leaves, a few on the middle ones and most on the upper. As rubbing with both inocula will have broken equal numbers of trichomes and bruised the epidermal cells equally, it seems that wounds *per se* do not necessarily constitute infection points; the injured cells must be in a suitable receptive state and this state differs for different viruses.

It is generally accepted that uninjured cells are immune from infection, but this view has been questioned on several occasions. DUGGAR and JOHNSON (1933) obtained infections with tobacco mosaic virus by spraying plants with infective sap through a fine atomiser, and concluded that the virus was entering through the stomata. SMITH and BALD (1935) also obtained infections with a tobacco necrosis virus merely by spraying infective sap over leaves believed to be undamaged, and SMITH (1935) states that large numbers of local lesions were produced on tobacco seedlings when they were sprayed with suspensions of potato virus X. On the other hand, CALDWELL (1931, 1932) states that he injected the intercellular spaces of *Nicotiana glutinosa* with extracts of tomato aucuba mosaic virus and never obtained infections unless cells were punctured, and that the exudates from hydathodes of infected tomato plants do not contain virus. Similarly, JOHNSON (1936, 1937) failed to demonstrate stomatal infection of a highly susceptible tobacco hybrid by spraying with tobacco mosaic virus, and SHEFFIELD (1936) obtained no infections when uninjured *Nicotiana glutinosa* were sprayed with aucuba mosaic virus. PRICE (1938) has repeated spraying experiments with tobacco necrosis on a number of different hosts. After they were sprayed, the plants were divided into two lots. The leaves of one lot were rubbed while the others were untouched. The leaves of the rubbed plants developed some hundreds of lesions, whereas most of those unrubbed remained healthy and the others developed only a few lesions. Significantly more lesions developed on unrubbed tobacco than on other plants. PRICE suggests that this is not because tobacco plants are more liable to stomatal infection, but because they have a compact type of growth and their leaves have more chances of rubbing together during natural growth movements. When tobacco plants were trimmed so that their leaves no longer rubbed together, no lesions developed on sprayed plants. With such conflicting evidence the possible occurrence of stomatal infection in the absence of wounds cannot be ruled out. But if it occurs it must be seldom, and it seems that for all practical purposes it can be ignored.

Although mechanical transmission is most important in experimental work, it is by no means negligible in the spread of virus diseases in nature. Some of the viruses occurring most commonly, such as tobacco mosaic and potato X, do not appear to be insect-transmitted. These are transmitted extremely easily by inoculation of sap and it is highly probable that they are transmitted in nature mechanically. Tobacco mosaic virus is so stable that it is not completely inactivated by the processes of curing tobacco. Its introduction into growing crops of tobacco and tomato can result from workmen handling the plants after using contaminated tobacco, or from infected debris in the soil. Once established in the crop, the virus spreads rapidly by the rubbing together of diseased and healthy plants and by cultural operations involving the repeated handling of plants. LOUGHNANE and MURPHY (1938) have shown that potato viruses X and F also spread between plants whose haulms are in contact, the rate of spread being increased when the rubbing of infected and healthy leaves is increased by allowing a fan to play on them. Potato spindle tuber is transmitted between tubers by the knife used for cutting the seed sets but potato virus X does not seem to be transmitted in this way. Tulip breaking virus also is readily carried from infected to healthy plants on knives used for cutting blooms, but narcissus mosaic virus apparently is not (VAN SLOGTEREN and OUBOTER 1941*a*, and 1941*b*).

The possibility of the transmission of viruses below ground-level by the infection of roots has been frequently raised. There is a good deal of evidence that it does occur, but again the results of different workers are conflicting. That the soil is a common source of primary infection for tobacco mosaic virus has been shown repeatedly (MCKINNEY, 1927; LEHMAN, 1934; JOHNSON, 1937). It has usually been assumed that infection from the soil takes place through the roots, although this has not been satisfactorily demonstrated and various workers have found it difficult to infect plants by inoculating roots. JOHNSON (1937) states that infection through the roots does not take place commonly even if the roots are deliberately wounded. He considers that infection of plants from infested soil takes place only through contact between the leaves or stems and the soil. Some other work would also suggest that if root infection occurs it is probably not of great economic importance.

SMITH (1937*a*, 1937*b*) has found that the roots of tobacco plants grown in soil contaminated with tobacco necrosis viruses become infected but the tops, except for occasional leaves touching the soil, remain free from virus. PRICE (1938) has confirmed this. He has also found that tobacco and *Nicotiana glutinosa* grown in soil contaminated with tobacco mosaic virus behave similarly. The roots of many plants become infected, but the leaves usually remain free from virus. Roots grown in water culture solutions containing the virus do not become infected unless they are rubbed. The virus is therefore unable to enter uninjured roots and presumably enters those grown in soil or sand through abrasions made by the growth movements of the roots. This work provides a definite demonstration of root infection, but the most interesting and surprising result is the limitation of infection to the roots. With tobacco necrosis viruses, or with tobacco mosaic virus in *Nicotiana glutinosa*, this is not so surprising, for when leaves are inoculated with these viruses only local lesions are produced. But the infection of a tobacco leaf with tobacco mosaic virus is followed by the spread of the virus to all parts of the plant including the roots. It seems, therefore, that although the virus is able to move freely from the stem down into the roots, it is unable to move from the roots to the stem. Unfortunately, PRICE kept his plants for only three weeks and it is therefore not possible to say whether the virus would ultimately pass into the tops, but the results support the statement of JOHNSON (1937) that root infection is of little economic importance.

At Rothamsted we have grown tobacco and tomato plants in soil contaminated with tobacco mosaic virus and kept such plants for months. Often the roots of these plants contained much virus, extracts producing hundreds of lesions per leaf when tested on *N. glutinosa*, although the leaves looked healthy and were virus-free. Occasionally, however, individual plants kept for long periods did develop symptoms in the tops. We attempted to encourage infection by cutting off the tops to force out new side shoots and by putting plants in the dark for some days to prevent the flow of food-stuffs from the leaves to the roots.

Both treatments slightly increased the number of plants that became infected in the leaves, but neither consistently caused the virus to move from the roots. FULTON (1941) infected the roots of tobacco and tomato plants with several different viruses by direct inoculation, but with no virus did he get any infection of aerial parts. The downward movement from the point of inoculation was as rapid as when stems were inoculated, but the upward movement in the roots was slow and restricted. This reluctance to invade stems from the roots, however may not be typical of all viruses. ROBERTS (1946) has found that potato virus X can spread from infected to healthy plants which are in contact only below ground and tomato plants infected through their roots showed leaf symptoms.

SMITH (1937*b* and *c*) found that if he atomised suspensions of tobacco necrosis viruses above tobacco plants the roots became infected. He also reports frequent natural spread among the plants in his glasshouses when growing in conditions usually regarded as safe against accidental virus infections. To account for these facts he suggests that the virus is air-borne, implying that it is normally carried in the air, from which it falls on to the soil and then enters the roots. SMITH's results show clearly that plant roots become infected with these viruses when the soil is watered with infective suspensions, and they have been confirmed by PRICE (1938). It is, therefore, to be expected that infections would also be obtained when suspensions are sprayed into the air, for the spray will settle on to the soil and the final result will be equivalent to adding a small volume directly to the soil. The successful demonstration that a virus is air-borne requires more than this. It is necessary to show not only that virus deliberately placed in the air can infect a plant, but that the virus can actually get into the air. There is at present no evidence for this either with the tobacco necrosis viruses or any others. PRICE observed no natural infections with tobacco necrosis in glasshouses containing large numbers of infected plants. Infected leaves could, of course, be blown from one plant on to the soil in the neighborhood of another, and infections may arise in this manner. The natural infections with this virus probably result from growing plants in infested soils, using contaminated water, or from the accidental mixture of infective plant debris with the soil. This seems especially likely as the natural infections are confined to the roots and to occasional leaves touching the soil, for if the virus were present in the air it would not all be expected to settle on to the soil. Some would fall on the leaves and the mere rubbing of healthy leaves would be expected to produce lesions, but this does not happen.

In the Rothamsted glasshouses natural infections occur, but these are often caused by viruses or virus-strains distinct from those being worked with in large quantities. They also happen as frequently in houses containing no artificially infected plants as in those containing large numbers. Hence it seems improbable that artificially infected leaves supply the sources of many of these natural root infections, and more likely that some soil factor is responsible.

**Transmission by Insects:** — Some viruses, such as potato X and tobacco mosaic, seem to spread largely, if not exclusively, without the aid of any specific vectors, but most depend for transmission on the activity of insects. Also, although the conditions under which plants are grown may influence susceptibility to infection and so play some part in determining the regional or seasonal occurrence of some diseases, there is little doubt that environmental conditions that affect the multiplication and movement of insects are usually most important in determining distribution. How insects transmit viruses is uncertain, but the process does not seem to be a mechanical carriage of infective sap on the exterior of the insect. Transmission depends on the insects feeding

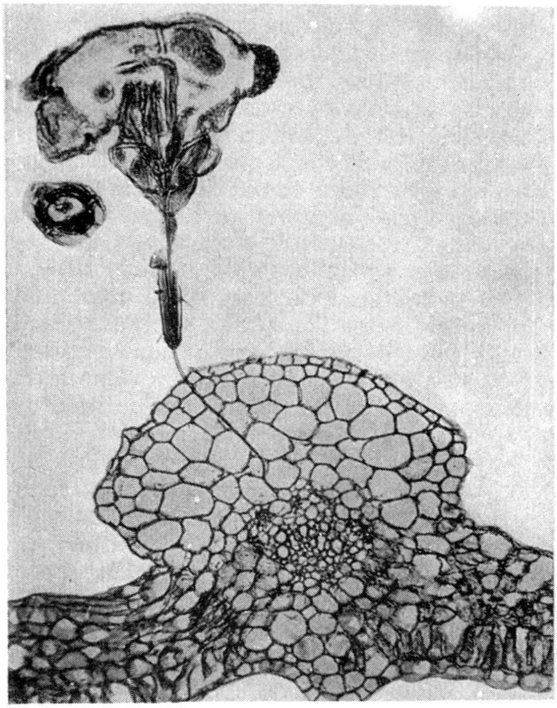

FIG. 19. — Section of potato leaf showing the path of the stylet of *Myzus pseudosolani*. Note the path is intercellular through the cortex and that the aphid is feeding in the phloem. × 80. (DYKSTRA, T. P. and WHITAKER, W. C., 1938, J. agr. Res. *57*, 319).

first on an infected plant, where they acquire virus, and then moving to and feeding on healthy plants. Suggestions have been made that insects infect plants in other ways, for example, by defaecation and mechanical damage, but these remain unconfirmed. On the other hand, it has been shown conclusively that insects act as vectors when deprived of all contact with plants except by their mouth-parts, and it seems likely that insects become infective and cause infections only while feeding.

The great majority of insects identified as vectors feed by sucking plant juice rather than by biting, and there are few authenticated records of insects with biting mouthparts acting as vectors. An exceptional virus in this respect is turnip yellow mosaic, which MARKHAM and SMITH (1949) have found to be transmitted by two species of flea-beetle, the mustard beetle, an earwig and a grasshopper, but not by any sucking insects tested. Other viruses which biting insects have been stated to transmit, also have sucking insects as vectors, and most of the statements have gone unconfirmed. The reasons for the failure of biting insects to make efficient vectors for most viruses are unknown, but it may be that these insects do too much damage to the cells on which they feed. However, there are probably additional and more specific reasons, for among insects possessing sucking mouthparts, and which appear to feed similarly, some are vectors and others not. Also, individual viruses are usually transmitted by one species of insect, or by a few closely related species, but not by other species, although these in turn may be vectors of other viruses. There are exceptions to this specificity of virus-vector relationship, of which the most striking is potato spindle tuber virus, which has been found to be transmitted by aphids (SCHULTZ and FOLSOM, 1925), grasshoppers, the tarnished plant bug, flea beetles, the leaf beetle and the larvae of the Colorado beetle (GOSS, 1931). A virus causing a mosaic of squash also has been found to be transmitted by beetles as well as by aphids (FREITAG, 1941), and onion yellow dwarf by more than 50 different aphids (DRAKE, HARRIS, and TATE, 1933).

By contrast with these examples, some other viruses show such great vector specificity that different strains of one virus have different vectors. Celery yellows virus, for example, is transmitted by *Thamnotettix montanus* and *T. geminatus,* whereas the related strain that causes aster yellows is not, but the two strains have a common vector in *Cicadula sexnotata* (SEVERIN, 1934). The vector-specificity of aster yellows virus, however, is less than previously thought, for SEVERIN (1947) has now transmitted it with 22 different species of leafhopper, belonging to twelve genera. What are considered to be related strains of potato yellow dwarf virus also have distinct vectors, for BLACK (1941) states that the New York strain is transmitted only by *Aceratagallia sanguinolenta* and the New Jersey strain only by *Agallia constricta.* Similarly, with what are thought to be strains of beet curly top virus, the one found in the Argentine is transmitted by *Agalliana ensigera* and not by *Eutettix tenellus,* the vector of the strain found in North America (BENNETT, 1946), and neither *Myzus persicae* nor any of the other aphids that transmit potato virus Y seems to be a vector of the serologically related potato virus C (BAWDEN and KASSANIS, 1947).

New vectors of plant viruses are constantly being discovered and there is no doubt that many viruses for which the only methods of transmission now known are grafting or sap inoculation are also transmitted by certain insects. Recent work has revealed vectors among new groups of insects, mealy-bugs, frog-hoppers, and those with biting mouthparts, and it is more than probable that other groups of insects not yet studied also contain vectors. The difficulties of discovering vectors do not arise simply because of the many species that infest a host, but also because certain conditions must often be fulfilled before an insect that can transmit will do so. In general, success is more likely to be achieved in transmission experiments when plants showing optimal symptoms are used as sources of infection and the insects are transferred to rapidly growing young test plants. Tests should also be made with insects fed for various periods on both the source of infection and the healthy test plants, for, as is shown in Chapter 5, variations in these respects affect the transmission of different viruses differently. In addition to the length of feeding periods, the condition of the host tissues on which insects feed may determine whether or not they act as vectors. The white fly that transmit cassava mosaic can live successfully on old leaves, and become infective from them, but they can only cause infection when they feed on young leaves less than one-quarter fully grown (STOREY and NICHOLS, 1938). Also, transmission may occur more easily with one host species than with another, for hosts vary in their susceptibility to infection and in the concentration of virus reached in their tissues, and the latter will affect the ability of vectors to become infective while feeding on them. Dandelion yellow mosaic virus, for example, is readily transmitted by *Myzus ornatus* to and from lettuce seedlings, but only with difficulty to and from dandelion (KASSANIS, 1947). An odd property of some viruses is that they are insect-transmitted when they occur in plants infected simultaneously with another virus for which the insect is a vector, but not otherwise. This phenomenon was first discovered with potato virus F, the cause of tuber blotch, which *Myzus persicae* does not transmit from infected plants unless they also contain potato virus A (CLINCH, LOUGHNANE and MURPHY, 1936). Another example is afforded by tobacco rosette, which SMITH (1945) has found to be caused by the joint action of two viruses, namely, vein-distorting and mottle, which differ strikingly in their methods of transmission. The vein-distorting virus is transmitted by *M. persicae*, but not by sap inoculation, whereas the mottle virus is sap-inoculable but on its own is not transmitted by the aphid. When present in plants simultaneously infected with the vein-distorting virus, that is, plants showing rosette symptoms, the mottle virus is transmissible by *M. persicae*. After feeding on rosette plants, individual aphids remain infective for some days and when transferred to a series of healthy plants at intervals may infect some with both viruses or with either one separately.

In the field different virus diseases spread at greatly different rates, some sufficiently rapidly to become epiphytotic in annual crops and others so slowly that in plantation crops it may take years for the numbers of infected plants to become doubled. Such differences may depend on differences between the susceptibility of different hosts to infection, on the virus content produced in infected plants, which together with other factors will affect the ease with which vectors become infective, and on differences between the prevalence and activities of the specific vectors. In searching for a vector of a particular virus, it is natural that common pests of the host should first receive attention, but the vector will often not be found among them. Diseases that spread widely may be transmitted by insects that are unimportant as pests and are never numerous in the crop. The devastation caused by peach yellows results from the activities of the leafhopper *Macropsis trimaculata*, which has only one brood a season, is active for only a few weeks during the summer and is never abundant (KUNKEL, 1933). Even when a common pest is found experimentally to be a vector, it may still be unimportant in the field. Experimentally *Aphis rhamni* is as efficient a vector of potato virus Y as is *Myzus persicae*, but in the field heavy infestations of *A. rhamni* occur commonly without a corresponding spread of the virus, whereas light infestations of *M. persicae* often lead to much spread. Similarly, both *Aphis fabae* and *M. persicae* are efficient vectors of sugar beet yellows virus in glasshouse tests (WATSON, 1946), but in the field spread depends largely on the activities of *M. persicae*, although it is rarely present in large numbers, and there may be little or no virus spread in crops blackened by heavy infestations of *A. fabae*. This varying efficiency of vectors in experiments and in the field probably reflects differences in the natural behaviour of the species, particularly differences in their movements. *M. persicae* is a relatively active species which moves freely and gives uniform infestations over a crop, whereas *A. rhamni* and *A. fabae* move less freely and build up dense, localised infestations. It is important to realise that two species may be equally capable of acting as vectors when they are transferred from infected to healthy plants, but until they move they are only potential vectors, and their behaviour in the field will depend on the frequency with which they move. A few individuals of an active species will do more transmitting than many sluggards. It seems likely that most transmissions of aphid-transmitted viruses are made by the relatively few winged migrants, whereas the much larger populations of wingless individuals, which are so damaging as direct pests, are of little consequence as vectors.

Before an insect-transmitted virus can spread, three conditions must be fulfilled: there must be a source of the virus, the specific vector must be present, and the vector must be moving about. Other things being equal, the number of new plants to become infected in a given time will be the product of these three factors, and a change in one may be compensated for by a change in another. For example in the south-east of Britain, the rate of increase of potato virus diseases is not correlated with the size of the *M. persicae* populations reached in the crops (DONCASTER and GREGORY, 1948), more spread sometimes occurring in fields with light infestations than in those with heavy ones. These discrepancies are probably largely accounted for by differences between conditions in the crops that determine the proportion of winged to wingless aphids and the amount of aphid movement. In dense crops where the stems of individual plants become entangled, wingless individuals will be afforded opportunities to move easily from plant to plant, whereas in more open crops only winged individuals are likely to do this. However, there are probably many other factors involved, of which the time of infestation may be important, for the susceptibility of plants to infection may alter with age. In north Wales, WHITEHEAD (1943) has noticed that spread of potato virus diseases depends as much on the time at which migrants first enter the crop as on the size of infestation ultimately achieved, more spread occurring the earlier the aphids arrive.

For an insect-transmitted virus to cause serious losses in an annual crop raised from seed, it is obvious that it must spread rapidly, a condition that can only be achieved if the crop is easily infected and if the vector is either very numerous or very active. On the other hand, in perennial crops, or in those propagated vegetatively, because the plants are exposed to infection over a long period, viruses may do serious damage though they are not easily transmissible and their vectors are relatively few and inactive. The losses caused by swollen shoot of cacao,

TABLE 4

INSECT VECTORS

| *Insects* | *Virus transmitted* |
|---|---|
| *Orthoptera* (Grasshoppers): | |
| *Chorthippus bicolor* (Charp.) | Turnip yellow mosaic |
| *Dermaptera* (Earwig): | |
| *Forficula auricularia* Linn. | Turnip yellow mosaic |
| *Thysanoptera* (Thrips): | |
| *Frankliniella insularis* Frankl. | Tomato spotted wilt |
| *Thrips tabaci* Lind. | —— —— —— |
| *Hemiptera* | |
| *Cercopidae* (Frog-hoppers): | |
| *Aphrophora angulata* Ball | Pierce's grapevine disease (Alfalfa dwarf) |
| *Aphrophora permutata* Uhler | Pierce's grapevine disease |
| *Clastoptera brunnea* Ball | —— —— —— |
| *Philaenus leucophthalmus* (Linn.) (several varieties) | —— —— —— |
| *Jassidae* (Leaf-hoppers): | |
| *Agallia constricta* Van Duzee | Potato yellow dwarf (New Jersey strain) |
| *Agallia quadripunctata* Prov. | Wound-tumour |
| *Aceratagallia sanguinolenta* (Prov.) | Potato yellow dwarf (New York strain) |
| *Agalliana ensigera* Oman | Sugar beet curly top (Argentine) |
| *Agalliopsis novella* Say. | Wound-tumour<br>Clover club-leaf |
| *Baldulus maidis* (DeL. and W.) | Corn stunt |
| *Carneocephala fulgida* Nott | Pierce's grapevine disease |
| *Cicadula sexnotata* Fall. | Aster yellows (Also transmitted by 20 other leaf-hoppers).<br>Celery yellows |
| *Cicadulina mbila* Naude | Maize streak<br>Maize mottle |
| *Delphax striatella* Fall. | Cereal mosaic |
| *Draeculacephala minerva* Ball | Pierce's grapevine disease |
| *Draeculacephala portola* Ball | Sugar cane chlorotic streak |
| *Empoasca devastans* Dist. | Brinjal mosaic |
| *Euscelis striatulus* Fall. | Cranberry false blossom |
| *Eutettix phycitis* Dis. | Little leaf of egg plant |
| *Eutettix tenellus* Baker | Sugar beet curly top |
| *Helochara delta* Oman | Pierce's grapevine disease |
| *Macropsis trimaculata* Fitch | Peach yellows |
| *Neokolla circellata* Baker | Pierce's grapevine disease |
| *Nephotettix apicalis* Motsch. | Rice dwarf (stunt) |
| *Peregrinus maidis* Ashm. | Maize stripe<br>Corn mosaic |
| *Perkinsiella saccharicida* Kirk. | Sugar cane Fiji disease |
| *Platymoideus acutus* Say. | Alfalfa witches' broom |
| *Thamnotettix argentata* Evans | Tobacco yellow dwarf |
| *Aleyrodidae* (White flies): | |
| *Bemisia gossypiperda* M. and L. | Cassava mosaic<br>Cotton leaf curl<br>Tobacco leaf curl |
| *Bemisia* sp. | Cassava brown streak |
| *Bemisia tabaci* Genn. | Abutilon variegation |

TABLE 4 *(Continued)*

INSECT VECTORS

| *Insects* | *Virus transmitted* |
|---|---|
| *Aphididae* (Aphids): | |
| *Acyrthosiphon pisum* (Harris) | Bean mosaic |
| | Alfalfa mosaic |
| | Narcissus mosaic |
| | Onion yellow dwarf |
| | Pea enation mosaic |
| | Pea mosaic |
| *Amphorophora rubi* (Kalt.) | Onion yellow dwarf |
| | Raspberry green mosaic |
| | Raspberry yellow mosaic |
| *Amphorophora sensoriata* Mason | Raspberry green mosaic |
| | Raspberry yellow mosaic |
| *Anuraphis padi* Linn. | Plum pox (mosaic) |
| *Aphis fabae* Scop. | Bean mosaic |
| | Narcissus mosaic |
| | Onion yellow dwarf |
| | Potato Y |
| | Soy bean mosaic |
| | Sugar beet mosaic |
| | Sugar beet yellows |
| | Tobacco etch |
| | Tulip break |
| *Aphis gossypii* Glover | Bean mosaic |
| | Cauliflower mosaic |
| | Cucumber mosaic |
| | Lily rosette |
| | Onion yellow dwarf |
| | Western celery mosaic |
| *Aphis idaei* v. d. Goot | Raspberry curly dwarf |
| *Aphis laburni* Kalt. | Groundnut rosette |
| *Aphis maidis* Fitch. | Onion yellow dwarf |
| | Sugar cane mosaic |
| *Aphis marutae* Oest. | Cineraria mosaic |
| *Aphis rhamni* Fonsc. | Potato A |
| | Potato Y |
| | Soy bean mosaic |
| | Tobacco etch |
| *Aphis spiraecola* Patch | Papaw mosaic |
| *Aulacorthum circumflexum* (Buckt.) | Cauliflower mosaic |
| | Cucumber mosaic |
| | Henbane mosaic |
| | Potato A |
| | Potato Y |
| | Potato leaf roll |
| | Soy bean mosaic |
| | Tobacco etch |
| *Aulacorthum solani* (Kalt.) | Cucumber mosaic |
| | Narcissus mosaic |
| | Potato leaf roll |
| | Soy bean mosaic |
| *Brevicoryne brassicae* (Linn.) | Bean mosaic |
| | Cabbage blackring |
| | Cauliflower mosaic |
| | Onion yellow dwarf |
| | Turnip mosaic |
| *Hysteroneura setariae* Thos. | Sugar cane mosaic |

TABLE 4 *(Continued)*
INSECT VECTORS

| *Insects* | *Virus transmitted* |
|---|---|
| *Macrosiphum euphorbiae* (Thomas) | Bean mosaic |
| | Cucumber mosaic |
| | Iris stripe |
| | Narcissus mosaic |
| | Onion yellow dwarf |
| | Pea enation mosaic |
| | Potato A |
| | Potato Y |
| | Potato leaf roll |
| | Potato spindle tuber |
| | Tobacco etch |
| *Macrosiphum rosae* (Linn.) | Narcissus mosaic |
| *Myzus ascalonicus* Doncaster | Cucumber mosaic |
| | Dandelion yellow mosaic |
| | Henbane mosaic |
| | Sugar beet yellows |
| | Tobacco etch |
| *Myzus cerasi* (Fab.) | Narcissus mosaic |
| *Myzus ornatus* Laing | Dandelion yellow mosaic |
| | Cucumber mosaic |
| | Potato Y |
| | Potato leaf roll |
| | Soy bean mosaic |
| *Myzus persicae* (Sulz.) | Bean mosaic |
| | Cabbage blackring |
| | Cauliflower mosaic |
| | Commelina mosaic |
| | Cucumber mosaic |
| | Dahlia mosaic |
| | Henbane mosaic |
| | Iris stripe |
| | Lettuce mosaic |
| | Onion yellow dwarf |
| | Pea enation mosaic |
| | Pea mosaic |
| | Potato A |
| | Potato F |
| | Potato Y |
| | Potato leaf roll |
| | Potato spindle tuber |
| | Potato unmottled curly dwarf |
| | Soy bean mosaic |
| | Sugar beet mosaic |
| | Sugar beet yellows |
| | Tobacco etch |
| | Tobacco mottle |
| | Tobacco vein-distorting |
| | Tulip break |
| | Turnip mosaic |
| | Western celery mosaic |
| *Pentalonia nigronervosa* Coq. | Banana bunchy top |
| *Pentatrichopus fragariae* (Theob.) | Strawberry crinkle |
| | Strawberry witches' broom |
| | Strawberry yellow edge |
| *Rhopalosiphum conii* Davids. | Celery yellow spot |
| *Rhopalosiphum pseudobrassicae* Davies | Bean mosaic |
| | Cauliflower mosaic |
| | Onion yellow dwarf |

TABLE 4 *(Continued)*

INSECT VECTORS

| *Insects* | *Virus transmitted* |
|---|---|
| *Yezabura malifoliae* Fitch. | Narcissus mosaic |
| *Yezabura tulipae* (Fonsc.) | Tulip break |
| *Coccidae* (Mealy-bugs): | |
| *Ferrisia virgata* Ckll. | Cacao swollen shoot (Strains A and B) |
| *Pseudococcus citri* (Risso) | —— —— —— (Strains A, C, and D) |
| *Pseudococcus njalensis* Laing | —— —— —— (Strains A, B and D) |
| *Coleoptera* (Beetles): | |
| *Diabrotica soror* Lec. | Squash mosaic<br>Cucurbit ring mosaic |
| *Diabrotica trivettata* Mann | Squash mosaic<br>Cucurbit ring mosaic |
| *Disonycha triagularis* Say. | Potato spindle tuber |
| *Epitrix cucumeris* Harris | Potato spindle tuber |
| *Leptinotarsa dicemlineata* Say. | Potato spindle tuber |
| *Phaedon cochleariae* (Fabr.) | Turnip yellow mosaic |
| *Phyllotreta cruciferae* (Goeze) | Turnip yellow mosaic |
| *Phyllotreta vittula* Redt. | Turnip yellow mosaic |
| *Systena toeniata* Say. | Potato spindle tuber |

for example, result from spread by vectors virtually incapable of independent movement. This disease does not spread readily unless infected and healthy trees are actually in direct contact with one another, and it seems probable that there are often two insects involved in transmission; mealy-bugs are the insects that acquire virus by feeding on diseased trees and infect healthy trees, but for their movement from tree to tree they largely depend on ants, which "farm" them for the honey-dew.

TABLE 4 gives a partial list of insect vectors and the viruses they have been found to transmit. It will be seen that some viruses have been transmitted by a number of different species of insect whereas others have been transmitted by only one or two. How far this difference in specificity of vectors is real, however, is impossible to say, for with many viruses only a few species have been tested as vectors. The table shows the predominance of leafhoppers and aphids as vectors, and that one insect may transmit many unrelated viruses. The pre-eminence of *M. persicae* as a vector may be less real than it appears; it is commonly the first insect to be tried as a vector of newly-discovered viruses and if it succeeds others are not tested, but its wide host range does make it peculiarly important as a field vector of many different viruses.

**Seed-Borne Viruses:** — Long before the existence of viruses was recognised, it was noticed that seedlings from "degenerate" stocks of plants such as the potato, which are normally propagated vegetatively, were more vigorous than plants raised from tubers. This observation led to the erroneous belief that continued asexual reproduction was in itself weakening and that vigour could be maintained only by frequent recourse to sexual reproduction. With the recognition of viruses, the beneficial effects of sexual reproduction could be re-interpreted as acting by eliminating viruses from the stock, for it was found that whereas viruses are almost invariably perpetuated when infected plants are propagated vegetatively, they rarely enter and survive in seed. The fact that seed from infected plants gives healthy progeny has been repeatedly confirmed with many unrelated viruses, differing as widely in their other properties as do tobacco mosaic and potato leaf roll or sugar beet curly top and potato X, and the wide measure of acceptance given to the belief in the deleterious effect of asexual reproduction itself implies the infrequency of seed-transmission. Consequently most viruses depend for their survival on infecting a continuous series of growing plants and their perpetuation from season to season depends largely on hosts that are biennials, perennials or that are vegetatively propagated. Such hosts usually provide the sources of infection for crops raised annually from seed, although contaminated soil or residues of infected plants may also be important sources of a few viruses. It is, too, largely by the transport of living plants, or of dormant vegetative organs of propagation, rather than of seeds, that most viruses become distributed over large distances. Nevertheless, although there is ample evidence that most viruses are rarely, if ever, seed-borne, there are some notable exceptions. A few viruses do invade and survive in the seeds set by infected plants, some doing so consistently and frequently and others being seed-borne either only occasionally or in certain hosts only.

Transmission through seed seems to be more characteristic of viruses of leguminous plants than of any others, for a number of these have been described as seed-borne. This was suspected by McCLINTOCK (1917) from field observations on the occurrence of a mosaic disease in Lima beans (*Phaseolus lunatus*) but it was first demonstrated unequivocally by REDDICK and STEWART (1919) with a virus causing a mosaic of *Phaseolus vulgaris*. Not all the seeds set by diseased plants are infected and the same pod may contain infected and virus-free seeds. The proportion of infected seeds depends on the length of time the parent plant has been infected; it is highest with plants which are themselves the progeny of infected seeds, and if plants become infected after the blossom has set the seed is all virus-free (FAJARDO, 1930; NELSON, 1932; HARRISON, 1935). The virus causing mosaic of soybean is another that is regularly seed-borne, from 10 to 25 per cent. of the seeds set by diseased plants being infected (GARDNER and KENDRICK, 1921). Seed transmission of a virus has also been reported in *Vigna sesquipedalis* (SNYDER, 1942), but other viruses of leguminous plants, such as pea mosaic, pea enation mosaic, groundnut rosette and lucerne (alfalfa) mosaic, seem rarely, if ever, to be seed-borne.

Lettuce mosaic virus is another that is regularly seed-transmitted, about 5 per cent. of the seed set by diseased plants producing infected seedlings. (AINSWORTH and OGILVIE, 1939). The large losses that occur from this disease are usually attributable to infected seed, for although the initial number of diseased plants is small, they supply early infection foci scattered through the crop, so that spread to adjacent plants can occur as soon as aphids arrive. Seed-transmission also seems to be more common in cucurbits than in most other plants. In squash, MIDDLETON (1944) has found seed-infection to be correlated with the grade of seed and that most infected seeds can be removed by careful winnowing; low-grade, light seed gave up to 37 per cent. infected seedlings, whereas heavy well-filled seed from the same lots gave only

2 per cent. He obtained as many infected plants from seed three years old as when it was freshly harvested; by contrast, with a seed-transmitted virus of muskmelon, RADER, FITZPATRICK and HILDEBRAND (1947) found that two batches of seed that produced 93 and 28 per cent. infected seedlings when sown soon after harvest produced only 3 and 6 per cent. infected seedlings after storage for three years. This virus was transmitted through the seed of *Cucumis melo, Cucurbita moschata, Cucurbita flexosus* and *Cucurbita pepo,* but not through the seed of *Cucumis sativa.*

A few other viruses also have been found to be seed-transmitted in some hosts, but not in others. BENNETT (1944) states that dodder latent mosaic virus is transmitted through about 5 per cent. of the seed set by *Cuscuta campestris,* but is not seed-borne in cantaloupe, pokeweed or buckwheat, all of which show symptoms when infected. In most hosts tomato spotted wilt virus also is not transmitted through the seed, but JONES (1944) states that it is in Cineraria as also is another virus causing mosaic symptoms. HENDERSON (1931) found that tobacco ringspot virus was frequently seed-borne in Petunia but not in tobacco. VALLEAU (1941), on the other hand, working with six different strains of this virus, found that all were seed-transmitted in tobacco. He gives no figures for frequency, but found that infected plants set many fewer seeds than uninfected ones and that much of the pollen set by infected plants was sterile. Many workers have raised large numbers of tomato and tobacco seedlings from seed taken from parents infected with tobacco mosaic virus and obtained only healthy plants, but a few have claimed a small number of seed-borne infections. These may be explained by AINSWORTH's (1933) observation that the virus occurs in the seed coats of tomato seed from infected parents and can be recovered from them even after germination. Although AINSWORTH got no infected seedlings in his tests, the seed coats obviously afford a source of virus from which young seedlings could become infected. Tobacco etch viruses also appear to enter the testa of the seed, for SHEFFIELD (1941) found the characteristic intracellular inclusion bodies there, but not in other parts of the seed. Inoculum taken from immature seed was infective, but no virus could be recovered from fully ripened seed and the seedlings raised from it were all healthy.

It is now clear that although seed-transmission is rare, it is far from negligible, and that because a virus is not seed-borne in one host it cannot necessarily be assumed it will not be in another. The ability to invade seed, like the ability to cause a certain clinical syndrome, is neither an intrinsic property of host nor pathogen, but a specific interaction between the two. In the same host, one virus may be seed-transmitted and another not, and the same virus may be seed-transmitted in one host but not in others.

The failure of most viruses to enter the seed of plants, of which they infect all other parts including the fruits, remains one of the most interesting unsolved problems of virus diseases. A number of theories has been advanced, such as the anatomical isolation of the embryo, the inactivation of the viruses by adsorption on to the seed proteins or by processes of maturation, but none has been supported by sufficient experimental evidence for it to be convincing. It may be that the reduction division is in some way concerned, for viruses that are not seed-borne also seem to fail to enter the pollen. CALDWELL (1934) found that the pollen from tomato plants suffering from aucuba mosaic was not infective; also no virus was demonstrable in pollen from plants suffering from sugar beet curly top (BENNETT and ESAU, 1936) or from plants infected with tobacco mosaic virus and potato virus X (GRATIA and MANIL, 1936). By contrast, REDDICK (1931) states that pollen from *Phaseolus vulgaris* infected with bean mosaic virus, which is regularly seed-transmitted, carries virus and can infect healthy plants by pollinating them.

**Soil-Borne Viruses:** — MAYER (1886) suggested that mosaic of tobacco could be contracted from the soil and BEIJERINCK (1898) showed that the virus could over-winter in infected plant residues in air-dried soil. Since then numerous other workers have confirmed that contaminated soil is an important reservoir of this virus, though the method by which infection occurs is still in doubt, for as shown earlier in this chapter many of the root infections that occur remain localised. However, only an occasional plant needs to become infected from soil-borne virus, because once infection of foliage occurs spread to adjacent plants will be rapid either by leaf contact or by workers acting as vectors. There are other sources of this virus, for example, processed tobacco and alternative hosts, but there is little doubt that many of the outbreaks in tobacco and tomato crops originate from virus surviving in the soil. The virus is so stable that it not only persists in plant roots and other debris remaining when the crop is harvested, but it also persists in composted tomato plants.

With tobacco mosaic virus there is no reason to assume that being soil-borne implies anything other than mere survival, for its properties are such that it would be expected to remain infective for long periods in soil. Other viruses, on the other hand, which lose infectivity much more rapidly, are also soil-borne and for these it would seem probable that there is some active phase of their existence in the soil. Examples of these are the wheat mosaic, lettuce big vein and tobacco necrosis viruses. MCKINNEY (1937) found that wheat mosaic occurred when seed was sown in contaminated soil and could be prevented by treating the soil with formalin. Clean soil can be infested by sprinkling it with infested soil, but not by mixing it with leaves from infected wheat plants (WEBB, 1927, 1928; MCKINNEY, 1946). The disease is more prevalent in heavy than in light soils; in some favourable soils the virus survives for more than 9 years, and MCKINNEY (1946) suggests that it does so in some soil-inhabiting organism that acts as a vector. JOHNSON (1945) finds that the disease is controlled by treating soil with insecticides, which he concludes act on a vector rather than directly on the virus itself, and he suggests the vector may be a nematode.

Big vein of lettuce in many ways resembles wheat mosaic; it is more common on heavy than light soils, its incidence increases with increasing soil moisture, the virus can persist for more than a year and infection is prevented by steam sterilisation (JAGGER and CHANDLER, 1934; THOMPSON, DOOLITTLE and SMITH, 1944; PRYOR, 1944). PRYOR (1946) found that diluting contaminated soil with 800 parts of sterile soil only slightly reduced the incidence of the disease, but he got only few infections when clean soil was watered with unfiltered leachates from infested soil and none with filtered leachates. The occurrence of tobacco necrosis also is only rationally explained by assuming the causative viruses are soil-borne, though these seem to have some more ready method than the other viruses of reaching uninfested soil. SMITH (1937*c*) found that tobacco plants growing in bacteriologically sterile conditions remain uninfected, as they do also in water-culture solutions, but infections are frequent in plants raised in soil that has been recently sterilised. This could be readily explained if the viruses have as hosts some members of the soil microflora, for the recolonisation of sterilised soil by such infected organisms would also re-establish the viruses. However, attempts to test this theory at Rothamsted have given it no support, for the viruses could not be obtained from mixed cultures of soil-inhabiting bacteria, nor could they be transmitted by such pathogenic fungi as *Thielaviopsis basicola* and *Rhizoctonia solani* (BAWDEN and KASSANIS, 1947*a*). Various other viruses, for example, potato X and tomato bushy stunt, give infections when suspensions of them are watered on to soil in which susceptible plants are growing, but there is no evidence to suggest that these are normally soil-borne in the sense that contaminated soil is important in deciding the occurrence of disease outbreaks. It is, however, clear that the question of soil-borne viruses warrants more attention than it has hitherto received and that there is much of interest to be discovered about the ways in which the viruses survive in the soil and infect the growing plants.

## *References:-*

AINSWORTH, G. C. (1933): Ann. Rep. Cheshunt Exp. Sta. *62.*
—— and OGILVIE, L. (1939): Ann. appl. Biol. *26*, 279.
BAWDEN, F. C. (1936): Ann. appl. Biol. *23*, 487.
—— and KASSANIS, B. (1947*a*): Ann. appl. Biol. *34*, 127.
—— and —— (1947*b*): *ibid. 34*, 503.
—— and KLECZOWSKI, A. (1945): J. Pomol. *21*, 2.
BEIJERINCK, M. W. (1898): Verh. Akad. Wet. Amsterdam, *6*, 1.
BENNETT, C. W. (1940): Phytopath. *30*, 2.
—— (1944): *ibid. 34*, 77.
—— (1946): Proc. 4th Meeting Amer. Soc. Sugar Beet Tech., 394.
—— and ESAU, K. (1936): J. Agric. Res. *53*, 595.
BLACK, L. M. (1941): Amer. Pot. J. *18*, 231.
BOYLE, L. M. and MCKINNEY, H. H. (1938): Phytopath. *28*, 114.
CALDWELL, J. (1931): Ann. appl. Biol. *18*, 279.
—— (1932): *ibid. 19*, 144.
—— (1934): *ibid. 21*, 191.
CLINCH, P., LOUGHNANE, J. B. and MURPHY, P. A. (1936): Sci. Proc. Roy. Dublin Soc. *21*, 431.
COCHRAN, G. W. (1946): Phytopath. *36*, 396.
COCHRAN, L. C. and LA RUE, J. L. (1944): Phytopath. *34*, 934.
COSTA, A. S. (1944): Phytopath. *34*, 288.
DONCASTER, J. P. and GREGORY, P. H. (1948): London H.M.S.O. A.R.C. Report Series No. 7.
DOOLITTLE, S. P. and WALKER, M. N. (1925): J. Agric. Res. *31*, 1.
DRAKE, C. J., HARRIS, H. and TATE, H. D. (1933): J. Econ. Ent. *26*, 841.
DUGGAR, B. M. and JOHNSON, B. (1933): Phytopath. *23*, 934.
FAJARDO, T. G. (1930): Phytopath. *20*, 469.
FREITAG, J. H. (1941): Phytopath. *31*, 8.
FULTON, R. W. (1941): Phytopath. *31*, 575.
GARDNER, M. W. and KENDRICK, J. B. (1921): J. Agric. Res. *22*, 111.
GOSS, R. W. (1931): Agric. Exp. Sta. Univ. Nebraska Res. Bull. *53*.
GRATIA, A. and MANIL, P. (1936): Compt. Rend. Soc. Biol. Paris *123*, 509.
HANSEN, H. P. (1947): 10th Meeting Inst. Intern. Rech. Betteravières, Brussels.
HARRIS, R. V. (1932): Journ. Pom. and Hort. Sci. *10*, 35.

HARRISON, A. L. (1935) : N. Y. State Agric. Exp. Sta. Tech. Bull. 236.
HENDERSON, R. G. (1931) : Phytopath. *21*, 225.
HOGGAN, I. A. (1933) : Phytopath. *23*, 446.
JAGGER, I. C. and CHANDLER, N. (1934) : Phytopath. *24*, 1253.
JOHNSON, F. (1941) : Phytopath. *31*, 649.
—— (1945) : Ohio J. Science *45*, 125.
—— and JONES, L. K. (1943) : Plant Dis. Rep. *27*, 656.
JOHNSON, J. (1936) : Amer. J. Bot. *23*, 40.
—— (1937) : J. Agric. Res. *54*, 239.
JONES, L. K. (1944) : Phytopath. *34*, 941.
KALMUS, H. and KASSANIS, B. (1945) : Ann. appl. Biol. *32*, 230.
KASSANIS, B. (1947) : Ann. appl. Biol. *34*, 412.
KUNKEL, L. O. (1933) : Contr. Boyce Thompson Inst. *5*, 19.
—— (1943) : In Virus Diseases, Cornell University Press, Ithaca.
—— (1945) : Phytopath. *35*, 805.
LEHMAN, S. G. (1934) : North Carolina Agric. Exp. Sta. Tech. Bull. *46*.
LOUGHNANE, J. B. and MURPHY, P. A. (1938) : Sci. Proc. Roy. Dublin Soc. *22*, 3.
MCCLINTOCK, J. A. (1917) : Phytopath. *7*, 60.
MCKINNEY, H. H. (1927) : J. Agric. Res. *35*, 13.
—— (1937) : U. S. Dept. Agric. Circ. 442.
—— (1946) : Soil Science *61*, 93.
MARKHAM, R. and SMITH, K. M. (1949) : Parasitology, *39*, 330.
MAYER, A. E. (1886) : Landw. Vers. Stat. *32*, 450.
MIDDLETON, J. T. (1944) : Phytopath. *34*, 405.
MURPHY, P. A. and MCKAY, R. (1926) : Sci. Proc. Roy. Dublin Soc. *18*, 169.
NELSON, R. (1932) : Agric. Exp. Sta. Mich. State Coll. Tech. Bull. *118*.
ORLANDO, A. and SILBERSCHMIDT, K. (1946) : Arq. Inst. biol. S. Paulo, *17*, 1.
PRICE, W. C. (1938) : Amer. J. Bot. *25*, 603.
PRYOR, D. E. (1944) : J. Agric. Res. *68*, 1.
—— (1946) : Phytopath. *36*, 264.
RADER, W. E., FITZPATRICK, H. F. and HILDEBRAND, E. M. (1947) : Phytopath. *37*, 809.
RAWLINS, T. E. and TOMPKINS, C. M. (1934) : Phytopath. *24*, 1147.
—— and —— (1936) : *ibid. 26*, 578.
REDDICK, D. (1931) : Extr. Deux. Congr. Int. Path. Comp. 363.
—— and STEWART, V. B. (1919) : Phytopath. *9*, 445.
ROBERTS, F. M. (1946) : Nature *158*, 663.
SAMUEL, G. and BALD, J. G. (1933) : Ann. appl. Biol. *20*, 70.
SCHULTZ, E. S. and FOLSOM, D. (1925) : J. Agric. Res. *30*, 493.
SEVERIN, H. H. P. (1934) : Hilgardia *8*, 339.
—— (1947) : Phytopath. *37*, 364.
SHEFFIELD, F. M. L. (1936) : Ann. appl. Biol. *23*, 498.
—— (1941) : J. Roy. Micr. Soc. *61*, 30.
VAN SLOGTEREN, E. and OUBOTER, M. P. (1941*a*) : Publication No. 64, Laboratorium Bloembollenonderzoek, Lisse.
—— and —— (1941*b*) : Meded. der Landbouwhoogeschool, Deel 45.
SMITH, K. M. (1935) : Ann. appl. Biol. *22*, 731.
—— (1935) : Plant Viruses. Methuen, London.
—— (1937*a*) : Parasitology *29*, 70.
—— (1937*b*) : *ibid. 29*, 86.
—— (1937*c*) : Nature *139*, 370.
—— and BALD, J. G. (1935) : Parasitology *27*, 231.
SNYDER, W. C. (1942) : Phytopath. *32*, 518.
SREENIVASAYA, M. (1930) : J. Indian Inst. Sci. *13*, 113.
STOREY, H. H. and NICHOLS, R. F. W. (1938) : Ann. appl. Biol. *25*, 790.
THOMPSON, R. C., DOOLITTLE, S. P. and SMITH, F. F. (1944) : Phytopath. *34*, 900.
VALLEAU, W. D. (1941) : Phytopath. *31*, 522.
WALLACE, J. M. (1947) : Phytopath. *37*, 149.
WATSON, M. A. (1946) : Proc. Roy. Soc. B, *133*, 200.
WEBB, R. W. (1927) : J. Agric. Res. *35*, 587.
—— (1928) : *ibid. 36*, 53.
WHITEHEAD, T. (1943) : Ann. appl. Biol. *30*, 85.

*Chapter 5*

# THE RELATIONS BETWEEN VIRUSES AND THEIR VECTORS

The primary purpose of much entomological work with plant viruses, because of its obvious economic importance, has been to identify insect vectors, and with this aim achieved many workers have remained content without studying the factors that may influence the vector's ability to transmit. Nevertheless, enough different viruses have been studied in detail to show that there are at least two different types of relation between them and their insect vectors, and it is likely that among the many that remain unstudied other relations exist. The differences between the two types as yet recognised are almost certainly set by the viruses and not the insects, for the same species, *e.g. Myzus persicae*, may show both relations with different viruses, whereas the same virus when transmitted by several insects behaves similarly with all. The two types of virus differ both in the time that their vectors take to become infective after feeding on infected plants and in the time for which their vectors remain infective. Vectors of the first type may acquire virus by feeding for short periods on infected plants, but they cannot immediately transmit it to healthy plants; after once becoming infective, vectors of this type of virus may remain so for long periods, sometimes for the remainder of their lives. Vectors of the second type, by contrast, can infect healthy plants immediately they have acquired virus, and they soon cease to be able to do so, sometimes becoming non-infective within minutes and always within hours of leaving the infected plants.

The existence of two different types of insect-transmitted viruses was early noticed and has been confirmed by many workers, but there has been less agreement on interpreting the differences and on the most suitable criterion for separating the two. At first it was generally accepted that viruses of the first type have some obligate relationship with their vectors, perhaps multiplying in them or undergoing some necessary stage of development, and that those of the second type were transmitted in a purely mechanical manner as contaminants on the outsides of insects' mouthparts (DOOLITTLE and WALKER, 1928; HOGGAN, 1933). This view is, indeed, still widely held (*cf.* KUNKEL, 1938; LEACH, 1940; COOK, 1947), although there are many experimental results conflicting with the idea that viruses of the second type are so transmitted, and it is now evident that it was premature to separate viruses into types labelled biologically and mechanically transmitted. Clearly a division is better based on observed phenomena rather than on interpretations of observations, but it is far from clear which observations are most suitable and easily defined. In the first edition of this book, viruses were divided into groups depending on whether or not vectors could infect healthy plants immediately after leaving infected ones. At that time the delay in developing infectivity seemed to be a relatively fixed period characteristic of individual viruses, and comparable with the incubation period in plants between the time of infection and the appearance of symptoms. Further work showed that there is much variation in the time taken by individual insects to become infective with the same virus, and that the implications carried by the term incubation period were unjustified in the present state of our knowledge on the mechanism of transmission. Hence, in the second edition, the term "incubation period" was replaced by the less committal "latent period" to describe the delay in development of infectivity, which was abandoned as a basis of division in favour of the length of time for which vectors remain infective.

In making this change, we were largely influenced by STOREY (1939) and WATSON and ROBERTS (1939, 1940). STOREY attempted to combine both latent period and retention of infectivity in defining the two types as *(1)* "Transmission negative on the first hosts, positive on succeeding hosts" and *(2)* "Transmission positive on the first hosts, negative on succeeding hosts". Although this division has the obvious virtue of relying only on observable results, it has one serious drawback; it can only be used if the time intervals spent by vectors on each of a series of test plants are suitably chosen for each virus. For example, WATSON and ROBERTS (1940) and KASSANIS (1941), by using short intervals on the test plants, have obtained infections in several successive plants with viruses that STOREY included in the second group. Similarly, by prolonging the time spent by vectors on the first host, it will be infected by some of the viruses included in STOREY's first group. The main difficulty with this suggested division is that no single time to be spent on the first host can be set that will certainly distinguish between all members of the two groups.

WATSON and ROBERTS (1939, 1940) used as a basis for division the length of time for which vectors remain infective, calling viruses whose vectors continue to be infective for long periods "Persistent viruses", and those whose vectors rapidly lose infectivity "Non-persistent viruses". More recently, however, WATSON (1946) has concluded that no satisfactory basis for division exists in the time taken either for development or loss of infectivity by the vectors, for she has found both to vary widely with the treatments to which vectors are subjected. Instead of a time basis, she suggests that a better criterion is provided by determining the effect of a preliminary period of fasting on the vector's ability to transmit. All the non-persistent viruses studied by WATSON are transmitted much more rapidly if the vectors are first prevented from feeding and then transferred to healthy test plants after feeding for only a few minutes on an infected plant, than if they feed continuously for long periods before transfer. By contrast, the chances of transmission occurring with persistent viruses are greater the longer the vectors feed continuously on an infected plant. These differences are striking, but they have at least one serious disadvantage as a criterion for grouping insect-transmitted viruses; some viruses that fail to persist in their insect vectors, for example, dandelion yellow mosaic (KASSANIS, 1947), strawberry mild crinkle virus (PRENTICE and HARRIS, 1946) and cacao swollen shoot (POSNETTE, 1947), are not transmitted when their vectors have fed for only a few minutes on infected plants. Hence, in spite of the fact that the length of time for which a vector remains infective with a virus may vary depending on the treatment given, in the present state of our knowledge, persistence of infectivity still seems the best basis. No exact time can be given that will apply in all circumstances, but a convenient division is to regard viruses as non-persistent if their vectors usually cease to be able to transmit within a day of leaving infected plants and as persistent if their vectors normally remain infective for more than one day.

**Viruses Which Persist in Their Insect Vectors:**—Some of the viruses whose vectors remain infective for considerable periods are listed in TABLE 5. A further feature they have in common is that their vectors, having acquired virus, undergo a latent period during which they are unable to infect healthy plants. With any one virus this period can vary with individual insects, and the times shown in the table are the shortest that have been recorded for the listed viruses. Some of these times given are almost certainly greater than the minimum necessary, but the latent periods attributed to different viruses differ so much more than the variations found between individual insects transmitting one virus, and are so great, that they are probably significant.

Except for a latent period and ability to persist in their vectors, the viruses listed in TABLE 5 seem to have little in common. Most are transmitted by leafhoppers, but aphids, whitefly, thrips and beetles act as vectors of others. Also, most have not been transmitted by inoculation with sap from infected plants, but some have occasionally or by using special methods, and tomato spotted wilt and turnip yellow mosaic viruses are readily inoculable. Only a few of the viruses have had their *in vitro* properties examined, but those that have differ widely. Tomato spotted wilt and potato yellow dwarf viruses are inactivated in a few hours at room temperature or by heating for 10 minutes to 50°C., whereas sugar beet curly top virus remains infective for weeks and has a thermal inactivation point around 80°C., and turnip yellow mosaic virus is suffciently stable to be isolated as a crystalline nucleoprotein.

The failure of mechanical transmission with so many viruses that are readily transmitted by insects remains unexplained, though several alternative speculations can be offered. The simplest is a quantitative one, based on the assumption, which there is every reason to think is true, that insect vectors are more efficient than other methods of inoculation in causing infection with dilute inocula. The virus content of sap from plants infected with this type of virus may be below that necessary to give infection when rubbed on to healthy leaves, whereas it may be adequate when introduced by vectors, either because these concentrate the viruses while feeding on the diseased plants or because they deliver it to cells which are more readily infected than those injured by customary methods of inoculation. SMITH and LEA (1946) have suggested that the main differences between the persistent and non-persistent viruses are explicable on the basis that the two differ in the concentrations they reach in infected plants and in the rates at which they inactivate in their insect vectors. There is little doubt that many of the differences in behaviour between viruses, for which in the past qualitative differences have been sought, can be explained quantitatively, but there are some facts that seem incompatible with this suggestion. The failure of sugar beet yellows and maize streak viruses to be transmitted mechanically, for example, is difficult to attribute to a low virus content. Serological tests with sap from beet plants with yellows show the presence of a specific antigen, which by analogy with other diseases is reasonably regarded as the virus itself, and this occurs in greater quantity than do some non-persistent viruses that are readily inoculable. Also, the virus content of maize streak plants is sufficient for leafhoppers to become infective when feeding for less than a minute, suggesting a virus concentration in the plants similar to that reached by many non-persistent viruses.

TABLE 5. — *Viruses with latent periods in their vectors:—*

| *Virus* | *Insect vector* | *Latent Period* | *Reference* |
|---|---|---|---|
| Abaca bunchy top | *Pentalonia nigronervosa* Coq. | 24 hours | OCFEMIA and BUHAY (1934) |
| Aster yellows | *Cicadula sexnotata* Fall. | 10 days | KUNKEL (1926*b*) |
| Cereal mosaic | *Delphax striatella* Fall. | 6 days | SUKHOV and SUKHOVA (1940) |
| Corn stunt | *Baldulus maidis* (DeL. and W.) | 16 days | KUNKEL (1946) |
| Corn mosaic | *Perigrinus maidis* Ashm. | 4 days | CARTER (1941) |
| Cotton leaf curl | *Bemisia gossypiperda* M. and L. | 6 hours | KIRKPATRICK (1931) |
| Cranberry false-blossom | *Euscelis striatulus* Fall. | Undetermined | DOBROSCKY (1931) |
| Maize streak | *Cicadulina mbila* Naude | 6 hours | STOREY (1928) |
| Pea enation mosaic | *Myzus persicae* (Sulz.) | 12 hours | OSBORN (1938) |
| Peach yellows | *Macropsis trimaculata* Fitch | 8 days | HARTZELL (1936) |
| Pierce's grapevine disease | *Draeculacephalo minerva* Ball. | Less than 4 days | HEWITT *et al.* (1946) |
| Potato leaf roll | *Myzus persicae* (Sulz.) | 24 hours | ELZE (1927) |
| Potato yellow dwarf | *Aceratagallia sanguinolenta* Prov. | 9 days | BLACK (1938) |
| Rice stunt | *Nephotettix apicalis* Motsch | 3 days | FUKUSHI (1935) |
| Strawberry mild yellow edge | *Pentatrichorus fragariae* (Theob.) | 1 days | PRENTICE (1946) |
| Sugar beet curly top | *Eutettix tenellus* Baker | 4 hours | SEVERIN (1921) |
| Sugar beet yellows | *Myzus persicae* (Sulz.) | 30 mins. | WATSON (1940) |
| Tobacco rosette | *Myzus persicae* (Sulz.) | Less than 1 day | SMITH (1946) |
| Tomato spotted wilt | *Frankliniella insularis* Frankl. | 5 days | BALD and SAMUEL (1931) |
| Turnip yellow mosaic | *Phaedon cochleariae* (Fabr.) | 24 hours | MARKHAM and SMITH (1949) |

Hence, although a too small virus content may well explain the failure of mechanical transmission with some persistent viruses, with others additional explanations seem necessary. One possibility is that to become established in healthy plants virus particles may need to be introduced into certain specific tissues, and this is done by the feeding vectors but not by ordinary inoculation methods. If this is so, it seems likely that the phloem is the tissue most usually involved, for most of the vectors are phloem feeders and there is additional evidence connecting some of the viruses intimately with the phloem. In plants suffering from potato leaf roll and sugar beet curly top, for example, the phloem is necrotic whereas other, even nearby, tissues are not. BENNETT (1934) states that the virus occurs much more abundantly in the phloem of curly top plants than in other tissues, and suggests that the other tissues may contain substances that inactivate it. A few plants have been infected with this virus by methods of inoculation that are thought to have placed the virus directly into the phloem (SEVERIN, 1924). FIFE and FRAMPTON (1936) found that *Eutettix tenellus* cultured on beet plants treated with carbon dioxide rarely succeeded in feeding normally on the phloem, and infective vectors rarely caused infection in such plants. Similarly, an association with phloem feeding and infection has been noticed with some of the other viruses of this type. Although the vector of maize streak virus can acquire the virus from parenchymatous tissues of infected plants, STOREY (1938) states that insects infect healthy plants only if they feed long enough to reach the phloem; infective insects made to feed through a membrane that prevented their stylets from reaching to the phloem did not cause any infection however long they fed. With vectors of potato leaf roll also, DYKSTRA and WHITAKER (1938) found from a cytological study that those that always feed on the phloem transmit the virus with greater regularity than those that may feed on other tissues. In the same way ROBERTS (1940) has found that the chances of aphids causing infection with sugar beet yellows virus are increased proportionally as the conditions of feeding increase the chances of the vector feeding on the phloem. The phloem need not necessarily be the specific tissue for all the viruses of this type, though only with the one causing Pierce's grapevine disease is there any evidence of another being concerned; HOUSTON, ESAU and HEWITT (1947) state that the leafhoppers that transmit this virus do so, in both the vine and lucerne, only when they feed on the xylem.

No positive interpretations can be given for the apparent need to place these various viruses in specific cells. With sugar beet curly top virus, it seems that infection does not become wholly systemic, suggesting that the virus may be able to multiply only in the phloem, but there is no evidence of this with other viruses. Maize streak virus, for example, is known to occur in the leaf parenchyma. It may be that different tissues, although all capable of supporting virus multiplication when infected, differ in their susceptibility to infection, in the sense that different quantities of virus need to be introduced to ensure that infection occurs. Inoculum adequate to cause infection if placed in the phloem might fail to infect epidermal cells or the leaf parenchyma. Alternatively, it may be that such viruses are unable to become established in wounded cells and that they need to be placed in either phloem or xylem to be able to escape from the effects of wounding. The feeding vectors can be expected to cause less damage than mechanical methods of inoculation, but even their mouthparts may damage penetrated cells too severely for the viruses to become established. Virus particles placed in the phloem or xylem vessels, however, would be in a favourable position to be carried rapidly away from the area of damage to enter uninjured cells at a distance. The fact that the viruses have to be placed into the vessels is not necessarily an indication that phloem or xylem are the only susceptible tissues, but may merely be an essential preliminary to entering other susceptible cells. There is evidence that some of these viruses do move much more rapidly from their entry points than viruses introduced by rubbing. Curly top virus has been detected

36 cm. from the entry point only one hour after infection (SEVERIN, 1924), maize streak virus 20 cm. (STOREY, 1928), and Pierce's grapevine virus 25 cm. (HOUSTON *et al.*, 1947). STOREY (1938) has also found that cutting off the piece of leaf on which infective leafhoppers have fed does not prevent infection with maize streak virus, which again indicates a much more rapid movement of virus away from the entry point than occurs with viruses introduced to leaves by rubbing, for these take four to five days to move from the inoculated portion of leaf.

Two other possible explanations for the failure of mechanical transmission of viruses that are insect-transmitted deserve comment. One is that the viruses are unstable in expressed sap, in which they are too rapidly inactivated for transmission to occur, but that they are protected from inactivation when in the vector. The great instability of tomato spotted wilt virus *in vitro* compared with its persistence in infective thrips lends some support to this theory, but it is unlikely that it can account for the failure of mechanical transmission of viruses such as sugar beet curly top, maize streak and aster yellows. BENNETT (1935) has found that *Eutettix tenellus* will feed satisfactorily on sugar solutions and after feeding on extracts of beet plants suffering from curly top, or on extracts of macerated infective leafhoppers, the vector could infect healthy plants. Using this method of transmission BENNETT has studied the properties of the virus in some detail and found that it must be ranked among the more stable ones; it withstands drying, heating to 75°C., ageing for 28 days at room temperature, 90% alcohol for 2 hours, *p*H changes between 3 and 9, and possesses considerable resistance to disinfectants. Clearly it would be unreasonable to attribute lack of transmissibility with this virus to its inherent instability. Nor is maize streak virus immediately inactivated in expressed sap, for STOREY (1933) has made *Cicadulina mbila* infective by injecting them with sap from diseased maize plants, although the same inoculum failed to cause infection in plants. Similarly, BLACK (1941) has transmitted aster yellows virus by inoculating *Cicadula sexnotata* with extracts from infective hoppers, and his results show that this virus remains active for a day or so *in vitro*. These results again suggest that vectors are more efficient at infecting plants with small amounts of virus than are mechanical methods of inoculation, for they are most simply interpreted by assuming that quantities of virus too small to cause infection when rubbed over leaves are adequate when introduced by vectors. However, an alternative interpretation would be that passage through a vector alters the infecting powers of the viruses, in much the same way as rust fungi in alternate hosts produce spores with different capacities for infecting different hosts. Thus the virus particles in maize plants may be incapable of directly re-infecting maize, but inoculable to *Cicadulina mbila*, in which the particles are altered so that they become capable of establishing infection in healthy maize. This theory would also conveniently explain the latent period as the time necessary for the virus to undergo such a developmental change. If any such change occurred, it might be expected that plants could be infected by mechanical inoculation of extracts from macerated infective vectors, whereas insects could not, but this is not so. Using the blood of infective insects as inoculum, STOREY has transmitted the virus to other leafhoppers, but not to plants. The significance of the negative result with plants, however, is uncertain for extracts of insects contain inhibitors of infectivity that can prevent infection occurring when plants are rubbed with other viruses that are normally readily inoculable (BLACK, 1939).

Inhibitors of infectivity behaving in much the same manner as those in macerated insects also occur in extracts of some plants, and these provide yet another alternative explanation for transmissibility by insects but not by mechanical inoculation. When tissues from plants suffering from curly top, maize streak, aster yellows and other such diseases, are macerated, the virus particles may be brought into contact with and combine with substances that render them non-infective. This would prevent their transmission by inoculation, but insects may either acquire them free from such inhibitors or be capable of separating the inhibitors from virus particles if the two occur combined in infected plants. If the second occurs, the time taken for the destruction or separation of the inhibitors could also account for the latent period. Therefore, in summary, at least four different reasons can be offered for the failure of mechanical inoculation to transmit viruses that are transmitted by insects: the viruses may occur at too great a dilution in plant extracts, they may inactivate too rapidly, they may encounter substances that inhibit infectivity, or they may need to be introduced into some specific tissue. These theories are not mutually exclusive, and each one may apply to some but not to other viruses.

The sequence of events entailed in the transmission by vectors of persistent viruses remains obscure, and it probably differs with different viruses and vectors. Turnip yellow mosaic virus is anomalous in having as vectors insects with biting mouthparts, so that the mechanism of transmission involved with this virus is obviously likely to differ radically from that with others. All the insects found by MARKHAM and SMITH (1949) to act as vectors are those with no oesophageal valve, and it seems probable that the virus is taken into the stomach with the chewed leaf tissue where it remains until regurgitated when it is in a condition to cause infection. The persistent viruses that have vectors with sucking mouthparts are most probably taken into the gut, pass through the gut wall into the blood, from which the virus particles pass to the salivary glands, where they become mixed with the salivary secretions and are injected back into plants by the feeding insects. With maize streak and sugar beet curly top viruses there is positive evidence that the virus particles enter the blood of vectors.

STOREY (1932, 1933) has found that the ability of individual *Cicadulina mbila* to transmit maize streak virus is determined by the permeability of the gut wall to the virus. The insect occurs in two strains, which are indistinguishable morphologically and inter-breed freely, but one strain (active insects) is a vector and the other (inactive) is not. The ability to transmit is inherited as a simple Mendelian factor linked with sex, the male being heterozygous for sex. Individuals of both strains pick up the virus when feeding upon infected plants, but in the active insects the virus can be found in the contents of the gut and the blood, whereas in inactive insects it can be found only in the contents of the gut. When feeding naturally the inactive insects never transmit the virus, but by operating upon them STOREY has been able to turn them into vectors. If the gut wall of an inactive insect is punctured, either before or soon after it has fed upon an infected plant, it behaves like an active insect. Virus can be detected in the blood of such insects and they are also able to transmit the virus. Similarly, if the virus is introduced directly into the blood of an inactive insect, by means of needles dipped either in infective maize sap or in the blood of infective insects, it is then able to transmit the virus to healthy plants. Both strains of the insect, therefore, can pick up the virus from infected plants, and both can transmit it to healthy ones. The sole difference between them seems to lie in the permeability of their gut wall, that of the active strain allowing the passage of the virus and that of the inactive preventing it.

The blood is most likely the main reservoir of virus in the infective insect, from which it probably passes only slowly into the salivary glands. STOREY was unable to detect any virus in the salivary glands of *Cicadulina mbila*, and BENNETT and WALLACE (1938) found significantly less in the dissected salivary glands of infective *Eutettix tenellus* than in other tissues. However, these tests were of necessity made on resting insects, and it is possible that the virus content of the glands of feeding insects, when actively secreting saliva, might be higher. This possibility is strongly suggested by SMITH's (1941) results. He fed infective *E. tenellus* through membranes on drops of sugar solution for three hours, and then concentrated the resulting mixture of saliva and sugar solution before feeding it to non-infective *E. tenellus*. Of 40 insects fed on such salivary secretions, 29 became infective. Although the salivary glands obviously can contain virus, especially in the feeding vector, there is other indirect evidence that they have only a small content and are not the main reservoirs. If infective leafhoppers are fed for fixed periods on each of a series of healthy plants, they usually fail to produce infection in every plant. Some individual leafhoppers infect more plants than others, and with all individuals the number infected increases with the length of time spent on the plants. It seems probable, therefore, that the salivary glands are easily exhausted of virus and that there is not a steady flow of virus from the blood to the glands, but that it takes place irregularly. If an insect is feeding on a plant while virus passes into its saliva, then it is to be expected that the plant will be infected, but if it does not the plant will remain healthy. The probability of virus entering the saliva will increase with increasing time and will also be greater the higher the virus content of the blood. The variations in the ability of individual leafhoppers to cause infection in short feeding times, which have frequently been recorded, can be explained on this theory by variations in the virus content of their blood or in some other factor affecting the rate of entry of virus into the salivary glands.

Apparently there is little virus wasted by excretion in the faeces. STOREY (1932) found virus in the contents of the rectum of *Cicadulina mbila* if they had recently fed on infected plants, but none in the naturally voided faeces. SEVERIN (1931) also found no virus in the faeces of *Eutettix tenellus* carrying sugar beet curly top virus and BENNETT and WALLACE (1938) found only litte.

One curious feature in the relationships of tomato spotted wilt virus and its vectors *Frankliniella insularis* and *Thrips tabaci* has not been found with other viruses and may be relevant here. Both the larvae and the adults can transmit the virus to healthy plants, but to become infective the thrips must feed on infected plants when they are larvae. The adults are unable to become infective (BALD and SAMUEL, 1931). This is still unexplained, no differences in the feeding habits or anatomy having been found to account for it. But if STOREY's results with maize streak and *C. mbila* can be applied to spotted wilt virus, it seems possible that the permeability of the gut wall may differ in the adult and larval thrips. The virus may be able to pass through the gut wall of the larva but not that of the adult. If this were so, the virus picked up by the feeding adult would either remain in the intestine or be excreted, the insect remaining non-infective. Whereas the virus picked up by larva would penetrate the wall and enter the blood, and the insect would become infective. An adult from such a larva would already have the virus in its blood and so could continue to transmit, although no longer able to acquire further virus.

How viruses manage to pass through insects' gut walls is unknown, for the particles are too large to diffuse through semi-permeable membranes. No signs of any injury or lesion have been found in the gut walls of infective insects. The absence of a peritrophic membrane in many vector species may render penetration easier than in other species, but the possibility that the virus plays an active rather than a passive part cannot be ruled out. It is possible that the ability to pass through the gut wall determines the specific relationships between viruses and their vectors. For example, potato leaf roll virus may be able to penetrate the gut wall of *Myzus persicae* but not of *Eutettix tenellus*, whereas curly top virus may be able to penetrate the wall of the latter but not of the former. It has been definitely established that some non-vector species of insects pick up viruses while feeding upon infected plants. STOREY (1933) has found maize streak virus in the intestine, but not in the blood, of *Peregrinus maidis* and BENNETT and WALLACE (1938) have found curly top virus in a number of species of aphids, thrips and leafhoppers after feeding on infected plants. From some of these non-vectors appreciable quantities of virus were recovered as long as a fortnight after they had access to a source of virus. The inability to transmit therefore, cannot be explained on the assumption that non-vectors destroy the viruses. However, factors other than the permeability of the gut wall to the particular virus are probably involved, for STOREY was unable to convert *P. maidis* into a vector by puncturing its gut wall as he could the inactive strain of *Cicadulina mbila;* nor did *P. maidis* transmit when maize streak virus was injected directly into its blood.

**Latent Period and Virus Multiplication in Vectors: —** After feeding on an infected plant, the vectors of all persistent viruses apparently need to wait some time before they can infect a healthy plant. As indicated in TABLE 5 this period varies from less than an hour to more than a fortnight with different viruses. Hence it may well be that, with different viruses, there are different reasons for the delay in development of infectivity. No certain interpretation of the phenomenon can be given, but a number of alternative theories can be suggested. One of these, that the latent period is the time needed in the vector for the virus to undergo some developmental change that affects its inoculability, has already been discussed. Another, and perhaps more probable, explanation is that it represents the time taken for virus particles ingested by the vectors to be ejaculated. With turnip yellow mosaic virus, this might simply be the time required for the digestion and regurgitation of virus-containing leaf tissues. With the other viruses, which have as vectors insects with sucking mouthparts, the latent period could be the time required for the virus particles to penetrate the gut wall, and pass through the blood stream to the salivary glands. The main objection to this theory is to account for the wide differences between the duration of the latent period shown by different viruses. However, this could be done either by assuming that viruses pass through the gut wall in different ways, or on the simple quantitative explanation that while feeding on infected plants vectors acquire large quantities of some viruses and relatively little of others, so that the chances of particles soon passing into the salivary glands will be large with the first type and small with the second. A third, and probably the most widely accepted, explanation is that the vectors acquire from infected plants too little virus to make them infective and that during the latent period this initial virus multiplies to the amount required to cause infection in plants. There is no *a priori* reason why plant viruses should not multiply in insects, and the possibility certainly provides a convenient explanation of many phenomena, but it also conflicts with many others.

At first sight virus multiplication seems the most reasonable explanation of the fact that once vectors have become infective they may remain so for long periods, but were it correct it would be reasonable to expect that all vectors would remain infective for long periods and probably for the whole of their lives. However, this does not happen, for whenever the performance of individual vectors has been followed in detail, it has been found that different individuals remain infective for varying periods, some soon losing infectivity and others not. Some results obtained by FREITAG (1936) and BENNETT and WALLACE (1938) with sugar beet curly top, and by STOREY (1938) with maize streak, suggest an explanation for these variations. They have all found that the length of time for which leafhoppers remain infective depends on the length of time they have fed on the source of the viruses. Vectors may become infective after a few minutes' feeding but if so they do not remain infective for long. By contrast, if they have fed for hours or days they remain infective for long periods, often for the remainder of their lives. Working with sugar beet yellows virus, WATSON (1940) also found that both the efficiency of the vectors and the length of time for which they remain infective increase with increased feeding time on the infected plant. The chances of transmission were also increased by prolonging the period spent on the healthy test plants. She got occasional infections with insects given a total feeding time of 30 minutes on the diseased and healthy plants, but the numbers increased greatly when the times on either were increased. WATSON challenges the whole concept of a fixed latent period, and suggests in its place a time during which the chances of successful transmission increase until a maximum is reached at which all insects capable of transmitting will have done so. Whether insects can infect only one or many plants will depend on how long they have fed on the source of infection, but when they cause infections will depend solely on the chance ejection of virus into the healthy plants. The results with curly top virus also fit with this idea. Most workers have found a latent period of at least 6 hours for this virus, but SEVERIN (1931) got some transmissions with vectors fed for only 10 minutes successively on infected and healthy plants. He concluded that these resulted from some anomalous method of transmission, possibly mechanical transfer on the vectors' mouthparts. WATSON (1940) has pointed out that if two methods of transfer were involved, there should also be two optima in the curve obtained when number of transmissions is plotted against time. Instead, with both curly top and yellows viruses, the curves are smooth, rising steeply at first and then flattening out towards a maximum.

Although the length of time for which vectors feed on infected plants may determine for how long they remain infective, FREITAG (1936) and STOREY (1939) have found, with sugar beet curly top and maize streak viruses, that it has no influence on the duration of the latent period. Similarly, with pea enation mosaic virus (CHAUDHURI, 1948) and strawberry mild yellow edge viruses (PRENTICE, 1946) the same latent period of about a day seems to occur whether the aphids have fed previously on diseased plants for a few hours or a few days. Hence with these viruses, which have relatively short latent periods, the latent periods seem to begin from the time the vectors leave the infected plant rather than from when they start to feed on it. Were the latent period the time required for virus to multiply to an infective concentration, there would seem little reason why the whole of it should not be spent on the infected plants and why adult insects raised on diseased plants should not be able to infect healthy plants immediately. Unfortunately no specific experiments seem to have been made on the possible interactions between the durations of infection feeding and latent periods with viruses such as aster yellows, which have prolonged latent periods, but in published work there is no clear indication that these can infect healthy plants immediately after feeding for a prolonged time on diseased ones.

In reviewing the second edition of this book, KUNKEL (1944) adversely criticised the views expressed in the previous paragraph, describing them as "serious errors" and stating that examples of such immediate transmission are reported in the literature. Unfortunately, he gives no references to such statements, and we have searched the literature again without encountering them. This search served only to emphasise the idea that the latent period begins at the end rather than at the start of feeding time on the diseased plants. Were it the other way around, it is difficult to see how the existence of a latent period could have been discovered with some viruses, for workers have used many insects per test plant and these were usually fed on the diseased plant for longer than the claimed latent period. Some at least should have been able to transmit immediately, yet none did, or a latent period would not have been detected. Although not done for this purpose, the results of an experiment by KUNKEL (1926*b*) with aster yellows are relevant and illuminating. Thirty insects were used in the experiment and after feeding on a diseased aster these were transferred individually to 23 successive batches of test plants, on each of which they spent periods varying between 1 and 7 days. The insects used were from eggs laid on the diseased aster on February 19th., and were transferred to the first batch of test plants on March 21st., when KUNKEL states many were adults and that "it may be assumed that most of the insects were about three weeks old". In the paper recording this experiment,

KUNKEL gives the latent (incubation) period as "at least 10 days" and in an earlier note (KUNKEL, 1926*a*) as varying from 6 to 10 days in adult insects and up to a fortnight in nymphs. The number and percentage of test plants that became infected in the first 10 and last 2 batches of test plants are shown in TABLE 6. The insects spent six days on the first batch of test plants, which together with the three weeks spent on the diseased plant greatly exceeds the range given for variations in the latent period, yet it will be seen that only 2 insects managed to cause infection. In the next period of 5 days, 11 insects caused infection, and only in the third batch of test plants, that is, up to 17 days after leaving the source of infection, were the insects producing approximately the same number of infections as they did in later transfers. The performance records of the individual insects also show considerable variability, some remaining infective for long periods and infecting most of the plants on which they fed, whereas others infected only few plants and soon lost infectivity. After leaving the diseased plants there was a period of three weeks during which increasing numbers of leafhoppers became able to cause infection, their ability to do so remained approximately constant for about 5 weeks, and then declined fairly steadily. It is perhaps significant, too, that the vectors that first caused infection were those that also remained infective longest and caused the largest total number of infections.

TABLE 6. — *The infection of successive batches of test plants with aster yellows virus by* Cicadula sexnotata *after feeding for three weeks on a diseased aster*:—

| *Transfer No.* | *Infection of test plants* No. infected/ No. colonised | *% infected* | *Time spent on test plant (days)* | *Total time since leaving diseased aster (days)* |
|---|---|---|---|---|
| 1 | 2/30 | 6 | 6 | 6 |
| 2 | 11/30 | 36 | 6 | 12 |
| 3 | 18/30 | 60 | 5 | 17 |
| 4 | 27/29 | 93 | 6 | 23 |
| 5 | 28/29 | 96 | 7 | 30 |
| 6 | 23/25 | 92 | 1 | 31 |
| 7 | 20/26 | 77 | 1 | 32 |
| 8 | 17/21 | 81 | 2 | 34 |
| 9 | 17/22 | 77 | 3 | 37 |
| 10 | 12/21 | 57 | 4 | 41 |
| 22 | 3/7 | 43 | 7 | 100 |
| 23 | 3/6 | 50 | 7 | 107 |

(Data from Table 4 in KUNKEL, 1926*b*).

These facts are all difficult to reconcile with the view that viruses multiply in their vectors and that the latent period is the time taken for the viruses to reach an infective concentration. A plant inoculated with a minimum infecting dose of virus takes longer to show systemic symptoms than one inoculated with a massive dose, but ultimately the two plants come to have the same appearances and virus-contents. Therefore, if the viruses multiplied in their vectors as they do in susceptible plants, it might be expected that the latent period would be lengthened by short feeding periods on infected plants. But there is no obvious reason why an insect which has acquired sufficient virus to become infective should not ultimately have the same virus-content, and remain infective for as long, as one receiving a larger initial dose of virus. Also, were the virus multiplying it is difficult to see why vectors which have once become infective should later lose this power.

A more reasonable interpretation of the observed phenomena is that vectors contain no more virus than they take up while feeding on infected plants, and that this amount is either stored in some tissue or circulates in the blood, to be dissipated only slowly and irregularly through the salivary glands. The reasons for the latent period are obscure, but on this view it could be regarded as the time taken for the virus to reach the saliva in an infective form. If this time is independent of the length of the infection feeding, it is possible that the insects imbibe some inhibitor from the diseased plant whose influence has to be removed before infection of plants can occur. There may be some loss of virus by destruction in the vector or by diffusion into tissues other than the salivary glands, but the loss during feeding will probably be greatest. If passage into the saliva is gradual and discontinuous, the prolonged infective state of insects given long infection feeding is readily explained, for there is good experimental evidence that insects can acquire from infected plants sufficient virus to infect a large number of healthy plants. BENNETT (1935) has shown that *Eutettix tenellus* fed artificially on phloem exudate diluted to 1/10,000, picked up sufficient virus to infect healthy sugar beet with curly top virus. Had the insects been feeding on diseased plants for the same length of time they would have been imbibing undiluted phloem contents. It is not unreasonable to assume, therefore, that they would have picked up something like 10,000 times the minimum amount of virus necessary to render them infective. As there appears to be little loss of virus in the faeces from feeding insects, it seems that most of the virus imbibed must enter the blood; therefore, providing only that the loss of virus from the blood to the saliva is slow, in relatively short feeding periods it will be possible for the vectors to accumulate more than sufficient virus for them to continue infecting plants for the remainder of their lives.

BENNETT and WALLACE (1938) have provided further evidence that insects can acquire much more virus from infected plants than they need to reach their maximum infectivity. The ability of leafhoppers to infect large numbers of plants with curly top virus increases rapidly with the length of time they are allowed to feed on the sources of the virus, and reaches a maximum in about two days. But if they are allowed to continue feeding on the source of the virus, the virus-content of the leafhoppers (measured by the ability of other leafhoppers to become infective when fed upon extracts of the infective hoppers) goes on increasing until they have fed for a fortnight, although their ability to infect healthy plants is not increased. As this increase occurs only when the infective insects are continuously fed upon diseased plants, it is obviously the result of more virus taken in by the mouth and not of multiplication in the insects.

Workers with most viruses have found that the progeny of infective vectors are free from virus provided that they are moved from diseased plants before they have fed, or provided they are born on healthy plants. Two exceptions to this rule are known, namely, those causing dwarf disease (stunt) of rice, vector *Nephotettix apicalis,* and club leaf of clover, vector *Agalliopsis novella,* both of which may be transmitted through the eggs laid by infective leafhoppers. FUKUSHI (1934, 1935) has found that transmission of rice dwarf virus to the progeny is determined solely by the female, crosses between infective males and non-infective females always giving virus-free offspring. Not all the progeny of infective females are themselves infective, but sometimes the virus is handed down through several generations. The record seems to be transmission to the seventh generation, while more than 1,000 plants were infected by 26 different leafhoppers derived in five generations as the progeny from one egg. To explain these results, FUKUSHI (1939) postulates virus multiplication because "the amount of virus contained in one egg must be extremely small" and KUNKEL (1944) considers this as "overwhelming" evidence for multiplication. Multiplication may occur and may be the true explanation, but other facts seem incompatible with it. Infective leafhoppers not only often fail to pass the virus to their offspring, but themselves cease to be infective, suggesting that if multiplication occurs it is sporadic and certainly not to the same extent as in plants. All these results are readily explicable on the assumption that different vectors acquire different amounts of virus while feeding on diseased plants, and that insects that soon lose infectivity are those that acquired relatively little virus, whereas insects that remain infective and transmit to their progeny are those that acquired much virus. No attempt has been made to correlate the length of time for which *N. apicalis* remain infective with the time for which they have fed on diseased plants, but the feeding times mentioned by FUKUSHI are days or weeks. In this time, as we have already seen, *Eutettix tenellus* can acquire many times the amount of sugar beet curly top virus required to make it infective for its whole life, and this may also be true with rice dwarf virus. The assumption that the amount of virus in an egg must be too little to infect 1,000 plants would seem quite unwarranted. The injection into a plant of a single virus particle may be all that is needed to cause infection, so that there is no need to postulate that the original egg contained much more than 1,000 particles, and there is no apparent reason why a leafhopper should not imbibe such a quantity while feeding on a diseased plant. We do not know the size of rice dwarf virus particles, but on the assumption that they are similar to other viruses, 1,000 would weigh about $10^{-14}$g., a quantity that would be inconsiderable even in the minute egg of a leafhopper.

Some infective nymphs of *N. apicalis* can infect healthy plants immediately they emerge from the eggs, whereas others do so only after a variable waiting period. This at first sight again suggests multiplication of the virus in the insect, for the newly-hatched nymph may contain too little virus to cause infection, and the waiting period may be the time required for this virus to multiply. But there are other equally probable interpretations. Newly-hatched nymphs may contain different initial quantities of virus or the virus they contain may be distributed differently. A nymph born with a high virus content, or with virus already in the salivary glands, might be expected to transmit immediately, whereas it might be a considerable time before virus entered the salivary gland of an insect with a low virus-content.

Similarly, the preliminary account published by BLACK (1948) of the transmission of clover-club leaf virus through eggs does not seem necessarily to call for virus multiplication as an explanation. Twenty-four out of twenty-seven insects raised from eggs laid by infective

*Agalliopsis novella* infected healthy plants, but out of a total of 642 plants colonised only 68 were infected. No insect transmitted until 3 weeks after hatching, and most infections were obtained when the insects were between 6 and 12 weeks old. As the insects on the average infected less than three plants each, though each test plant was colonised for a week, it seems that they were only weakly infective and inherited only small quantities of virus from their parents.

KUNKEL (1937, 1938) has found that *Cicadula sexnotata* carrying aster yellows virus are unable to transmit if exposed to high temperatures. Leafhoppers exposed to a temperature of 32°C. for a day lost the ability to infect healthy plants, but on lowering the temperature to 24°C. they quickly regained the ability without again feeding on a source of the virus. Infective leafhoppers kept at the high temperature for a week regained the ability to transmit when subsequently kept at the lower temperature, but they regained it much more slowly than those heated for one day. Leafhoppers heated for one day regained the ability to transmit in a few hours at the lower temperature, those heated for a week required two days or longer, whereas those heated for more than 12 days never regained the ability to transmit.

KUNKEL regards these facts as strong evidence that the virus multiplies in the insect vector. He interprets his results as indicating that long heat treatments inactivate all the virus in the infective insect and that short treatments inactivate a part only. The time required for treated insects to regain the ability to transmit he regards as a latent period during which the non-inactivated virus multiplies sufficiently to render the insects infective. This interpretation seems to the writer to necessitate a number of assumptions for which there is no positive evidence. First, as the ability of the vector to infect plants is destroyed by a short period of heating and is regained in a few hours, destruction of a small amount of virus must be assumed to render the vector non-infective. Thus it appears that the fully infective insect can be carrying only the minimum charge of virus necessary to infect a plant, a conclusion widely differing from the results of BENNETT with *Eutettix tenellus* which show that this leafhopper can carry many times the charge of virus necessary for its maximum infectivity. Secondly, it seems necessary to assume that the virus multiplies much more readily in the insect than in susceptible plants. For KUNKEL's results show that insects treated for one day cease to transmit, and on his interpretation then contain too little virus to cause infection in a plant, yet after a week's treatment it seems there is still enough to multiply in the insects themselves.

KUNKEL found that the leafhoppers lived and bred for long periods at the high temperature without apparently being harmed, and because of this he states the effect of heat must be on the virus. Even if this is assumed, and the loss of infectivity is regarded solely as a result of virus inactivation, there would seem to be no reason for assuming that the recovery of infectivity is a result of virus multiplication in the vectors. The body of an insect is such a complex system and so little is known about it that there are many other equally probable explanations. Virus in or near the salivary gland may be inactivated by heat more rapidly than virus in the blood. Then the loss of infectivity after short heat exposure could be explained by the inactivation of the virus in the gland; and the waiting period as the time required for virus again to enter the saliva. If continuous heat inactivates the virus in the blood, then with decreasing virus concentration the probability of virus entering the saliva from the blood will be reduced, and the length of time required for the virus to enter the saliva, and for the insect to become infective, will be increased. But heating may also have effects other than inactivating the virus. The insect's blood or tissues may be affected so that the virus is more firmly absorbed, or precipitated. If this were so then enough time would be needed for the insect to return to normal before the virus was again liberated and free to enter the saliva. Or again, heating may make the salivary gland less permeable to the virus. Indeed, there are so many possible alternative explanations of these results that it would seem extremely premature to regard them as evidence for the multiplication of the virus in the vector.

BLACK (1941) has also described results of experiments which he considers provide evidence for the multiplication of aster yellows virus in its vector. *Cicadula sexnotata* were fed on diseased asters for one or two days and then transferred to healthy plants. At intervals, batches of insects selected at random were macerated, diluted with saline, and inoculated into non-viruliferous insects, which were later tested for their ability to infect healthy aster plants. Extracts from insects macerated within four days of feeding on the yellowed aster failed to render any inoculated insects infective, whereas extracts from those macerated after twelve days when diluted 1/1000 succeeded in causing some vectors to become infective. BLACK interprets this as demonstrating that the virus multiplied at least a hundred-fold during the latent period, but it would seem at least as probable that, during this period, the virus imbibed while the vector fed on the diseased aster moved from one tissue to another, or changed from an insoluble to a soluble state, so that at the end of the latent period it was more readily extracted from the macerated insects than at the start. For instance, it may be that early in the latent period, the virus particles are attached to the gut wall, while later they are free in the blood. When the results of these inoculations to insects are examined closely, it is also clear that they cannot be directly interpreted in terms of virus content, and that failure with one undiluted inoculum and success with another diluted 1/1000 is no positive evidence of a higher virus content in one than in the other. With many individual inocula, BLACK obtained greater success in rendering insects infective when the extracts were diluted 1/100 or 1/1000, than when they were undiluted or diluted only 1/10. Obviously, dilution can hardly be expected to increase virus concentration, and some other explanation is needed. There is no certain one, but from infective tests with plants it is known that there are many substances that act as inhibitors and that non-infective mixtures of viruses and such substances become infective when largely diluted. Such inhibitors have been found in extracts of insects by BLACK (1939), who showed that mixtures of tobacco mosaic virus and clover leafhopper juice increase in infectivity when diluted. Hence, if these substances also interfere with the successful inoculation of virus to insects, the correlation between transmission and virus content of the inoculum is far from sure.

In addition to macerating test insects that had fed for only a day or two on diseased asters, BLACK (1941) also macerated control insects that had fed for long periods, and the greater number of transmissions obtained with extracts of the controls is most striking. He attributes the relative lack of transmission with extracts of the test insects to the fact that they "did not pick up much virus", an interpretation hardly compatible with the idea of ready multiplication, for the virus content achieved in plants is not set by the size of the initial infecting dose. Thus with aster yellows virus, as with curly top virus, it seems that the virus content of the vector is largely determined by the feeding time on the diseased plant; if there is any multiplication during the latent period, it is less effective in increasing the virus content than continued feeding on the diseased plant.

With viruses such as maize streak and aster yellows, which can be mechanicallly transmitted to their vectors, experiments could be made to settle the problem of virus multiplication in the vector. Vectors could be inoculated with various dilutions of virus-containing extracts, and after a suitable lapse of time they could be macerated and their extracts inoculated at various dilutions to non-viruliferous vectors. If the vectors inoculated with the minimum concentration of virus ultimately came to have the same virus content as those inoculated with concentrated extracts, there would be good evidence of multiplication. Secondly, insects could be inoculated with dilute extracts, then macerated after a suitable time and their juice again diluted and inoculated to fresh insects. If this process succeeded in producing a long series of infective insects, as it undoubtedly does of infected plants, the evidence for multiplication in the vector would be unquestionable. Until such proof is forthcoming, the balance of evidence is against multiplication on any considerable scale, although the possibility of limited multiplication with some viruses cannot definitely be excluded. All the presumptive evidence offered in support of such a view, however, can be interpreted differently.

Most workers have failed to find any significant differences between infective and non-infective individuals of vector species. Infective individuals live as long and breed as freely as non-infective. No morphological changes or disease symptoms have been described. This, of course, cannot be taken as evidence that the viruses do not multiply, for viruses multiply freely in many plants (carriers) in which they cause no symptoms. HARTZELL (1936, 1937) claims to have found intracellular inclusions in *Macropsis trimaculata* carrying the virus causing peach yellows. He also described similar bodies in diseased trees, but they differ so widely from the inclusions found in other virus diseases that, until more evidence has been obtained about them, it would be premature to accept them as true inclusion bodies. A beneficial effect from feeding on celery and aster plants infected with aster yellows virus has been described by SEVERIN (1946), who found that several leafhopper species multiply more freely, pass through nymphal stages sooner, and live longer than those feeding on healthy plants. Whether this is a direct effect of the virus on the insects, or an indirect effect by altering the nutritive quality of the host plant, is uncertain.

**Viruses Whose Vectors Soon Lose Infectivity: —** The vectors of this second type of virus can infect healthy plants immediately after leaving a diseased one. Indeed, the less the delay in transferring insects from infected to healthy plants, the greater is the probability that they will cause infection. This difference from the behaviour of the persistent viruses often makes it possible, using only one vector, to isolate persistent and non-persistent viruses separately from plants infected with both; when transferred from the diseased plant to a succession of test plants, the vector will infect the first with the non-persistent virus only whereas it will infect later plants in the series with the persistent one only. In this way, using *Myzus persicae* as a vector, pure cultures of individual viruses can be isolated from plants infected with the following pairs, potato Y and leaf roll, sugar beet mosaic and yellows, pea mosaic and pea enation mosaic; similarly *Pentatrichopus fragariae,* when fed on strawberry plants infected with both mild crinkle and mild yellow edge viruses, will transmit only the first to plants on which it feeds immediately and only the second to plants on which it feeds one day after leaving the diseased plant.

Not all the known non-persistent viruses have had their properties *in vitro* examined, but those that have behave rather similarly (WATSON and ROBERTS, 1939). They are mostly inactivated by 10 minutes' heating around 60°C., by a few days' ageing at room temperature and they are easily inactivated by acid and alcohol. Except cacao swollen shoot viruses, which are transmitted by mealy-bugs, all those so far identified have aphids as vectors. The first non-persistent viruses to be studied were all readily transmissible by needle inoculation or rubbing methods, which, together with the fact that the whole process of becoming infective and infecting a healthy plant need occupy a vector no more than a few minutes, led many workers to conclude that the insects act simply mechanically, transferring the virus as a contaminant on the outsides of their stylets, as a needle might. The rapid development of infectivity by these vectors suggests that there is a different sequence of events from that involved with the persistent viruses, and the rapid loss of infectivity by the vectors clearly makes it unnecessary to discuss the possibility of these viruses multiplying in the insects. Nevertheless, there is much evidence that the vectors do not behave merely as needles.

If insects acted solely by getting their mouthparts contaminated externally with virus, it is difficult to see why all those with similar feeding habits should not be equally efficient vectors, except that those with larger stylets might perhaps be expected to carry a little more virus. However, this is not so, and insects with similar feeding habits will transmit some viruses but not others, and, with viruses that have a number of aphids as vectors, one species is often much more efficient than another. SEVERIN and FREITAG (1938) transmitted cucumber mosaic virus with eleven different aphid species, but the percentage infection obtained with different species varied from 14 to 84. Similarly WATSON (1938*b*) found that *Myzus persicae* was more efficient than *Aulacorthum circumflexum* in transmitting henbane mosaic virus (hyoscyamus virus 3), but that the two aphids were almost equally efficient in transmitting cucumber mosaic virus. The specific relationship between viruses and vectors can also be shown by feeding different aphids on plants infected with virus mixtures, when one species may transmit one virus and another a different one. For instance, if *Myzus persicae* and *Myzus ornatus* are fed on lettuce plants infected simultaneously with lettuce mosaic and dandelion mosaic viruses and are then transferred to healthy lettuce, *M. persicae* will transmit only the former and *M. ornatus* only the latter (KASSANIS, 1947). Likewise, from cruciferous plants infected with both cauliflower mosaic and cabbage blackring viruses, *M. ornatus* will transmit only the former, whereas *M. persicae* will transmit both (KVICALA, 1945). This specificity in the viruses that individual insects can transmit is further illustrated in TABLE 6A, where the abilities of two rather similar aphids to transmit seven non-persistent viruses, and a persistent one (sugar beet yellows), are compared (DONCASTER and KASSANIS, 1946).

Again, although *M. persicae* is listed in the table as a vector of potato virus Y, it does not transmit all the known strains. In many tests at Rothamsted, we have failed consistently to transmit one strain (potato virus C) with this aphid or with any of the eight other species that transmitted the other strains studied. The efficiency of the different species as vectors of virus Y varies greatly and the results with some, *e. g. Myzus ornatus,* suggest that occasional individuals are vectors whereas most of the population are not. (BAWDEN and KASSANIS, 1947). From such results as these, it appears that there is some kind of positive reaction between individual viruses and insects that determines whether the insects act as vectors or not.

TABLE 6A. — *Comparative transmission of different viruses by* Myzus persicae *and* Myzus ascalonicus:—

| *Virus* | *M. persicae* | *M. ascalonicus* |
|---|---|---|
| Potato Y | *19/20 | 0/47 |
| Cucumber mosaic | 18/20 | 41/50 |
| Henbane mosaic | 10/10 | 20/35 |
| Tobacco etch | 10/10 | 0/30 |
| Lettuce mosaic | 24/24 | 0/38 |
| Sugar beet mosaic | 10/10 | 0/10 |
| Dandelion yellow mosaic | 0/30 | 7/33 |
| Sugar beet yellows | 10/10 | 5/10 |

*Numerator is the number of plants infected, denominator number colonised. With each virus, the numbers of individual insects used per test plant was the same for both aphid species.

If transmission by insects were merely mechanical, it might be expected that viruses most readily transmitted by artificial inoculation would also be most readily transmitted by insects. And, if the viruses were merely acquired as a contaminant on the mouthparts, those occurring in infected plants in the greatest concentration might be expected to be most easily acquired and transmitted. But it is not so. Tobacco mosaic virus and potato virus X are more easily transmitted by rubbing or by needle inoculation, and occur in greater concentration in extracts of infected plants, than any of the viruses under discussion. Neither of these normally appears to be aphid-transmitted. HOGGAN (1931) has shown that *Myzus persicae* fed on tobacco infected with both tobacco mosaic and cucumber mosaic viruses transmit only cucumber mosaic virus, and SMITH (1931) could always get pure cultures of potato virus Y by feeding *M. persicae* on plants infected with both this and potato virus X. By contrast, if inoculations were made with needles by first pricking plants infected with these pairs of viruses and then pricking healthy ones, the viruses most likely to be transmitted are certainly tobacco mosaic and potato X.

These two viruses have been studied in detail by many different workers, but there is only one serious claim for transmission by insects. HOGGAN (1931) found that a number of aphid species all failed to transmit tobacco mosaic virus from infected tobacco plants. However, when fed on infected tomato plants infections were obtained with *Myzus pseudosolani* and *Macrosiphum gei,* but not with *Myzus persicae. M. pseudosolani* gave consistently more infections than the other aphids. These rather curious results have not been confirmed, but if the transmissions are accepted as true insect-transmissions and not accidental contaminations, again they do not suggest that the virus is carried mechanically on the mouthparts. On the contrary, the varying ability of the different aphids to transmit, together with the fact that the insects do not become infective when fed on tobacco, although this host has a higher virus-content than tomato, suggest some relationship between the virus and vectors. In these experiments very large numbers of aphids were transferred from infected to the healthy test plants. In spite of this, all the test plants rarely became infected with tobacco mosaic virus as they did in similar experiments with cucumber mosaic virus. A possible explanation for these results is that most individuals of the vector species are unable to transmit tobacco mosaic virus, or in STOREY's terminology are "inactive", whereas occasional individuals may be "active". If this were so, increasing the number of aphids transferred would increase the probability of including an active individual and increase the probability of getting infection. The failure of workers other than HOGGAN to transmit tobacco mosaic virus with aphids may be due to the smaller number of insects used or to their cultures containing no active individuals.

If insects transmit viruses merely by adsorbing infective plant sap on to their mouthparts, it is difficult to understand why tobacco mosaic virus should not be transmitted by any insect. On the other hand, if the viruses must be sucked up and then ejected possible explanations can be offered for the fact that insects do not normally transmit this most infectious virus. It is possible that the virus is quickly inactivated in the intestinal tract of aphids. It is unlikely to be destroyed, for *in vitro* tobacco mosaic virus withstands treatments that rapidly destroy most insect-transmitted viruses, but it is possible that the secretions of aphids contain substances that specifically inhibit its infectivity. Another possible explanation is that vectors of other viruses are unable to ingest tobacco mosaic virus. BENNETT and WALLACE (1938) have shown that non-vector leafhopper species and aphids contain sugar beet curly top virus after feeding on diseased beet, and SMITH (1941) has found that tobacco mosaic virus passes through the alimentary canal of caterpillars, but it still remains to be shown that sucking insects fed on plants with tobacco mosaic contain any virus.

Also, if aphids are only transferring sap mechanically, it is difficult to account for their success as vectors, because they are much more efficient than any mechanical methods of inoculation. With viruses such as henbane mosaic or potato Y, using sensitive test plants and inoculation methods, about $10^{-4}$ ml. of sap from an infected leaf is needed to cause infection. Yet aphids that have fed for only a minute on an infected plant and for another on the test plant can transmit, and there is no reason to suppose they transfer anything like this quantity, although it must be admitted our knowledge on this point is inadequate. However, WATSON (HAMILTON, 1935) found that after 6 hours' feeding a previously starved aphid gained less than 0.00001 g. in weight, equivalent to $10^{-5}$ ml. of sap, so that in one minute rather less than $10^{-6}$ ml. would probably be imbibed. By feeding aphids on radio-active materials, she found that less than 10% of the material imbibed by aphids was ejected into plants they fed on later, so that it seems likely that vectors can cause infection with rather less than one-thousandth the amount of virus needed by ordinary inoculation methods.

Considerable evidence for the view that non-persistent viruses have more than a mechanical relationship with their vectors has been provided by WATSON (1936, 1938*a* and *b*) and WATSON and ROBERTS (1939, 1940). This work was done chiefly with henbane mosaic virus and *Myzus persicae*, but similar results have since been obtained with potato Y, cucumber mosaic, tobacco etch, lettuce mosaic, pea mosaic, cabbage blackring and cauliflower mosaic viruses with *Myzus persicae* and other aphids. After feeding on infected plants, aphids were transferred singly to the healthy test plants, only one aphid being placed on each plant, and the efficiency of the vectors was measured by the number of healthy plants out of a hundred infected by single aphids. Working with single aphids in this way it was possible to control the length of time they fed on both the sources of the virus and on the healthy test plants. The results obtained are most unexpected and contrast remarkably with those obtained with viruses having latent periods.

The efficiency of insects as vectors of all these viruses is much increased if they are prevented from feeding for some time before they are fed on the source of the virus. As short a fasting period as 15 minutes has a pronounced effect on the percentage of transmissions. Little further increase in efficiency occurs after one hour's fasting, but this treatment may result in from 7 to 9 times the number of infections that would be produced by unstarved aphids. This remarkable effect of fasting is obtained only if the insects are fed on the source of the virus for short periods before being transferred to healthy test plants. The optimum conditions for transmission are to use insects that have previously fasted for at least one hour, feed them on the diseased plant for about 2 minutes and then transfer them immediately to the healthy plants. Increasing the length of time the aphids fed on the infected plant greatly reduces the number of infections obtained, and if they are allowed to feed for as long as an hour, insects starved before use give no more infections than those unstarved. The reduction in the efficiency of the vectors, therefore, is produced by continuous feeding and it occurs whether the aphids feed continuously on infected or on healthy plants. The optimum conditions for transfer are such that can occur seldom in nature with wingless insects, and even with winged insects it must be uncommon for an insect that has not fed for an hour to alight on an infected plant, feed for a few minutes only and then fly to a further plant. The fact that conditions in nature normally reduce the efficiency of aphids as vectors may explain the relatively slow spread of this type of virus even when the vector is present in large numbers.

The results with insect-transmitted viruses such as henbane mosaic virus are directly opposite to those with curly top of sugar beet and maize streak. The efficiency of starved vectors of the first decreases rapidly with the length of feeding time on the source of the virus, whereas the longer the vectors of the second feed the more efficient they become. It seems improbable, therefore, that the ability of *Myzus persicae* to transmit henbane mosaic virus can depend solely on the amount of virus obtained while feeding. It is apparent that in long feeding periods the aphids will have chances to acquire more virus than in short periods, but after long periods they are less efficient vectors. That they do acquire more virus by long feeding is shown by the fact that unstarved aphids do transmit a little better if fed on a diseased plant for 12 hours than if fed for less than an hour.

Infective aphids can infect several healthy plants in succession without again having access to a source of the virus, provided that they feed for only short periods on each plant (WATSON and ROBERTS, 1940; KASSANIS, 1941). The rate at which they lose their ability to transmit is also determined by the feeding. Aphids fed for a few minutes on the source of the virus and starved before and after this feeding period may remain infective for 12 hours, whereas those allowed to feed continuously either before or after feeding on the source of the virus cease to be infective within an hour. Starving the vectors after the infection feeding prolongs the time for which they remain infective, an effect not found with persistent viruses like sugar beet yellows, whose vectors seem to lose infectivity more rapidly when fasting than when feeding (WATSON, 1946).

It seems impossible to reconcile WATSON's results with the view that the viruses are merely carried as contaminants on the outside of the insect's mouthparts. For there is no apparent reason why the mouthparts of starved aphids should absorb more virus than those of unstarved. Neither is there any obvious reason why the stylets of starved aphids should absorb from a plant more virus in 2 minutes than in 1 hour, nor why virus should disappear from the stylets of unstarved insects in an hour and take 12 hours to disappear from those of starved insects. To explain these results it seems necessary to assume that the virus transmitted by aphids is taken in by the mouthparts and then injected back into plants. Then, as WATSON has shown, the results can be explained by postulating something in the aphid that inactivates the virus, the something being present in relatively large quantities in aphids which have fed continuously for some time but absent from, or present in smaller quantities in, aphids which have been starved. The something need not necessarily destroy the viruses, for it might behave like trypsin and certain other proteins, merely reducing infectivity by acting on the plant or by forming a non-infective complex with the viruses. Varying amounts of such substances in different aphids, or the presence of different substances that act specifically on individual viruses, could also account for the varying abilities of different species as vectors of different viruses.

Such inhibitors do occur in the bodies of insects, both vector and non-vector species. WATSON showed that juice from crushed *Myzus persicae* inactivated henbane mosaic virus and SMITH (1941) found that extracts from caterpillars neutralised the infectivity of tobacco mosaic and tobacco necrosis viruses. BLACK (1938) too has shown that extracts of macerated leafhoppers and aphids inhibit tobacco mosaic virus and that the leafhopper extracts also inhibit several other viruses. There is no evidence that in uninjured vectors feeding normally the viruses ever come into contact with these inhibitors, but their occurrence in extracts lends some support to WATSON's suggestion.

There is further evidence that continuous feeding reduces the vector's efficiency, although it must increase the amount of virus acquired by the vector. WATSON has shown that the longer previously-starved aphids feed continuously on diseased plants, the less likely are they to transmit. On the other hand, if they are given two or more short infection feedings and are starved before the first and between the subsequent feedings, their ability to transmit is increased. From this it appears that the aphids have acquired more than their usual dose of virus by having the short interrupted feedings, but have avoided the deleterious effects produced by continuous feeding. Aphids which have once been infective can become so again if given a second feeding on infected plants (KASSANIS, 1941), so that the loss of infectivity is not the result of any permanent change in the vector.

Not all non-persistent viruses are transmitted more readily by vectors that have been first starved and then fed for only a few minutes on the diseased plants. Vectors of dandelion yellow mosaic virus, for example, cease to be infective within an hour of leaving a diseased plant, yet they produce no more infections if starved before their infection feeding. This is probably because the virus content of diseased plants is too low for the vectors to become infective during a short feeding period, an interpretation indicated by the difficulty of mechanical inoculation. The vector has to feed on a diseased plant for at least an hour before it becomes infective, probably because it needs this period to acquire the virus from the host plant, and during this time any beneficial effects of a preliminary fast have been destroyed by the continuous feeding on the infected plant. Thus, despite the apparently large differences, the transmission of dandelion yellow mosaic virus may not differ qualitatively from that of viruses whose vectors respond to a preliminary starving. The difference may merely be set by virus concentration in the host plant, and were a host known in which the dandelion virus occurred as concentrated as, say, potato virus Y, the two might well behave similarly.

The plant-tissue relationships of the non-persistent viruses during transmission may also differ from some of those that have incubation periods in their vectors. The aphids can become infective by feeding on a diseased plant, and infect a healthy plant, in feeding times too short for their stylets to reach the phloem. It might be suggested, therefore, that the decrease in vector efficiency produced by long infection feedings occurs because aphids at first have their stylets in tissues more favourable for picking up the viruses than are those fed on later. But if this were so, the decrease would be expected with all insects and not only with those previously starved. Also, the effect of continuous feeding on the efficiency of the vectors would be determined solely by the time they fed on the infected plant, whereas it is also determined by continuous feeding on healthy plants before access to infected plants.

The mechanism of transmission of this type of virus is still obscure. The evidence suggests that the viruses must be sucked up through the vector's mouthparts and not just carried on the outside, but how they get into a position from which they can be ejected into another plant is unknown. The short time elapsing between feeding on an infected plant and infecting a healthy one would seem to preclude the penetration of the gut wall, circulation in the blood and entry into the salivary glands as was suggested for persistent viruses.

**Effect of the Number of Insects in Transmitting:** — Many workers have found that they get transmission of viruses fairly readily when large numbers of insects are transferred from diseased to healthy plants, but get few or none when single insects are used. This has often led to the suggestion that insects can inject sub-minimal infective doses of viruses into plants, the effects being cumulative so that a plant receiving sufficient of these doses will become infected. And it is implied that the infection produced by a group is contributed to by all the members, none of which could have produced it alone.

There is strong evidence against this view. WATSON (1936) finds with henbane mosaic virus and *Myzus persicae* that the percentage of plants infected increases regularly with the number of aphids placed on each healthy plant. In experiments in which 1, 5, 10 and 20 aphids were used per plant the numbers of plants infected agreed well with the numbers expected on the basis that each infection was local and independent. There was no large increase when large numbers of aphids were placed on each plant as would occur if the effects of individuals were cumulative. Indeed, the number of infections obtained when large numbers of aphids were placed on each plant was rather smaller than would be expected from the number produced by single aphids, although not significantly so.

With maize streak virus and *Cicadulina mbila* STOREY (1938) also finds that the probability of infection by a group of leafhoppers is the probability that one or other of the members of the group would alone cause infection. From experiments in which individual infective leafhoppers were allowed to feed several times, he further concludes that the probability of a single leafhopper causing infection in a number of distinct feeding periods is the probability that it would cause infection in one or other of those feeding periods. These results seem to support the suggestion made earlier in the chapter that the passage of the virus into the saliva takes place irregularly; they imply that the insect injects virus into the plant in distinct doses, the effect of each being independent of any other dose that may be injected by the same or another leafhopper. If the virus were continually present in the saliva and being injected into the plant at a steady rate, whether or not a leafhopper caused infection in a single feeding period would depend solely on the length of the feeding period. All infective leafhoppers cause infection if allowed to feed on healthy plants for extended periods. If single leafhoppers are given a number of short infection feedings of different lengths, however, infection does not necessarily occur in the longer periods. Infection often occurs in some feeding periods of about ten minutes and not in others of from four to five times as long. From this it seems that the ability of insects to infect when their stylets are in suitable positions does not depend merely on the amount of saliva ejected, but on the chance of some of the saliva containing virus.

It is probable that the mass-action effect suggested by earlier workers is not a real effect, but was obtained because experimental conditions were unfavourable for large numbers of transmissions. If conditions were such that only ten per cent of the insects could transmit, then most of the healthy plants colonised with a hundred insects would become infected, but only ten per cent of those colonised with single insects would become infected.

*Addendum.* — Since this chapter was written, several papers have appeared that bear on problems discussed in the sections on persistent viruses. With three such viruses, wound-tumour transmitted by *Agallia* sp. (MARAMOROSCH, K., 1950, Phytopathology *40*, 1071), strawberry virus 4 transmitted by *Pentatrichopus fragariae* (PRENTICE, I. W. and WOOLLCOMBE, T. M., 1951, Ann. appl. Biol. *38*, 389) and potato leaf roll transmitted by *Myzus persicae* (KASSANIS, B., 1952, Ann. appl. Biol. *39*, 157), evidence has been published that vectors can complete the latent period on infected plants and, provided they have fed there for long enough, can transmit immediately they are transferred to healthy plants. This is also true of aster yellows virus (MARAMOROSCH, private communication), so that some explanation other than that the latent period begins at the end of the infection feeding is needed to explain the results shown in TABLE 6; the differences in the lengths of time required by different insects to become infective possibly reflect differences in the lengths of time they took to acquire virus after first starting to feed on the infected plants. Whether it is true with all persistent viruses is uncertain, but experiments with pea enation mosaic virus (CHAUDHURI, R. P., 1950, Ann. appl. Biol. *37*, 342) and a virus from strawberry with yellows (MELLOR, F. C. and FITZPATRICK, R. E., 1951, Canad. J. Bot. *29*, 411), suggest that the time their aphis vectors take to become infective after leaving infected plants is unaffected by the length of the infection-feeding period.

There is now, too, direct evidence that some leafhopper-transmitted viruses multiply in their insect vectors. From one infective leafhopper, clover club-leaf virus was found to be transmitted through eggs for more than 20 successive generations (BLACK, L. M., 1950, Nature *166*, 852), a result incompatible with the hypothesis that the original insect contained all the virus detected in later generations. Also, experiments of the type outlined on p. 96 of this book, have been made with aster yellows virus (MARAMOROSCH, K., 1952, Phytopathology *42*, 59) and have shown that this virus can be transmitted serially by inoculation through 10 successive leafhoppers. This probably means that some other viruses with latent periods in their vectors also multiply in them, but it would be premature to assume that all do. Indeed, there is evidence that vectors of sugar beet curly top and potato leaf roll viruses contain only as much virus as they acquire while feeding on infected plants. Thus, different explanations may be needed to account for the latent periods of different viruses.

## *References:-*


BALD, J. G. and SAMUEL, G. (1931): Council Sci. Ind. Res. Austral. Bull. *54*.
BAWDEN, F. C. and KASSANIS, B. (1947): Ann. appl. Biol. *34*, 503.
BENNETT, C. W. (1934): J. Agric. Res. *48*, 665.
—— (1935): *ibid. 50*, 211.
—— and WALLACE, H. E. (1938): *ibid. 56*, 31.
BLACK, L. M. (1937): Cornell University Memoir *209*.
—— (1938): Phytopath. *28*, 863.
—— (1939): *ibid. 29*, 321.
—— (1941): *ibid. 31*, 120.
—— (1948): *ibid. 38*, 2.
CARTER, W. (1941): Ann. Entom. Soc. Amer. *34*, 551.
CHAUDHURI, R. P. (1948): (Unpublished data).
COOK, M. T. (1947): Viruses and Virus Diseases of Plants, Minneapolis, Burgess Publishing Co.
DOBROSCKY, I. D. (1931): Contrib. Boyce Thompson Inst. *3*, 59.
DONCASTER, J. P. and KASSANIS, B. (1946): Ann. appl. Biol. *33*, 66.
DOOLITTLE, S. P. and WALKER, M. N. (1928): Phytopath. *18*, 143.
DYKSTRA, T. P. and WHITAKER, W. C. (1938): J. Agric. Res. *57*, 319.
ELZE, D. C. (1927): Meded. Landbouwhoogesch. Wageningen *31*, 1.
FIFE, J. M. and FRAMPTON, V. L. (1936): J. Agric. Res. *53*, 557.
FREITAG, J. H. (1936): Hilgardia *10*, 305.
FUKUSHI, T. (1934): J. Fac. Agric. Hokkaido Imp. Univ. *37*, 41.
—— (1935): Proc. Imp. Acad. Japan *11*, 301.
—— (1939): *ibid. 15*, 142.

Hamilton, M. A. (1935) : Ann. appl. Biol. *22*, 243.
Hartzell, A. (1936) : Contrib. Boyce Thompson Inst. *8*, 113.
—— (1937) : *ibid. 8*, 375.
Hewitt, W. B., Byron, R. H., Frazier, N. W. and Freitag, J. H. (1946) : Phytopath. *36*, 117.
Hoggan, I. A. (1931) : Phytopath. *21*, 199.
—— (1933) : *ibid. 23*, 446.
Houston, B. R., Esau, K. and Hewitt, W. B. (1947) : Phytopath. *37*, 247.
Kassanis, B. (1941) : Ann. appl. Biol. *28*, 238.
—— (1947) : *ibid. 34*, 412.
Kirkpatrick, T. W. (1931) : Bull. Entom. Res. *22*, 323.
Kunkel, L. O. (1926*a*) : Phytopath. *16*, 67.
—— (1926*b*) : Amer. J. Bot. *13*, 646.
—— (1937) : *ibid. 24*, 316.
—— (1938) : J. Econ. Entom. *31*, 20.
—— (1944) : Science *99*, 450.
—— (1946) : Amer. J. Bot. *39*, 831.
Kvicala, B. (1945) : Nature *155*, 174.
Leach, G. L. (1940) : Insect Transmission of Plant Diseases. McGraw-Hill, New York.
Markham, R. and Smith, K. M. (1949) : Parasitology *39*, 330.
Ocfemia, G. O. and Buhay, G. C. (1934) : Philipp. Agriculturist *22*, 56.
Osborn, H. T. (1938) : Phytopath. *28*, 749.
Posnette, A. F. (1947) : Ann. appl. Biol. *34*, 388.
Prentice, I. W. (1946) : Nature *158*, 24.
—— and Harris, R. V. (1946) : Ann. appl. Biol. *33*, 50.
Roberts, F. M. (1940) : Ann. appl. Biol. *27*, 348.
Severin, H. H. P. (1921) : Phytopath. *11*, 424.
—— (1924) : *ibid. 14*, 80.
—— (1931) : Hilgardia *6*, 253.
—— (1946) : *ibid. 17*, 121.
—— and Freitag, J. H. (1938) : Hilgardia *11*, 495.
Smith, K. M. (1931) : Proc. Roy. Soc. B. *109*, 251.
—— (1941) : Parasitology *33*, 110
—— (1946) : *ibid. 37*, 131.
—— and Lea, D. E. (1946) : Parasitology *37*, 25.
Storey, H. H. (1928) : Ann. appl. Biol. *15*, 1.
—— (1932) : Proc. Roy. Soc. B. *112*, 46.
—— (1933) : *ibid. 113*, 463.
—— (1938) : *ibid. 125*, 455.
—— (1939) : Bot. Rev. *5*, 240.
Sukhov, K. S. and Sukhova, M. N. (1940) : C. R. Acad. Sci. U. S. S. R. *26*, 479.
Watson, M. A. (1936) : Phil. Trans. Roy. Soc. B. *226*, 457.
—— (1938*a*) : Proc. Roy. Soc. B. *125*, 144.
—— (1938*b*) : *ibid. 125*, 305.
—— (1940) : *ibid. 128*, 535.
—— (1946) : *ibid. 133*, 199.
Watson, M. A. and Roberts, F. M. (1939) : Proc. Roy. Soc. B. *127*, 544.
—— and —— (1940) : Ann. appl. Biol. *27*, 227.

*Chapter 6*

# MUTATION, INTERFERENCE BETWEEN STRAINS, AND RECOVERY FROM DISEASE

The fact has long been known that persons who have once suffered from smallpox rarely contract the disease a second time, even though exposed to it, and it has also been appreciated that men and animals could, as a result of a non-fatal attack, acquire a lasting immunity against certain other virus diseases. From JENNER's pioneer work in protecting people against smallpox by vaccination with cowpox, it was also clear that the benefits of resistance might be acquired without the risks attendant on infection with a virulent pathogen. It is natural, then, that much of the work on animal virus diseases should have been devoted to a search for relatively innocuous viruses suitable for vaccines and should have led early to the discovery that viruses are not stable entities, but can undergo variations and exist in strains of different virulence. More than 60 years ago, PASTEUR did the first experimental work on changing virulence, when he showed that the pathogenicity of rabies was reduced by repeated passage in the rabbit's central nervous system, and from which he produced the attenuated or "fixed" form used as a vaccine.

No such incentives existed for seeking variants of plant viruses. There was no evidence that plants possess anything equivalent to the antibody-forming mechanisms to which animals owe their recovery and subsequent immunity, and it was only too evident that most virus-diseased plants did not recover but remained infected until they died. Nor were there any comparable immunological techniques by which the workers on plant viruses could relate viruses that caused different symptoms. It is not surprising, therefore, that clinically distinct plant viruses were for long tacitly assumed to be unrelated, and that the occurrence of some in strains of varying virulence went unrecognized. Now, however, the mutability of plant viruses is generally appreciated and the study of virus strains is being increasingly pursued. There are various reasons for this change of emphasis. The application of serological methods to plant viruses showed that some which were regarded as distinct were related, and this pointed the need, in any attempt at classification, for grouping strains around a named type. Also, the study of virus strains was encouraged by the discovery in plants of two phenomena that simulate the acquired immunity and vaccination against severe diseases so well established with animals. The first phenomenon was observed with plants infected with tobacco ringspot virus, which after first suffering an acutely necrotic disease recover and never again succumb (WINGARD, 1928). The second came from the observation that plants infected with tobacco mosaic virus were protected against the effects of a virus that caused a severe "white" mosaic when inoculated to healthy plants (THUNG, 1931). The mechanisms underlying these two phenomena are still obscure, but there are differences from the superficially similar behaviour in animals in which the immunity is sterile and results from antibody production, for the plants are resistant only as long as they contain active virus. Both phenomena have since been encountered with additional viruses, and the protection afforded by one strain against infection with another has been found to apply generally to the many viruses that occur in strains of different virulence.

It is now apparent that it is usual rather than exceptional for plant viruses to vary and that few, if any, occur in single immutable forms. So many viruses have been found to exist in strains of different virulence, and so many strains of some individual viruses have been identified, that no attempt can be made to describe them in detail. Over fifty strains of tobacco mosaic virus have already been recognised, and, as there is good reason to consider that new strains are constantly arising,

FIG. 20. — Tobacco leaf showing symptoms of tobacco mosaic. In addition to the general mild type of mottling, the leaf shows a bright yellow spot. Inoculum taken from this spot causes a general yellow chlorosis. (KUNKEL, L. O., 1934, Ann. Rep. Quebec Soc. Prot. Plants *25-26*, 23.)

the number that could be identified is probably limited more by experimental techniques and investigators' perseverance than by the total number possible. Many variants of some viruses occur naturally, but with others most have been found during the course of experimental work designed for the purpose. Many of the latter type would have little chance of occurring commonly in the field, where the principle of natural selection can be expected to operate and determine the prevalent types. Related strains compete with one another, because the presence of one interferes with the multiplication of another, and the strains that occur most commonly are the most infective ones, that is, those most easily transmitted and which invade plants most rapidly and multiply most extensively. Unless a newly-produced strain is more infective than its progenitor, its chances of survival under natural conditions are small, whereas in experimental work poorly infective strains can be isolated and transmitted separately to test plants in which they then can multiply free from competition with more active strains. Some such strains have been described as "unstable", for it is difficult to perpetuate them even experimentally, and they are often regarded as varying more frequently than the common, naturally-occurring strains. In the absence of any quantitative data, however, this seems an unnecessary assumption, for a mutant is more likely to make its presence obvious from a poorly infective parent than if it comes from a highly infective one. There is little doubt that most new strains that arise from virulent viruses pass unnoticed, whereas, for obvious reasons, changes involving increased virulence are much more likely to be observed.

The type of change most frequently recorded is in pathogenicity, affecting the symptoms caused or the host range, and examples of the range of symptoms produced by related strains of a few viruses have been indicated in Chapter 2. Here an attempt is made to discuss the general problem of variation, using specific examples as illustrations, rather than to describe in any detail the individual characters of strains.

**Kinds of Variation:** — The first suggestions that plant viruses might alter in virulence occur in the work of CARSNER (1925), JOHNSON (1925), and MCKINNEY (1926, 1929). CARSNER showed that sugar beet curly top virus was altered by passage through *Chenopodium murale*, because when returned to beet it produced only mild symptoms. The change was apparently permanent, for propagating the attenuated virus continuously in beet did not increase its virulence, but LACKEY (1932) found that passage through *Stellaria media* could restore it to a condition in which it again produced a serious disease in beet.

JOHNSON found that tobacco mosaic virus was altered by keeping recently infected plants for some days at temperatures between 35° and 37°C. The plants showed milder symptoms than those grown at lower temperatures, and continued to do so when later grown under more normal conditions. That the change was in the virus and not in the host was shown by transmission to other plants grown continuously at lower temperatures, which in turn also showed only mild symptoms. HOLMES (1934), by incubating pieces of inoculated tomato stems for 15 days at 34.6°C., carried the process of attenuation still further, and obtained a culture of tobacco mosaic virus that produced no clear mosaic symptoms in tobacco. To retain cultures of this "masked" strain in plants grown at normal temperatures, new isolations from single lesions are needed at intervals; otherwise, infected plants begin to show symptoms because more virulent forms develop and tend to dominate the masked strain. A similar effect of attenuation by exposure to heat has been described with aster yellows virus by KUNKEL (1937), who got an avirulent strain by keeping infective leafhoppers for some days at 32°C., and SALAMAN (1937) found that incubating tobacco roots infected with potato virus Y for a week at various temperatures above 22°C. produced less virulent forms. Strains of tobacco mosaic virus more virulent than the normal were obtained by MCKINNEY (1926, 1929) by the simple method of selecting inoculum from different leaf parts. Bright yellow spots (Fig. 20) often occur in leaves and inoculum from these caused a bright yellow mottling, very different from the normal symptoms of mosaic. MCKINNEY suggested that the virus in the yellowed areas was not a contaminant in the customary use of the word but that it was derived as a mutant from the ordinary type. He also noticed that plants infected with inocula containing a mixture of the yellow and the ordinary mosaic viruses, showed symptoms intermediate between those of plants inoculated with either separately, and that the smaller the proportion of one in the inoculum, the less were the symptoms characteristic of it and the more slowly did these symptoms develop.

JENSEN (1933, 1936, 1937) has isolated over fifty strains of tobacco mosaic virus from the yellow spots. Some of these are apparently similar to tomato aucuba mosaic virus, a strain commonly found in nature, but others differ from any strains previously described. JENSEN distinguishes between the various strains by the ease with which they infect,

FIG. 21. — Comparable leaves from two shoots of one potato plant showing segregation of strains of potato virus X. Most of the plant showed only mild symptoms and contained an avirulent strain (left-hand leaf), but one shoot showed a severe mosaic with much necrosis and contained a virulent strain.

and by the symptoms they cause in, tobacco, tomato, *Nicotiana sylvestris* and *N. glutinosa*. All of them at normal temperatures produce only necrotic lesions on *N. glutinosa*, but some strains produce smaller lesions than others. All cause local lesions on *N. sylvestris*, but some cause necrotic lesions and others chlorotic. Also, some give systemic infection and others do not. In tobacco and tomato, the variants differ most in the severity of the mottling or necrosis and in the degree of distortion and stunting produced, but they also differ in the ease with which they can be transmitted by pricking with needles and in the rapidity with which they move from the entry point in the plant. Making transfers by single pin-puncture inoculations, some strains gave infection in over fifty per cent of the inoculations, whereas others were transmitted in less than one per cent of the attempts. JENSEN has transmitted some of these variant strains continuously for a period of more than two years and many of them retain their original characteristics. Sometimes, however, especially with strains that move slowly in the infected plants, reversion seems to occur. Then the young leaves of inoculated plants develop either symptoms similar to ordinary tobacco mosaic or else a systemic infection with a rapidly-moving strain of the tomato aucuba mosaic virus type that produces a general chlorosis.

Similar, though less numerous, variants occur in cucumber mosaic virus and potato virus X. The commonly occurring cucumber mosaic virus causes in tobacco a rather mild green type of mottling. PRICE (1934) observed that infected plants sometimes produced bright yellow spots, like those noticed by MCKINNEY with tobacco mosaic. The spots varied in size, shape and intensity of colour, and occurred more frequently in plants growing rapidly than in those growing slowly. PRICE

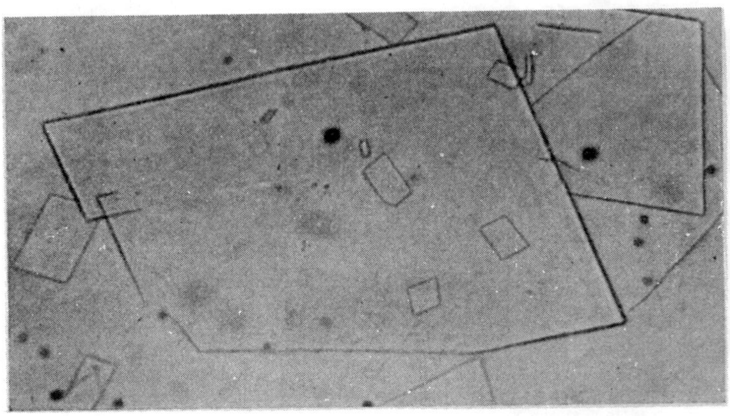

FIG. 22*a* and *b*. — Serologically related strains of a tobacco necrosis virus which consistently crystallise in different ways. *a:* Thin lozenge-shaped plates formed by one strain, and *b:* Hexagonal prisms formed by the other. ×200.

cut out areas containing these yellow spots, macerated them and rubbed the extracts over leaves of healthy tobacco plants. After a few days yellow local lesions developed and pin-puncture inoculations were made from these local lesions to further healthy tobacco seedlings. Only 14 infections were obtained out of 956 inoculations, but from these, five variants were distinguished by the intensity of the mottling and the degree of stunting produced. One strain produced necrotic symptoms in tobacco and moved only slowly from the entry point, but this was unstable, and the young leaves of plants inoculated with it always contained a more rapidly moving strain. The other strains remained constant and produced their characteristic symptoms when transferred serially through numbers of different susceptible species. All the strains isolated from tobacco produced black necrotic local lesions on *Vigna sinensis* (cowpea) and failed to give systemic infections. But among the necrotic lesions produced on cowpea by the normal strain of cucumber mosaic virus, PRICE noticed two chlorotic lesions. From these he isolated a further strain sharply differing from all the others because it infected cowpea systemically, producing a mosaic. During work on insect transmission of the common strain of cucumber mosaic virus, HOGGAN (1935) found occasional plants showing bright yellow mosaics instead of the usual symptoms. From these plants she obtained a strain of the virus which apparently remained unchanged and produced a bright yellow mosaic in tobacco over a period of five years. It is probable that the insects had fed on the occasional yellow areas before they were transferred to healthy plants, so effecting the same separation as if these areas had been cut out, macerated and inoculated separately to healthy plants.

Cucumber mosaic virus has a wide host range and there is little doubt that many viruses given different names in the past are really variants of it. PRICE (1935) showed that southern celery mosaic virus was a strain of cucumber mosaic virus that was rather more infective and had a slightly different host range from the usual type. He also

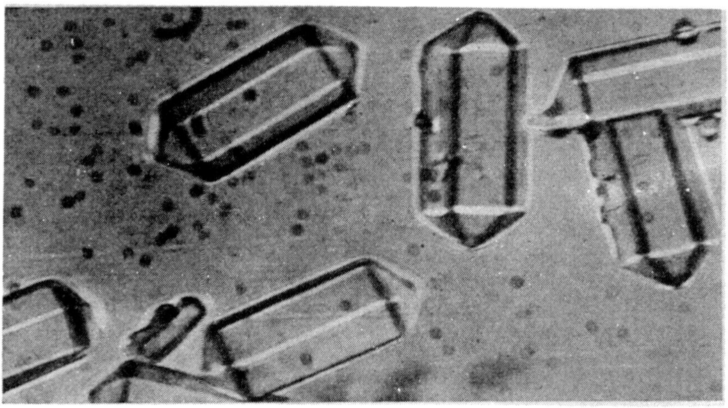

FIGURE 22*b* (*see* p. 110).

found that lily mosaic virus was another strain, which when first transferred to tobacco usually gave local lesions only, but when repeatedly cultured in this host changed to a form that became systemic and gave a mottle (PRICE, 1937). BHARGAVA (1948) has found that, in addition to differing in host range and type of symptoms caused, different strains differ in their transmissibility by aphids and in the concentration they attain in infected plants, which in turn affects such properties as thermal inactivation and dilution end-point.

What are now recognised as strains of potato virus X were early distinguished under such names as potato mottle or ringspot viruses, depending on the type of symptoms caused in tobacco. SMITH (1929) suggested that the mottle form gradually passed into the ringspot form as a result of continued passage through tobacco, but it is more likely that his original plants contained a mixture of strains; the gradual increase in virulence of the culture would result from increasing the proportion of the ringspot form in the inoculum, which would occur every time the most severely affected leaves were used for sub-culturing. SALAMAN (1933, 1938) described six strains of virus X and several more have now been identified, all readily differentiable by the kind of symptoms they cause in tobacco and a range of potato varieties. SALAMAN found that tobacco plants inoculated from naturally-infected potatoes usually contain more than one strain. The most common reaction in tobacco is a mild mottle, but often the leaves also have a few scattered yellow or necrotic spots. Unselected inoculum reproduces the original clinical picture, but inoculum taken exclusively from the yellow or necrotic spots produces a general yellow mottle or ringspot. Virulent cultures of virus X can be attenuated by passage through sugar beet, for SALAMAN (1938) found that a culture that previously gave ringspot symptoms in tobacco gave only a mottle after causing local lesions in beet. Sudden changes in virulence sometimes appear in potatoes; tubers saved from infected parents that showed only a faint mosaic, or were symptomless carriers, occasionally give progeny with severe mosaic or necrotic symptoms, which sometimes affect the whole plant and sometimes are restricted to one shoot only (Fig. 21).

Variations in virulence or in host range are the simplest types to detect, and most of the strain differences so far recognised lie in these two features. There is, however, ample evidence that variation in other characters also occurs. Different strains of aster yellows (SEVERIN, 1934) and potato yellow dwarf (BLACK, 1941) viruses have different leafhoppers as vectors, and one strain of potato virus Y does not appear to be transmitted by any of the aphids that are vectors of other strains. Within one of the tobacco necrosis viruses, strains have been detected by their varying behaviour when precipitated in the same conditions from ammonium sulphate solutions; one consistently crystallises as thin, lozenge-shaped plates, the second as hexagonal prisms, whereas the third has produced no recognisable crystals (Fig. 22) (BAWDEN and PIRIE, 1942). Hence it is probable that many viruses, now regarded as identical because they are clinically indistinguishable, will become differentiated as strains when their properties are more fully determined. Many would probably become so differentiated merely by extending studies of host plant reactions, for mutation may affect the reactions caused in one host but not in another, and in the same host different strains may produce the same symptoms in one set of environmental conditions but not in another. There are, for example, strains of tobacco mosaic virus that produce identical mosaic symptoms in some types of White Burley tobacco, but in other types they are clearly different for some cause systemic mosaics and others give necrotic local lesions or a lethal systemic necrosis (KASSANIS and SELMAN, 1947). The effect of environment on strain identification was indicated by MCKINNEY (1935), who found two strains of tobacco mosaic virus that are clinically indistinguishable in tobacco plants grown at 24°C., but which cause diseases of different severity if the plants are kept at 15°C. An opposite effect of temperature has been described with cabbage blackring virus, strains of which are differentiable by the symptoms caused in *Nicotiana glutinosa* at 24-28°C., but not at 16°C. (POUND and WALKER, 1945).

**Origin of Variation:** — As we have seen there is now conclusive evidence that plant viruses vary and it has been shown repeatedly that passage through a new host, or subjecting infected plants to some unusual environment, often causes a change in behaviour. Such changes appear to occur suddenly and to be permanent, so that it is perhaps admissible to apply the word mutation, although the application to viruses of words coined to express changes in higher organisms is not without dangers. There is, however, less reason to believe that the treatment applied has actually caused a mutation. It may have acted simply as a selective mechanism that specifically encouraged a previously existing strain at the expense of others. In other words mutations may be occurring spontaneously regardless of any applied treatment, but the newly-produced forms normally remain undetected and make their presence obvious only when selected by the treatment. It is much more difficult to detect the fact that a plant contains a mixture of related virus strains than that it is infected with a mixture of viruses. Two unrelated viruses will not interfere with one another and both can become established throughout the plant and cause their specific effects. But conditions are very different with two strains of one virus. Each interferes with the multiplication of the other, neither becomes fully systemic, and the symptoms shown will be largely those characteristic of the predominating strain. Hence the presence of new mutants in small quantities will pass unnoticed unless some treatment is applied that specifically encourages them. Such selection may be supplied by transmission to a new host, by changing environmental conditions, or by selecting as inoculum small pieces of leaf that show unusual symptoms.

Thus the attenuation of sugar beet curly top virus by passage through *Chenopodium murale,* and the recovery of virulence by a subsequent passage through *Stellaria media,* has two equally plausible explanations. *C. murale* may cause the virus to mutate, and *S. media* may either reverse this change or produce another mutation towards increased virulence. On the other hand, both hosts may be acting simply as filter plants and specifically separating strains from mixtures, for curly top virus may mutate spontaneously and infected plants may usually contain strains that vary in virulence. If these strains all multiply equally in sugar beet, then their presence would be undetectable and symptoms over many transmissions would remain those of the predominant type. If *C. murale* is susceptible to strains avirulent for sugar beet, but is resistant to, or immune from, virulent ones, it will select the former, whereas if *S. media* is susceptible to virulent strains and resistant to avirulent ones, it will reverse the procedure. Similarly with potato virus X, the change from a ringspot to a mottle form by passage in sugar beet may occur because the new host caused a mutation, but it is equally likely that the original stock contained a mixture of strains, in which ringspot forms predominated, and that the new host favoured the mottle forms at the expense of the others. No final judgement can be made between these alternative hypotheses, but the existing evidence favours the idea of random mutation with the treatments that seem to cause changes merely selecting previously produced variants.

The best evidence for a new host acting as a filter rather than causing changes has been provided by JOHNSON (1947), working with tobacco mosaic virus and sea-holly *(Eryngium aquaticum).* He found that sea-holly naturally infected with tobacco mosaic virus always contained strains that were avirulent towards tobacco, and when he inoculated healthy sea-holly with virulent stocks from tobacco, they regularly became infected systemically with avirulent strains. When he used cultures of virulent strains recently derived from single local lesions, however, he got no systemic infection in sea-holly, but when such cultures were propagated without further selection, they again produced systemic infection in sea-holly, from which again avirulent strains were recovered. Hence it seems that forms which are incapable of infecting sea-holly systemically are, when multiplying in tobacco, producing variants that can, and these, unless frequently eliminated by recourse to single-lesion isolates, gradually increase until they occur abundantly enough to be inoculable to sea-holly. JOHNSON found that the constancy of his single-lesion isolates varied with the temperature at which infected tobacco plants were grown, more variants being noted at 35°C. than at normal glasshouse temperatures.

From the different results obtained when inoculum is taken from parts of leaves showing different symptoms, it is clear that plants infected with some viruses do normally contain mixtures of strains, for the selection of inoculum cannot cause changes but can only separate previously formed mutants. Thus it is likely that treatments other than transmission to new hosts also produce changes by selecting existing variants rather than by themselves causing mutation. For example, HOLMES (1934) has shown that the masked strain of tobacco mosaic virus which he obtained by heating infected tissues multiplies more rapidly than mosaic-producing strains in plants grown at high temperatures. Hence there is no reason to assume that the heat-treatment caused the mutant; it is more likely that it merely allowed to become predominant an avirulent strain that previously constituted only a minor part of the stock.

The occurrence of some viruses in so many strains is in itself indisputable evidence that they mutate, and the only debatable points seem the frequency of mutation and whether any applied treatment that changes behaviour causes mutation or exerts a selective action on existing mutants. Unequivocal evidence on these points could be obtained only by knowing that virus cultures used in experiments are pure lines, that is, contain one strain only. Unfortunately, with present techniques, this is impossible. With viruses the nearest equivalent to a single-spore fungus culture is a stock derived from a single local lesion, but it remains unproven that a local lesion is caused by a single virus particle. Inoculum must contain many millions of tobacco mosaic virus particles before it will cause a local lesion in *Nicotiana glutinosa.* Most of the particles are certainly not concerned in the infection, but it cannot be assumed than only one is. However, KUNKEL (1934), working with mixtures of tobacco mosaic virus strains, found that single lesions usually contained one strain only, although rarely two were present. Using highly diluted inocula, the chances of different strains being concerned in causing a single infection are obviously reduced, yet JENSEN (1936) showed that variants causing yellow mosaics could be obtained from such single-lesion isolates of tobacco mosaic virus when propagated in tobacco. Some of these strains are so difficult to transmit, and have such a low dilution end-point, that it seems unlikely that they were carried as contaminants through single-lesion transfers, and more likely that they originated in the plants in which they were detected.

Further evidence for the frequent mutation of tobacco mosaic virus has been provided by MCKINNEY (1935), who found that *Nicotiana glauca* becomes systemically infected with the common strains but gives only local lesions with strains that cause yellow mottles in tobacco. He showed that tobacco plants inoculated from systemically infected *N. glauca* first showed a mild green mosaic, but they later developed yellow spots from which he isolated strains which failed to infect *N. glauca* systemically. It is conceivable that such strains might infect *N. glauca* systemically when inoculated together with other strains although they do not when alone, but there is circumstantial evidence against this. McKINNEY has found one strain that causes a green mottling in tobacco and has never produced any yellow spots. When this was mixed with variants causing yellow spots and inoculated to *N. glauca,* subsequent transfers from systemically infected leaves produced the normal green mottling with no yellow spots in tobacco.

The common strains of tobacco mosaic virus are more infective, *i. e.,* are more readily transmitted, invade tobacco plants more rapidly and reach a higher concentration than most of those that have been derived experimentally. If mixtures of tobacco mosaic virus and other strains are inoculated to the same plant, unless the other strains are greatly in excess, they tend to be overrun and their effects suppressed. By contrast, the variants described give reasonably constant reactions. McKINNEY (1937) states that those he has isolated occasionally give rise to further strains, but that none has ever reverted to tobacco mosaic virus. This is strong evidence that his cultures were not originally mixtures. If some strains can be obtained free from the parent type, there would seem to be no obvious reason why tobacco mosaic virus itself should not also be obtained pure, except that mutations occur in most infected plants.

At first sight the frequency with which variants can be detected may seem too high for the production of spontaneous mutants. However, if the mutation rate is thought of in terms of virus particles instead of host plants, the number of variants that can be detected is by no means excessive. A tobacco plant can be infected by rubbing with about $10^{-10}$ of a gram of tobacco mosaic virus. Two months after infection at least 50 mg of virus can be isolated from such a plant. Even if it is assumed that all the virus in the inoculum entered the plant, which it certainly did not, there has been a virus increase of approximately a hundred million times, and the plant will contain at least a million million ($10^{12}$) virus particles. In the fruit-fly *Drosophila melanogaster,* mutation rates have been studied in detail, and in a normal environment a mutation can be expected in from one hundred thousand to a million individuals. Therefore, if the mutation rate in tobacco mosaic virus is of the same order as in *Drosophila,* it is obvious that a large number of variants could be expected in each infected plant.

In view of these figures it is not the number of mutations that is surprising, but rather the constancy of the symptoms produced by tobacco mosaic virus. This can probably be explained by the intense competition between a parent strain and its variants. The variant will only be able to multiply and produce its characteristic symptoms if it can get into tissues containing little or none of the parent strain, for the presence of one strain will prevent another from developing. To become evident it must either be more invasive than the parent strain and be able to move into newly formed tissues more rapidly, or it must be fortunate enough to find some cells not fully infected with the parent strain, as seems to happen in the yellow areas that regularly appear in tobacco mosaic plants. The commonly occurring strains of tobacco mosaic virus are more invasive and occur in infected plants in greater quantities than any of the experimentally produced variants yet described, and if mixtures are inoculated to healthy plants, the symptoms of tobacco mosaic soon predominate. Thus most of the mutants produced from tobacco mosaic virus seem unable to compete with it, and remain undetected because they do not multiply sufficiently to produce any characteristic effects. Most mutations that have been produced experimentally in plants and animals have been deleterious in the sense that the changes make them less likely to survive in nature than the parent. This also appears to be true with tobacco mosaic virus. Virus strains best fitted to survive in nature are those most easily transmitted, and which rapidly invade and reach a high concentration in the plant without causing too serious a disease. The commonly occurring tobacco mosaic virus possesses all these properties, but the derived mutants mostly lack one or more. The constancy of the symptoms produced by tobacco mosaic virus, in spite of the numerous mutations, can probably be explained on the Darwinian theory of natural selection and survival of the fittest. But, of course, which are the fittest may well vary from time to time as conditions change. The introduction into culture of new plant varieties that resist the currently common strains but are susceptible to others, will favour the latter and change the relative proportions in which different strains occur. Other practices may also have such effects. The strains of virus X now prevalent in British potato stocks are not those that spread most rapidly and multiply most extensively, but are those that cause no clear leaf symptoms. This is because commercial stocks are inspected and all plants showing symptoms are removed, so that only strains that produce no definite leaf symptoms are able to survive and accumulate.

The mutants derived from tobacco mosaic virus can, in their turn, also mutate (JENSEN, 1936; HOLMES, 1936; NORVAL, 1938). Studies of these mutants suggest that individual characters of the virus change independently of one another and that one character may remain constant while another varies. For example, HOLMES finds that strains causing yellow mottles in tobacco may possess other different characters which are linked with those of their immediate progenitors. Strains derived from an invasive parent were themselves mostly of the type that rapidly became systemic in tobacco, whereas most of those from the weakly invasive "masked" strain lacked this property. Similarly, NORVAL found that twelve clinically distinct strains derived from one that caused necrosis in tomato all possessed this property, whereas strains derived from other sources rarely did. With potato virus X, too, SALAMAN (1938) described five characters which he considers vary independently. The ability to inherit different combinations of characters suggests unit differences in the structure of virus particles analogous to those attributed to genes in organisms. The comparison, so often made, between virus particles and genes seems less appropriate, for the particles apparently carry the equivalent of many genes and so are more logically compared with portions of chromosomes.

We have no idea of the types of change within the virus particles that lead to mutations. This is not surprising when the size and complexity of the particles is considered, for many separate changes could occur without being detectable by current techniques. Variations in active side groups, loss or gain of different groups, or rearrangement of the constituent units, many such changes might alter biological properties without affecting a large enough proportion of the whole particle to be chemically indentifiable. In the synthesis of such large particles there is ample opportunity for different structural variations, and not all the completed particles could be expected to follow the parent's pattern precisely. Some may be changed in such a manner that they are no longer capable of acting as viruses and causing infection. This could account for the presence in sap from tobacco mosaic plants of non-infective particles which nevertheless possess the general chemical, physical and serological attributes of virus particles (BAWDEN and PIRIE, 1945). In genetical terminology, such particles would be equivalent to mutants with lethal genes, and the same term might also be applied to such changes as those caused by ultra-violet or X-ray irradiation, which have no detected effects on chemical, physical or serological behaviour, but which destroy infectivity.

While changes leading to loss of infectivity can occur for unknown causes, it might seem premature to seek causes for relatively minor changes in virulence. Nevertheless, there are detectable differences between strains of one virus, the differences becoming more striking the larger the biological differences between the strains. Between strains recently derived from the same stock, the only differences so far detected lie in their antigenic structure, and, though slight, they are sufficient to show that biologically-distinguishable strains are structurally not identical, for such strains share most of their antigens but they have a few specific ones. Between some naturally-occurring strains of tobacco mosaic virus, which differ widely in their pathogenicity, much greater differences have been found. Tomato aucuba mosaic, cucumber 3, and the type form of tobacco mosaic virus, for example, have different isoelectric points, and tobacco mosaic virus, cucumber 3 and the rib-grass strain (HOLMES, 1941) share some antigens, but many of their antigenic groups are different (BAWDEN and PIRIE, 1937; KNIGHT and STANLEY, 1941). Strains that differ to this degree also have demonstrably different chemical constitutions. The detailed amino-acid analyses of purified preparations of different strains that have been obtained by KNIGHT (1947) are given in TABLE 17 (Chapter 10), but warrant comment here. They show that strains which differ greatly in their pathogenicity may also differ in the quantities of certain amino-acids they contain, and that amino-acids present in one may be absent from another. The strains are so alike in many of their properties that it can be assumed they have a common ancestor and that, in the course of time, they have either dropped or acquired certain amino-acids. The relevance of this to the spontaneous mutations that seem to occur frequently, however, is uncertain, for the clinically distinct strains that arise in this manner have not yet been found to have different amino-acid constitutions. It may be that such mutants also vary because of amino-acid differences which are too slight to be detected by present techniques, but it is equally likely that other differences are involved.

The reactions between viruses and their hosts that are responsible for the development of symptoms are unknown, but severity and type of symptom is not necessarily correlated with quantity of virus produced. Strains of viruses such as tobacco mosaic and potato X that produce mild symptoms often occur in higher concentration than those that produce severe diseases. Symptoms do not, therefore, seem to occur merely because virus multiplication directly weakens the host, but seem more likely to result from secondary effects, reactions between host constituents and specifically active groups on the virus particles. Hence changes in such groups, or the gain or loss of a certain group, might lead to the formation of strains capable of causing different symptoms, but otherwise being indistinguishable from the parent type.

Attempts have been made (ANSON and STANLEY, 1941; MILLER and STANLEY, 1941, 1942) to produce mutations in tobacco mosaic virus by treating it with substances that react with the amino, carboxyl, indolic, phenolic and sulphydryl groups. A large proportion of these groups could be changed without affecting the infectivity of the virus, but none of these changes simulated mutation, for all the derivatives produced the usual symptoms and virus isolated from plants inoculated with them was the normal type and not like the chemically prepared derivatives. However, MILLER and STANLEY (1941) noticed one interesting effect with the phenylureido derivatives of the virus, for although these produced as many lesions as untreated virus in *Nicotiana glutinosa* they were less infective when tested in *Phaseolus vulgaris.* Hence it seems that chemical changes in some groups may affect ability to infect different plants differently, although with such treatments it may be that the host cells merely reverse the changes caused in the laboratory and that different hosts vary in their ability to do this.

No quantitative work has yet been done on the relative mutation rates in different viruses, but qualitative observations suggest that there may be considerable differences. We have already seen that mutant forms can be found in most plants infected with tobacco mosaic virus, and with potato virus X mutation also seems to be frequent. By contrast, thousands of different plants have been infected with tomato bushy stunt virus without any variations being detected. This difference may in part have a simple quantitative explanation, for tobacco mosaic virus and potato virus X reach higher populations in infected plants than most other viruses, and, other things being equal, the more particles there are produced, obviously the more chances there are of mutants occurring. But it is possible that the different chemical constitution and structure of individual viruses may make some more liable to variation than others. Potato virus Y, for example, which has particles similar in gross structure to tobacco mosaic virus and potato virus X, reaches a much lower concentration in infected plants than tomato bushy stunt virus, yet many variants of virus Y have been distinguished and none of bushy stunt. Too few viruses have yet been studied for any generalisations to be safe, but on existing evidence there is a suggestion that those which contain relatively little nucleic acid, and whose particles aggregate to form long rods, tend to produce more variants than those which contain more nucleic acid and have spherical particles.

**Acquired Immunity:**—The use of the term acquired immunity in plant pathology is open to grave objections, but as it has been widely applied, both for plants that recover from an acute disease and for the resistance conferred on plants against virulent strains by the presence of avirulent ones, it will be continued here for want of a better. There are two principal objections to its use. First, the resistance acquired is rarely absolute, and hence does not justify the word immunity, and the resistance is against a set of symptoms rather than against a pathogen. Secondly, the term is being applied to phenomena that differ fundamentally from those in animals for which it was first coined. In animals acquired immunity can be either active or passive, but both are usually of the sterile type, resistance to infection being conferred by the presence of antibodies. The acquired immunity is active if the animal has itself produced antibodies against the pathogen and passive if they have been obtained ready-made from the blood of another animal. In some diseases, for example, the salivary gland disease of guinea pigs (COLE and KUTTNER, 1926) and infectious anemia of horses (KOCK, 1924), animals seem to contain virus long after they have recovered and become immune, but this is exceptional. With the phenomena in plants that have been described under the term acquired immunity, on the other hand, the persistence of active virus in the recovered or protected plants is an invariable and essential part of the mechanism. Freedom from a second attack of an acute disease, or protection from the effects of virulent strains, persists only as long as tissues are infected. What is acquired is really a resistance to a set of symptoms and, as the recovered or protected plants are rarely wholly normal, the resistance to severe diseases is acquired only at the cost of continually suffering lesser ones.

Recovery from an acute disease was first described by WINGARD (1928) with plants infected with tobacco ringspot virus, which after reacting initially with an acutely necrotic disease later produce new leaves that may bear no distinct lesions. WINGARD found that suckers or cuttings taken from recovered plants produced plants that looked healthy and that such plants developed no symptoms when inoculated with sap taken from necrotic leaves. He suggested that infected plants acquire an immunity to the disease, a view that PRICE (1932, 1936*a*, 1936*b*) has vigorously supported. VALLEAU (1941) and MCKINNEY (1941) have opposed the application of the term acquired immunity to the phenomenon and state that the disease simply has two phases, first acute and later chronic. It is certainly true that the recovered plants do not acquire an immunity from the pathogen. PRICE has propagated infected plants by means of cuttings through ten generations and the progeny all contained virus which caused typical ringspot symptoms when transferred to healthy seedlings. None of the vegetatively-propagated plants showed any leaf symptoms and they did not react when inoculated with the ringspot virus, whereas similar cuttings from virus-free plants reacted severely. Hence, although the recovered plants are undoubtedly more accurately described as tolerating the pathogen, the fact that they have acquired a lasting immunity from the ringspot symptoms can hardly be denied.

VALLEAU (1941) finds that the degree of recovery depends on the environment, and is never total. Plants grown at 20°C. continue to show leaf symptoms, though not of the ringspot type, and only at 26°C. do the recovered plants grow normally. Even then much of the pollen is sterile and the plants set few seeds. MCKINNEY (1941) also considers that recovery is incomplete and states that infected plants yield less than uninfected ones. VALLEAU (1941) and MCKINNEY and CLAYTON

(1944) suggest that only leaves that are well differentiated at the time they become infected are in a physiological state to react acutely, and that infection of embryonic tissues so alters their metabolism that they are no longer in a suitable condition to produce ringspot symptoms. In this connection it is worth noting that the few examples of seed-transmission of ringspot virus recorded gave seedlings that showed no ringspot symptoms (VALLEAU, 1932), showing that the ability to tolerate the virus does not necessitate having the acute disease, but depends only on the early presence of virus in the cells. The virus content of recovered leaves is less than that of leaves showing ringspot symptoms. PRICE (1936*a*) found that sap from diseased leaves was five times as infective as that from recovered leaves, and STANLEY (1939) obtained approximately six times the yield of virus from necrotic as from symptomless leaves. No other differences were noticed between the virus purified from the two types of leaf. Hence it seems that leaves which have become well differentiated before they become infected contain materials that stimulate virus multiplication and that react to infection by showing acute symptoms, whereas the presence of virus early in the life of cells prevents or reduces the formation of such substances.

The sequence of an acute initial reaction followed by a later chronic stage is common to a number of virus diseases, especially those with symptoms of a systemic necrosis, but the recovery is rarely so complete as with tobacco ringspot. The leaf-drop streak caused by potato virus Y in many potato varieties is a transient symptom that is succeeded by mosaic, and the necrotic reaction caused by one strain of this virus in tobacco plants is also only temporary. Plants infected with dodder latent mosaic virus (BENNETT, 1944) at first show severe symptoms from which they later partly or wholly recover, and the ringspot symptoms caused in tobacco by some strains of potato virus X and cucumber mosaic virus are usually succeeded by milder mosaic lesions. In some of these infections the change in symptoms is also accompanied by a reduction of the virus content of sap. A similar recovery also occurs in tobacco plants infected with sugar beet curly top virus; studies of this has led WALLACE (1938, 1940, 1944) to suggest that the plants acquire an immunity because they develop "protective substances", which he considers can be transferred to other plants and confer on them a form of passive acquired immunity. WALLACE finds that Turkish tobacco plants first react severely when infected with curly top virus but then recover and may become almost symptomless. The initial reactions of tobacco plants are said to be less if they are grafted with scions from recovered plants than if they are infected by leafhoppers. Infected tomato plants suffer severely and, unlike tobacco plants, do not normally recover, but WALLACE (1944) states that if they are infected by grafting with scions from recovered tobacco plants "the tomato plants acquire an immunity similar to that observed in recovered tobacco plants". WALLACE suggests that the phenomena involved differ fundamentally from those described for other plant viruses and seem closely comparable to reactions known for animal viruses. However, PRICE (1943) could find no evidence to support the antibody theory; he confirmed the recovery of infected tobacco plants, but suggested that the variations in severity of symptoms depended merely on the site of infection and amount of inoculum. Another possibility, which does not seem to have been excluded, is that the curly top cultures were mixed and that propagation in tobacco specifically encourages the multiplication of avirulent strains, which are also less readily transmitted by leafhoppers than the virulent ones. If this is so, then antagonism between strains could account for the whole phenomenon. WALLACE's results suggest that curly top is a complex of different viruses as well as related strains, for he finds that recovered plants are protected against some of his isolates but not against others. This may also be true of tobacco ringspot, for PRICE (1936*b*) has found that plants which have recovered from yellow ringspot also resist ordinary and green ringspot viruses, whereas plants recovered from ringspot No. 2 develop typical symptoms when reinoculated with any of the others.

FIG. 23. — Two leaves of *Nicotiana sylvestris* as they appeared 5 days after both had been inoculated with tomato aucuba mosaic virus. The leaf on the left was healthy whereas that on the right was suffering from tobacco mosaic at the time of inoculation with aucuba mosaic virus. The local lesions appear only on the leaf which was healthy at the time of inoculation. (KUNKEL, L. O., 1934, Phytopath. *24*, 437).

The term acquired immunity has been applied to the phenomenon of interference between virus strains because it has been largely studied by using the ability of avirulent strains to protect plants from the effects of virulent ones. The phenomenon was indicated by MCKINNEY in 1929, but was first clearly demonstrated by THUNG (1931) and SALAMAN (1933). From a tobacco mosaic plant, THUNG isolated a strain that caused white mosaic when transferred to healthy plants but which produced no extra symptoms when inoculated to plants already suffering from mosaic. He suggested that one strain was antagonistic to another and that the presence of one in a cell prevented the entrance of the other. SALAMAN similarly found that tobacco plants infected with strains of potato virus X that cause mild mosaic developed no further symptoms when inoculated with strains that alone cause ringspot or severe mosaic.

Since then many workers have confirmed these results and obtained similar ones with other viruses.

The mechanism underlying the protection remains undetermined. Strains have no direct inactivating effect on one another, for if two enter a healthy plant together both multiply. However, they tend to settle into separate parts of the same leaves, so that by selecting inoculum each can be regained in apparently pure culture. Their antagonism is probably restricted to competition for suitable substrates. Some strains that reach relatively low virus concentrations protect completely against those that reach higher concentrations, so that protection does not seem to occur simply because there is a limited amount of a particular virus a cell can produce. It is more likely, perhaps, that, as a preliminary to multiplication, virus particles must become attached to some cell constituents and that there are only limited numbers of such sites in each cell. Then if the sites are occupied by one strain another would be unable to multiply in that cell. Protection extends only to cells that are actually infected and in plants in which systemic infection is incomplete, tissues that are uninvaded retain their normal susceptibility to other strains.

Infection with one strain does not prevent another from entering a plant but prevents the second from multiplying sufficiently to produce its normal effects. The extent to which this occurs depends on how thoroughly the first strain is established. KUNKEL's (1934) experiments with tobacco mosaic and aucuba mosaic viruses, which produce respectively a systemic mottling and necrotic local lesions in *Nicotiana sylvestris*, show this relation clearly. *N. sylvestris* becomes resistant to aucuba mosaic virus when infected with tobacco mosaic virus (Fig. 23), resistance being conferred either by directly inoculating individual leaves or by systemic invasion. In inoculated leaves, the resistance to aucuba mosaic virus is restricted to the rubbed areas, in which it increases as the tobacco mosaic virus multiplies, but is usually only complete if a concentrated inoculum was used. The inverse relationship between content of tobacco mosaic virus and number of lesions caused by aucuba mosaic virus is also clearly shown by the quantitative studies made by SADASIVAN (1940). Similarly, with potato virus X, SALAMAN (1938) has shown that the protection afforded to tobacco plants against virulent strains by infection with avirulent ones depends on the extent to which the latter have multiplied and systemically invaded the plants.

The degree of protection conferred on a plant by infection with one strain against the effects of another will depend on many factors. It will vary with the length of time for which the plant has been infected with the first strain, the extent to which this has invaded all the susceptible tissues, and the method whereby the second strain is introduced. For instance, strains of potato virus X that completely protect potato plants from the effects of other strains when the latter are introduced by leaf-inoculation, may give only incomplete protection if they are introduced by grafting (BAWDEN, 1934; KÖHLER, 1935). Likewise some strains may give complete protection against others in one host but not in another, a result interpretable on the assumption that different strains vary in their ability to invade different plants systemically. This may explain the observations that different strains of potato virus X give varying degrees of protection against other strains (KÖHLER, 1935).

It is easily demonstrated that avirulent virus strains protect plants from the effects of virulent ones, but it is less easy to show the converse. However, using strains that are avirulent on one host but that have a characteristic effect on another, it can be shown that the effect is reciprocal. That is, if an avirulent strain prevents the establishment of a virulent one, the presence of the virulent one will also interfere with the avirulent. Although the antagonism is mutual, there may be quantitative differences in the degree to which different strains protect plants against each other. Strains that multiply rapidly and move readily will give complete protection against less aggressive strains, whereas the latter would be unlikely to protect completely against the former, though they would delay the establishment of the aggressive strains and reduce their effects. For example, McKinney (1935) found that plants infected with the commonly-occurring strains of tobacco mosaic virus were invariably protected against strains that cause yellow mosaics, but the latter gave only partial protection against the former. When plants with yellow mosaic were reinoculated with tobacco mosaic virus, the symptoms produced on newly-formed leaves became progressively less severe. When kept for long periods, or when the apical buds were removed and new side shoots forced out, the young leaves showed predominantly the green mottling of tobacco mosaic virus with only a few scattered yellow spots. Similarly, plants infected with aggressive strains of potato virus X prevent the development of the less aggressive strain that causes foliar necrosis (Bawden, 1934), but the latter only partially protects against the former. Such results show that the protection does not work by preventing a second strain from entering a plant but by interfering with its multiplication. The extent to which strains prevent one another from becoming established will depend on their relative abilities to multiply and move readily; once infected, individual cells are probably immune from infection with another strain, which to become established at all must encounter uninfected cells. Hence the more completely a strain becomes systemic, and the more rapidly it invades newly-produced tissues, the more effective it will be in preventing the establishment of other strains.

The protection acquired by an infected plant is usually specific to serologically-related virus strains, infected plants still being susceptible to unrelated viruses. However, there are exceptions to this, of which one is the resistance to potato virus Y conferred on plants by infection with tobacco etch virus (Bawden and Kassanis, 1941, 1945). This is probably a different phenomenon from that already described, for the protection is not reciprocal, tobacco plants infected with virus Y being as susceptible to severe etch virus as healthy plants. If the two viruses are inoculated simultaneously to healthy plants, the symptoms produced are identical with those in plants inoculated with severe etch virus alone, and the systemically-infected leaves seem free from virus Y. Similarly, when plants fully infected with virus Y are infected with severe etch virus, the final symptoms are those of severe etch and such plants contain no detectable virus Y. Hence, severe etch virus not only protects plants from infection with virus Y but also suppresses it in tissues in which it was already well established. Mild etch virus, which occurs in lower concentrations than either severe etch or virus Y, completely protects plants against severe etch, but only partially suppresses virus Y.

It seems probable that host metabolism is altered by the etch viruses so that substances which are essential for the multiplication of virus Y cease to be formed in cells infected with severe etch virus and are reduced in those infected with mild etch virus. To account for the disappearance of virus Y from tissues in which it was previously present in considerable quantities, there is no need to postulate that the etch viruses inactivate it. It is more likely that the virus is continually denaturing *in vivo* and the virus content is normally kept reasonably constant only because there is continual production of new virus. The rate at which virus Y disappears from leaves infected with severe etch virus is approximately the rate at which it denatures *in vitro,* and only viruses that are stable *in vitro* reach and maintain high concentrations in infected plants.

In concluding this section, it is worth summarising the various effects that may be obtained when a plant already infected with one virus becomes inoculated with another. If the two are related strains, the multiplication of the second will be inhibited and the plant will be protected against it either wholly or partially. If the two are unrelated, one of three things may occur. Both may multiply readily, when they may be synergistic and produce a different and more severe effect than either does alone, or the result may be merely the summation of the separate effects, or the symptoms may be characteristic of one only, because this suppresses and supplants the other.

### *References:-*


ANSON, M. L. and STANLEY, W. M. (1941) : J. Gen. Physiol. *24,* 679.
BAWDEN, F. C. (1934) : Proc. Roy. Soc. B. *116,* 375.
—— and KASSANIS, B. (1941) : Ann. appl. Biol. *28,* 107.
—— and —— (1945) : *ibid. 32,* 52.
—— and PIRIE, N. W. (1937) : Brit. J. exp. Path. *18,* 275.
—— and —— (1942) : *ibid. 23,* 314.
—— and —— (1945) : *ibid. 26,* 294.
BENNETT, C. W. (1944) : Phytopath. *34,* 77.
BHARGAVA, K. S. (1948) : (unpublished).
BLACK, L. M. (1941) : Amer. Pot. J. *18,* 231.
CARSNER, E. (1925) : Phytopath. *15,* 745.
COLE, R. and KUTTNER, A. G. (1926) : J. Exp. Med. *44,* 855.
HOGGAN, I. A. (1935) : Ann. appl. Biol. *22,* 27.
HOLMES, F. O. (1934) : Phytopath. *24,* 845.
—— (1936) : *ibid. 26,* 896.
JENSEN, J. H. (1933) : Phytopath. *23,* 964.
—— (1936) : *ibid. 26,* 266.
—— (1937) : *ibid. 27,* 69.
JOHNSON, J. (1925) : Science *64,* 210.
—— (1947) : Phytopath. *37,* 822.
KASSANIS, B. and SELMAN, I. W. (1947) : J. Pomol. *23,* 167.
KNIGHT, C. A. (1947) : J. Biol. Chem. *171,* 297.
—— and STANLEY, W. M. (1941) : J. Biol. Chem. *141,* 39.
KOCK, G. (1924) : Un. S. Africa Dep. Agr. 9-10th Rep. 253.
KÖHLER, E. (1935) : Angew. Bot. *17,* 61.
KUNKEL, L. O. (1934) : Phytopath. *24,* 437.
—— (1937) : Amer. J. Bot. *24,* 316.
LACKEY, C. F. (1932) : J. Agric. Res. *44,* 755.
MCKINNEY, H. H. (1926) : Phytopath. *16,* 893.
—— (1929) : J. Agr. Res. *39,* 557.
—— (1935) : *ibid. 51,* 951.
—— (1937) : J. Heredity *28,* 51.
—— (1941) : Phytopath. *31,* 1059.
—— and CLAYTON, E. E. (1944) : Phytopath. *34,* 60.
MILLER, G. L. and STANLEY, W. M. (1941) : J. Biol. Chem. *141,* 905.
—— and —— (1942) : *ibid. 146,* 331.
NORVAL, I. P. (1938) : Phytopath. *28,* 675.

POUND, G. S. and WALKER, J. C. (1945) : J. Agr. Res. *71*, 255.
PRICE, W. C. (1932) : Contrib. Boyce Thompson Inst. *4*, 359.
—— (1934) : Phytopath. *24*, 743.
—— (1935) : *ibid. 25*, 947.
—— (1936*a*) : *ibid. 26*, 503.
—— (1936*b*) : *ibid. 26*, 665.
—— (1937) : *ibid. 27*, 561.
—— (1943) : *ibid. 33*, 586.
SADASIVAN, T. S. (1940) : Ann. appl. Biol. *27*, 359.
SALAMAN, R. N. (1933) : Nature *131*, 468.
—— (1937) : *ibid. 139*, 924.
—— (1938) : Phil. Trans. Roy. Soc. B. *229*, 137.
SEVERIN, H. H. P. (1934) : Hilgardia *8*, 305.
SMITH, K. M. (1929) : Ann. appl. Biol. *16*, 1.
STANLEY, W. M. (1939) : J. Biol. Chem. *129*, 429.
THUNG, T. H. (1931) : Z. Ned.-Indisch Natuurwetensch. Congr. Bandoeng, Java 450.
VALLEAU, W. D. (1932) : Ky. Agr. Exp. Sta. Bull. *327*, 43.
—— (1941) : Phytopath. *31*, 522.
WALLACE, J. M. (1938) : Phytopath. *28*, 674.
—— (1940) : *ibid. 30*, 26.
—— (1944) : J. Agr. Res. *69*, 187.
WINGARD, S. A. (1928) : J. Agr. Res. *37*, 127.

*Chapter 7*

# SEROLOGICAL REACTIONS OF PLANT VIRUSES

It is impossible here to give any details of the theory and practice of serology. For a full account of general serology the reader is referred to TOPLEY and WILSON (1937) and MARRACK (1938), and for plant serology to CHESTER (1937*a*).

TOPLEY and WILSON define an antigen as any substance which when introduced parenterally into the animal tissues, stimulates the production of an antibody, and which, when mixed with that antibody, reacts with it in some observable way. An antibody is defined as any substance which makes its appearance in the blood serum or body fluids of an animal, in response to the stimulus provided by the parenteral introduction of antigen into the tissues. The term parenteral is used to emphasise the fact that the antigen must reach the animal tissues in an unaltered state. It will be noted that two properties are attributed to the antigen; it must have the power of stimulating the production of an antibody, and it must react specifically with that antibody. Both of these properties are essential, as there are substances which react specifically with antibodies, without possessing the power of stimulating their formation. Such incomplete antigens are known as haptens.

Substances with widely different properties can act as antigens, but antigens are most commonly large molecules or molecular aggregates, and they are usually, though not invariably, composed wholly or partly of protein. They have on their surfaces so-called determinant groups, which may have various structures and are not necessarily chemically complex, and which determine the specificity of the serological reactions. How the antibodies are produced in response to the injection of antigens is unknown. Except for the specific reactions with their antigens, they behave like serum globulins. The union of antibody and antigen *in vitro* can be detected and measured by observing various consequences. The most obvious is the separation of a visible precipitate, and the precipitin reaction has been the chief one employed with plant viruses, but complement fixation, neutralisation of infectivity, and the production of anaphylaxis have also been used.

**Preparation of Antisera:** — Antisera against plant viruses are most conveniently prepared by injecting animals with purified preparations. Animals can safely be immunised by injecting such preparations into the veins, a method that usually produces a high antibody content more rapidly than injecting into the peritoneal cavity. The use of purified virus preparations also ensures that the antiserum does not contain antibodies to normal plant constituents. In rabbits, the animals most often used, a single injection of 1 mg of purified virus will sometimes give antisera with precipitin titres of greater than 1 in 100, but to ensure a good titre it is better to give a series of injections at intervals of a few days.

Although it is preferable to use purified virus preparations for immunisation, this is not always feasible and specific antisera to many viruses can be made using clarified infective sap. To avoid undesirable reactions in the animals, this should be injected intraperitoneally, from 5 to 10 injections, each of about 5 ml. of sap, being given at intervals of 5 days. The total number of injections, and the volume of sap, needed to produce a good antibody response varies with different viruses, depending on their stability and the concentration at which they occur in sap. The animals are bled 10 days after the last injection, and the blood allowed to clot when the serum is separated. In addition to rabbits, antisera against plant viruses have been produced in guinea pigs (CHESTER, 1936*a*) and fowls (NEWTON and EDWARDS, 1936).

Viruses are more highly antigenic than normal plant constituents, but when antisera have been prepared against clarified sap or partially purified virus preparations, they should be tested against extracts of healthy plants to determine whether they also react with them. If they do, the antibodies to normal plant constituents should be removed. This is done by absorbing the antisera with healthy plant sap; the serum is mixed with 2 to 5 times its volume of the sap and allowed to stand over-night, when the precipitate is removed by centrifugation. The supernatant fluid is tested again against normal plant sap, and the absorption is repeated until no reaction is obtained. DOUNIN and POPOVA (1938) described an alternative method of obtaining the antisera free from antibodies to normal plant constituents. An antiserum is first formed by injecting animals with sap from healthy plants. Infective sap from the same plant species is then allowed to react fully with this antiserum, so that all the antigenic constituents of normal plants are precipitated. After centrifugation, the supernatant fluid is used as an immunising antigen.

The time for which antisera remain active depends on their original strength, sera with low antibody-content becoming inactive sooner than those with initially high contents, and on the type of virus. Antisera to viruses with spherical particles lose their precipitating power more rapidly than those to viruses with elongated particles. For instance, if antisera with the same initial precipitin titres are stored in similar conditions, those to tomato bushy stunt virus may cease to precipitate within three to four years while those to tobacco mosaic virus remain almost unchanged. Antisera to tobacco mosaic virus and potato virus X, when kept at 1°C. and sterile, retain high precipitin titres for more than 10 years.

**Methods of Testing:** — Precipitation as a result of combination between soluble antigens and their antibodies occurs when the two are mixed in suitable proportions in the presence of electrolytes. The rate of precipitation is increased by treatments that increase the frequency of impact between antigen and antibody, for example, by increasing the concentrations of antigen and antiserum, by heating or by ensuring continuous mixing either by shaking or by convection currents. Precipitation occurs optimally at a definite ratio of antigen to antibody; departures from this ratio delay precipitation, and if there is a great excess of one, particularly an excess of antigen, precipitation may be wholly inhibited. Hence, unless the optimal proportions are known, tests should be made over a range of antibody/antigen ratios, for otherwise negative results lose significance. The range is simply covered by keeping the concentration of one, usually antiserum, constant in all the tubes, while individual tubes contain different amounts of antigen. The following method can be used either for qualitative tests or for quantitative measurements and has the advantage that the reactions can be watched continuously without interfering with the mixing of the reagents. Tests are made in thick-walled test-tubes, 7 mm in diameter, 1 ml. of diluted antiserum being added to 1 ml. of antigen solution, all dilutions being made in 0.85% NaCl solution. Immediately antigen and antibody are mixed, the tubes are placed in a water-bath at 50°C. with the fluid columns half immersed so that convection currents keep the fluids continuously mixing. The water-bath is illuminated from behind by a strip-light and has a glass front so that the tubes can be observed. By this method the optimal precipitation ratios, and precipitation end-points (of either antisera or virus solutions) are readily determinable under constant conditions.

The method is obviously applicable only to clear, heat-stable antigen solutions, and the difficulties of preparing such stable preparations of plant antigens have been long recognised (WELLS and OSBORN, 1911). The sap of many plants, whether healthy or virus infected, after clarification by centrifugation or by filtration, forms spontaneous precipitates after standing for some time at room temperature. This precipitation occurs even more rapidly in a heated water-bath. In experiments in which clarified sap is used as the test antigen, therefore, a rigid system of controls, in which the sap is mixed with normal serum and saline, must be used to differentiate between the specific and the spontaneous precipitates. This difficulty can be overcome, in part at least, by subjecting sap to treatments which destroy unstable components. If the virus is heat-stable, the sap can be heated for a few minutes at about 60°C., when a bulky green coagulum separates which is readily removed by filtration or centrifugation and leaves a clear, stable supernatant fluid. If the virus is heat-labile, the stability of the clarified sap can be increased by freezing, as many of the unstable components are thereby made insoluble. Alternatively, the sap can be clarified by the addition of $Na_2HPO_4$ and centrifuging, when the virus can be precipitated from the supernatant fluid with ammonium sulphate and redissolved in water. Provided that the precipitation tests are not carried out at too high temperatures, antigens prepared in this way are sufficiently free from spontaneous precipitates to give reliable results.

Provided they are rigidly controlled, reliable tests can be made with clarified sap or with crude extracts obtained by squeezing macerated leaves in cloth. Indeed, specific reactions are sometimes obtained with unclarified and not with clarified sap. To obtain a positive precipitin reaction, there must be enough antigen present to produce visible floccules after combining with antibody. With sap from some infected plants, clarification may reduce the virus content below the essential minimum. Also, using crude sap the bulk of the precipitate is increased because plastids and other suspended material flocculate together with the virus-antibody precipitate. Tests with crude sap can be made by mixing 0.5 ml. of the sap at room temperature with an equal volume of suitably diluted antiserum in Wassermann tubes, as described by CHESTER (1937*b*), or by mixing drops of sap and antiserum on slides, as in the agglutination tests described by VAN SLOGTEREN (1945). The last are particularly suitable for such work as testing potato stocks for the presence of virus X, for with a suitable tool for crushing tissues and extracting small volumes of sap, large numbers of plants can be examined in a short time.

The complement fixation reaction is more sensitive than precipitation and might well succeed in demonstrating the union of virus and antibody with extracts in which there is too little virus for precipitation to be seen. However, it has not yet been used widely in plant virus work, probably because the technique is rather difficult and laborious. The result of the combination between antigen and antibody is observed indirectly by its effect in inhibiting another reaction. The reaction inhibited is the lysis (dissolution) of sheep-blood corpuscles by the serum of rabbits immunised by injection with sheep corpuscles. Two constituents are necessary for the lysis of the corpuscles. One is the antibody

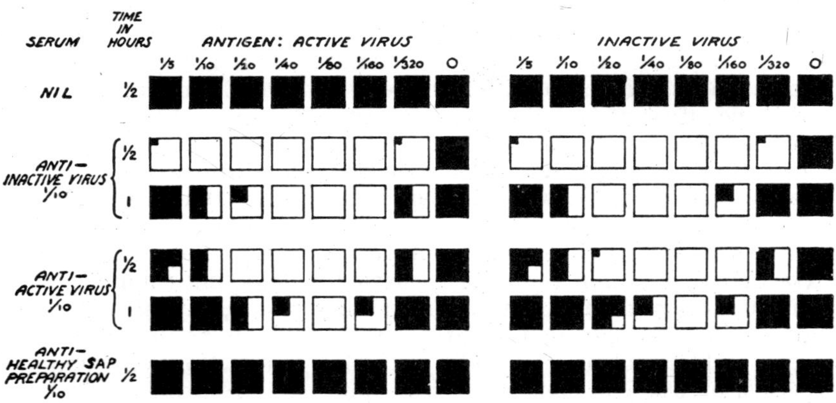


FIG. 24. — Potato virus X. Complement fixation by active virus preparations and preparations inactivated with nitrous acid, with antisera prepared against each, and against a healthy plant preparation. The haemolytic amboceptor was used at three times its minimal strength with the dose of complement used. Fixation was allowed to proceed for 1 hour before the sheep-blood corpuscles were added. The degree of heamolysis is indicated by the blackening of the squares: a totally black square indicates a tube in which the haemolysis was complete, a white square one in which there was no haemolysis. Note the optimal zone of fixation.

produced in the rabbit in response to the injection of corpuscles. The other is a thermo-labile, non-specific constituent of normal serum, known as complement. When antigens and antibodies unite, complement is used up or fixed. The amount fixed varies with the extent to which antigen and antibody react, and is measured by observing whether or not there is still sufficient free complement to allow the lysis of the sheep-blood corpuscles in the presence of anti-sheep-cell rabbit serum. This serum, known as the haemolytic amboceptor, is freed from its complement by heating to 56°C., the standardised complement used in the tests being provided by fresh guinea pig serum.

Fixation experiments are carried out in two parts. First, the antigen and antibody under test are mixed and complement added. The mixtures are allowed to stand for the complement to be fixed. After an hour, the washed sheep corpuscles and the haemolytic amboceptor are added, when the fluids are mixed and kept at 37°C., readings being made at intervals on the degree of lysis of the blood cells. The reactions taking place are summarised below:

1) Haemolytic amboceptor + Complement + Corpuscles = Lysis
2) Antigen + Antibody + Complement = Fixation
3) 2 + Haemolytic amboceptor + Corpuscles = No lysis.

It will be noticed that the lysis of the corpuscles has nothing to do with the union of the antigen and antibody under examination. It is purely an indicator reaction, showing whether or not the antigen and antibody have united and fixed the complement. As in the precipitation tests, the complement fixation experiments should be carried out over a range of antigen and antibody dilutions, for zones of optimal reaction are obtained, excess of one or the other inhibiting the fixation of complement (See Fig. 24).

The union of a toxic antigen with its antibody is usually accompanied by a loss of toxicity. Similarly, the mixing of viruses with their antisera is accompanied by a loss of infectivity. At first sight this would seem to be the simplest serological reaction to use with viruses, for all that appears to be needed is to mix infective sap with antiserum and determine the result on infectivity. Unfortunately, it is not quite so simple, for normal sera and heterologous antisera also greatly inhibit infectivity. The specific effect of homologous sera can be taken as the difference between the reduction in infectivity caused by homologous and other sera, but this is far from exact as the unspecific neutralisation varies considerably with different sera; therefore, to ensure that a specific serological reaction has occurred and reduced infectivity, accurate infectivity tests are needed using adequate controls with normal serum and saline.

The fourth reaction that can be used for detecting combination between antigens and their antibodies is anaphylaxis, in which the union is indicated by changes in animal tissues. In the living animal, there may be more or less violent spasmodic muscular contractions, inflammation of tissues, or other abnormal effects, sometimes resulting in death. *In vitro,* the most sensitive form of anaphylactic shock is the Schultz-Dale (DALE, 1913) technique. Virgin guinea pigs are immunised by injecting the antigen. After three weeks, they are killed and the two horns of the uterus are removed. To each end of the horns is attached a thread. The uterine horn is then placed in an aerated Ringer's solution, kept at 37°C. The lower end is tied rigidly to the bath, the upper end is attached to a kymograph needle. A small quantity of the antigen is then introduced into the Ringer's solution, and a positive reaction is shown by a rapid contraction of the uterine horn. This is followed by a slow relaxation, both contraction and relaxation being recorded on the kymograph. A guinea pig can be sensitised simultaneously against several antigens, and the uterine horns will react successively with each antigen. But each horn will react only once with one antigen, as one reaction completely de-sensitises it. CHESTER (1936*a*) has used this method with great success in differentiating between the antigenic constituents of healthy and virus-infected plants. Care must be used in the interpretation of the results of this technique with plant antigens. Extracts of plants, especially of members of the *Solanaceae,* contain substances causing non-specific muscular contractions. These can be distinguished from the specific serological reactions as they are given by the uterine horns of unsensitised guinea pigs; also, the muscles will react to the non-specific toxins several times in succession, provided they are washed and allowed to relax after each contraction. CHESTER has found that these toxic substances can be removed from plant extracts by a few hours' dialysis against Ringer's solution.

**The Antigenicity of Plant Viruses:** — The first indication that plant viruses might be antigenic is found in the work of DVORAK (1927), who prepared antisera separately against healthy and mottled Triumph potatoes. Precipitin tests showed that extracts of both potatoes had antigens in common, but each antiserum gave a higher titre with its homologous than with its heterologous antigen. That is, the antiserum prepared against sap from mottled potatoes reacted more strongly with sap from mottled potatoes than with sap from healthy potatoes, and *vice versa.* By precipitating the sap with ammonium sulphate, DVORAK showed that the difference lay in the globulin fraction of the sap, and the explanation offered was that the disease had altered the precipitability of the globulins. As it is now known that all Triumph potatoes, whether obviously diseased or apparently healthy, are virus-infected, these tests were probably distinguishing between plants infected with different mixtures of viruses, rather than between virus-free and virus-infected plants. This fact would also explain the higher titre obtained by DVORAK for antiserum prepared against "healthy" potatoes than that obtained by subsequent workers using virus-free plants.

Proof that a virus-infected plant contained a specific antigen was provided by PURDY BEALE (1928, 1929, 1931). She produced antisera separately against extracts of healthy tobacco and tobacco infected with tobacco mosaic virus. Some antigenic substances were found to be common to both extracts, but after antiserum to infective sap was completely absorbed with healthy sap it still precipitated and fixed complement with infective sap. Extracts of plants other than tobacco infected with tobacco mosaic virus also reacted with this antiserum, whereas extracts of plants infected with other viruses did not.

GRATIA (1933*a*, 1933*b*) first showed that plants infected with different viruses contain specific antigens. He prepared antisera separately against extracts of plants infected with a potato mosaic virus and with tobacco mosaic virus. Each antiserum precipitated strongly with the extract used to produce it, but not with the other. BIRKELAND (1934) obtained similar results, but further showed that extracts from plants infected with viruses believed to be related strains contained common antigens. He prepared sera against sap from plants infected separately with tobacco mosaic virus, attenuated tobacco mosaic virus, two strains of spot necrosis (probably a mixture of potato viruses X and Y), potato ringspot, and cucumber mosaic virus. The serum prepared against one strain of tobacco mosaic virus reacted with extracts of plants infected with the other, but not with the other viruses. Similarly, serum prepared against either strain of spot necrosis precipitated with the other and with potato ringspot, but not with tobacco mosaic virus or cucumber mosaic virus. And serum prepared against cucumber mosaic virus reacted only with sap from plants infected with this virus. Similar results have since been found with other viruses, extracts of plants infected with related strains containing common antigens, whereas those of plants infected with distinct viruses are serologically unrelated.

Although these tests showed conclusively that virus-infected plants contain specific antigens, and that serological reactions could be used to identify the virus, they did not indicate the nature of the specific antigens. Most workers considered that the viruses were themselves antigenic, but it was equally possible that the specific antigens were products of the host plant produced by infection. A good deal of work has been done in an attempt to decide between the two possibilities, and the evidence now overwhelmingly favours the view that the specific antigens are the viruses themselves.

MATSUMOTO and SOMAZAWA (1932, 1933) examined tobacco mosaic plants in some detail by serological methods, and found that the specific antigen was always associated with the presence of the virus, as indicated by infectivity tests. If virus was present, so was the antigen regardless of whether symptoms could be seen. BIRKELAND (1934) showed that the passage of sap from tobacco mosaic plants through filters retaining the virus also removed the specific precipitating antigen. SPOONER and BAWDEN (1935) obtained similar results with potato virus X; filtrates through collodion membranes fixed complement and precipitated with antiserum if they were infective, but not if they were virus-free. Experiments were also made to determine the effects on the serological reactions of inactivating the viruses. Most treatments that destroyed infectivity also destroyed the serological reactions, the antigen specific to an infected plant possessing the stability characteristic of the infecting virus. CHESTER (1935) showed that when tobacco mosaic virus, tobacco ringspot virus and potato viruses X and Y, are inactivated by heating, or by progressive strengths of $KMnO_4$, $AgNO_3$ and chloramine-T, the serological reactions are retained as long as the viruses are present and active; they diminish in strength in direct proportion to the loss of infectivity and they disappear when the viruses cease to be infective. Similarly, when potato virus X is inactivated by the proteolytic enzymes, trypsin, pepsin and papain, the reduction in serological activity is directly proportional to the reduction in infectivity (BAWDEN and PIRIE, 1936). Some lesions are occasionally obtained from preparations which give no precipitate with antiserum, but this merely indicates the greater sensitivity of the infectivity test. A similar effect is obtained by dilution, the precipitation end-point being lower than the infection end-point.

The early workers studying the relation between infectivity and serological activity were fortunate in their choice of viruses and treatments used, for had they selected others they might have reached different conclusions. Sap from plants with tobacco necrosis and tomato bushy stunt, for example, can lose infectivity as a result of heating or ageing without suffering any loss of serological activity. Similarly, with all viruses yet tested, there are some treatments that destroy infectivity without affecting the ability to react with antisera. Typical of these are treating with formaldehyde, dilute hydrogen peroxide, or nitrous acid, and irradiation with X-rays or ultra-violet light (BAWDEN, 1935; STANLEY, 1936; BAWDEN and PIRIE, 1936, 1937*a*, 1937*b*, 1938*a*, 1938*b*). It is probable that these treatments leave the antigenicity of the virus unimpaired, although only non-infective preparations of potato virus X have been shown to be fully antigenic, *i. e.*, capable of stimulating the formation of antibodies *in vivo* in addition to reacting with them *in vitro*. BAWDEN, PIRIE and SPOONER (1936) showed that preparations lost their infectivity if treated with nitrous acid at $p$H 4 and 0°C., but they still reacted with virus antiserum. Also, when injected into rabbits they produced an antiserum indistinguishable from that produced by infective preparations. Both antisera fixed complement and precipitated with infective virus preparations, but not with the sap of healthy plants, and both were equally effective in neutralising the infectivity of virus X when mixed with it *in vitro*. The results of complement fixation and precipitation tests with antisera

TABLE 7.—*Potato virus X. Precipitation test with active virus, virus inactivated with nitrous acid, and healthy plant preparation, and their respective antisera:—*

| | | *Antigens* | | |
|---|---|---|---|---|
| *Serum prepared against —* | *Dilution* | Active virus suspension 1/20 (1-hour reading) | Nitrite-treated inactive virus suspension 1/20 (1-hour reading) | Healthy sap preparation 1/20 (24-hour reading) |
| Active virus suspension | 1/4 | ++++ | ++++ | — |
| | 1/8 | ++++ | ++++ | — |
| | 1/16 | +++ | +++ | — |
| | 1/32 | ++ | ++ | — |
| | 1/64 | + | + | — |
| | 1/128 | — | — | — |
| | 0 | — | — | — |
| Nitrite-treated inactive virus suspension | 1/4 | ++++ | ++++ | — |
| | 1/8 | ++++ | ++++ | — |
| | 1/16 | ++++ | ++++ | — |
| | 1/32 | +++ | +++ | — |
| | 1/64 | ++ | ++ | — |
| | 1/128 | + | + | — |
| | 0 | — | — | — |
| Healthy sap preparation (all 24-hour readings) | 1/4 | — | — | — |
| | 1/8 | — | — | — |
| | 1/16 | — | — | — |
| | 1/32 | — | — | — |
| | 1/64 | — | — | — |
| | 1/128 | — | — | — |
| | 0 | — | — | — |

+ signs indicate the degree of precipitation; ++++ a bulky precipitate; + a small precipitate; — no precipitation.

prepared against active virus, virus inactivated by nitrous acid, and healthy plant extracts, with the three antigens are shown in Fig. 24 and TABLE 7.

At first sight the inactivation of viruses by such treatments, without affecting the serological reactions, suggests that the specific antigens are not the viruses. However, some of these treatments also destroy the pathogenicity of bacteria without destroying their antigenicity. Also, it will be shown in later chapters that these treatments have no effect on other characteristic properties of the viruses. It seems that they change the viruses enough to render them non-infective, but not enough to disrupt the particles or change the antigenic groups.

If the antigens specific to virus-infected plants are merely diseased-host products, antigenically they differ remarkably from normal plant constituents. Animals injected with tobacco mosaic sap readily produce antisera that precipitate and fix complement with their homologous antigen when they are diluted 1/1,000 or more. By contrast, some workers have failed to demonstrate any specific precipitating antibodies in the serum of animals injected with healthy tobacco sap, and when they have been demonstrated the titre of the serum has always been low. It will be noticed in Fig. 24 and TABLE 7, that no reactions were obtained using healthy plant extracts (or the serum prepared against them) treated in the same way as extracts from infected plants. On the other hand, CHESTER (1936*a*) has shown that some constituents of healthy solanaceous plants are active antigenically in producing anaphylaxis, as indicated by the Schultz-Dale technique, whereas the antigen specific to tobacco mosaic plants is inactive. This was shown in the following manner. Guinea pigs immunised with either healthy or tobacco mosaic sap give anaphylactic shock with both. But if the uterine horns of pigs immunised with infective sap are allowed to react with healthy sap they are de-sensitised and give no further reaction when tested against infective sap or against concentrated virus preparations. Hence the antigens responsible for this anaphylactic reaction must be all constituents of normal plants. In sharp contrast to the results of the anaphylaxis absorption experiments, the serum of guinea pigs immunised against infective sap still precipitated and fixed complement strongly after it had been allowed to react fully with healthy plant sap. In the same animal, therefore, some normal plant constituents can be active as producers of anaphylaxis but inert as producers of precipitating antibodies, whereas the antigen specific to infected plants produces a good precipitating antiserum but fails to cause anaphylaxis in the excised uterine horns. CHESTER suggests that normal plant constituents and the virus may behave differently in the Schultz-Dale reaction because of their different sizes. The first are probably small particles or molecules which readily diffuse into the uterine muscles, whereas the virus particles are too large to do this. SEASTONE, LORING and CHESTER (1937) have shown that tobacco mosaic virus is not inert as a producer of anaphylaxis. They confirmed CHESTER's earlier results that the virus did not give the Schultz-Dale reaction, but found that the injection of infective sap, or of purified virus, into living guinea pigs which had been immunised against infective sap and then completely de-sensitised with healthy plant sap, frequently produced fatal anaphylactic shock. Thus it appears that the virus does produce antibodies with which it can unite to cause anaphylaxis, but the union is prevented in the excised uterus and occurs only in the living animal. BEALE and SEEGAL (1941) have also found that purified tobacco mosaic virus produces anaphylactic shock *in vivo*, although the virus and normal tobacco proteins were much less anaphylactogenic than animal proteins.

If the antigens specific to infected plants were merely host reaction products, it also becomes necessary to assume that every susceptible species can produce exactly the same product. Healthy plants of tobacco and phlox contain no demonstrable amounts of serologically related substances, but both are susceptible to tobacco mosaic virus. When they are infected, both contain a common antigen, for antiserum prepared against one reacts strongly with sap from the other. It is, of course, possible that in addition to multiplying in these two unrelated plants, the virus also causes the production of this common antigen, but at present there is no reason to doubt that it is the virus itself.

The fact that antiserum prepared against infective sap specifically neutralises infectivity when mixed with the virus *in vitro* again suggests a close relationship between the antigen and the virus. This is further indicated by the fact that the serological reactions of infective sap are directly proportional to virus content as measured by infectivity tests (BEALE, 1934; BAWDEN, 1935). However, by far the strongest evidence in favour of the viruses themselves being the specific antigens comes from the work on the purification of the viruses. The purest virus preparations are the most active serologically, *i. e.* with increasing purity less solid material is necessary to give a visible reaction with antiserum, and the liquid crystalline and crystalline nucleoprotein preparations give precipitin titres of the same order as other antigens that have been carefully purified. There is no doubt that these nucleoproteins are the antigens specific to virus-infected plants. Although it cannot be proved that they are the viruses (they may be host-reaction products contaminated with active virus), all the evidence suggests that they are. And for the present there would seem to be no reason to doubt that the viruses themselves are the exceedingly active antigens present in infective sap.

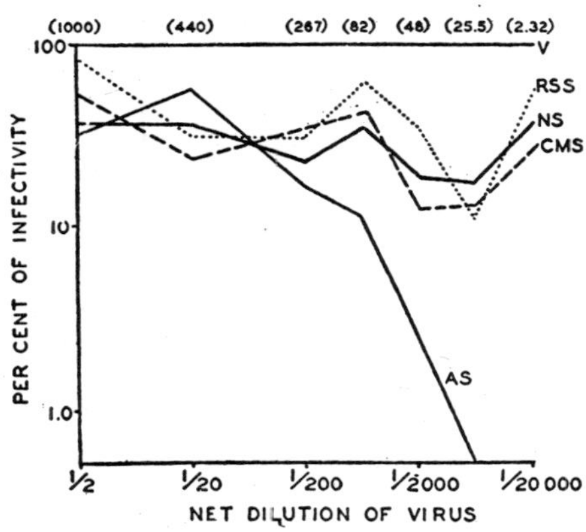


FIG. 25. — The inactivation of tobacco mosaic virus with various sera. Inhibition of infectivity of tobacco mosaic virus at various dilutions by normal serum (NS), tobacco ringspot virus antiserum (RSS), cucumber mosaic virus antiserum (CMS), and tobacco mosaic virus antiserum (AS), all at dilutions of 1/200. Infectivity expressed as percentage of serum-free control (V); numbers in parenthesis above each point in the dilution scale are the actual number of lesions formed by the control on bean leaves. (CHESTER, K. S., 1934, Phytopath. *24*, 1180).

**Neutralisation of Infectivity:** — MULVANIA (1926) found that the addition of fresh, normal rabbit serum to tobacco mosaic sap caused a decrease in the infectivity. Other workers have confirmed this, but have found that the serum of rabbits immunised with tobacco mosaic sap is more effective than normal serum in inhibiting infectivity. (PURDY BEALE, 1928; MATSUMOTO, 1930; SILBERSCHMIDT, 1933). Similar results have also been obtained with potato virus X (SPOONER and BAWDEN, 1935). The most detailed work on inhibition of infectivity has been done by CHESTER (1934) and KASSANIS (1943).

CHESTER prepared antisera to tobacco mosaic virus, tobacco ringspot virus, and cucumber mosaic virus, and found that each serum and normal rabbit serum reduced the infectivity of any of the three viruses when mixed with them. This property was also shared by other fluids containing proteins, such as healthy tobacco sap, milk and solutions of ovalbumin. In addition to this non-specific effect, however, each antiserum specifically neutralised the infectivity of the particular virus used to produce it, but not that of the other two viruses. For example, normal rabbit serum, tobacco ringspot virus antiserum, and cucumber mosaic virus antiserum, all reduced the infectivity of tobacco mosaic sap equally, but tobacco mosaic virus antiserum reduced it more (Fig. 25).

CHESTER (1934) states that the non-specific and the specific effects of antiserum are produced differently. He claims that the non-specific effect decreases the susceptibility of the host plant, whereas the specific one acts directly on the virus. Two facts were given as evidence for this claim. First, the full effect of normal serum occurred immediately the serum and virus were mixed, whereas the effect of antiserum increased with the length of time the mixtures were allowed to stand before being inoculated to plants. Secondly, the percentage reduction in infectivity caused by adding a given amount of normal serum was approximately the same at all virus dilutions, whereas the specific effect of antiserum was greater at high dilutions than at low. The effect of various antisera on the infectivity of tobacco mosaic sap is shown in Fig. 25. That the homologous antiserum has the greatest effect in reducing infectivity cannot be disputed. But that the specific effect of tobacco mosaic virus antiserum is qualitatively as well as quantitatively different from the effect of normal serum cannot be taken as proved. YOUDEN, BEALE and GUTHRIE (1935) have fitted CHESTER's results to the dilution curve $y = N(1-e^{-ax})$ (*see* p. 157), and find that they are better suited by a reduction in the concentration of active virus rather than a reduction in the susceptibility of the host plant, indicating that the normal serum is acting directly on the virus. CALDWELL (1936), from a series of dilution experiments with normal serum, also concluded that the effect was on the virus and not the host. KASSANIS (1943) in tests with several viruses found no qualitative differences between the neutralising action of homologous and of other sera. Heterologous sera reduced infectivity rather more than normal sera, and the specific neutralisation by homologous antisera could be used for demonstrating serological relationships only if sera of the same age and stored similarly were compared. The unspecific neutralisation of infectivity by freshly prepared sera was so large that the specific effect of homologous sera was difficult to detect. The unspecific action of sera decreased rapidly unless the sera were kept frozen, and in old sera the specific action predominated. The relationship indicated by neutralisation tests were the same as those found from precipitin tests, but neutralisation and precipitation seemed to be caused by different antibodies. There was no correlation between precipitin titre and neutralising power, and removal of precipitating antibodies did not remove the specific neutralisating action.

TABLE 8. — *Effect of dilution on the infectivity of mixtures of tobacco mosaic virus and sera:—*

| *Dilution* | Average number of lesions per leaf | | |
|---|---|---|---|
| | Saline control | Normal serum | Tobacco mosaic virus antiserum |
| 1/1 | 240 | 24 | 2 |
| 1/5 | 210 | 40 | 4 |
| 1/25 | 52 | 38 | 9 |
| 1/125 | 29 | 18 | 9 |
| 1/625 | 13 | 5 | 6 |

It seems probable that both the specific and the non-specific effect of antiserum in reducing infectivity is a direct result of the serum on the virus, and that the viruses from non-infective complexes by uniting with the serum proteins. The union with the antibody is a firmer one than with the ordinary serum proteins. But union with the antibody does not destroy the virus. The infectivity is merely neutralised, and the effect is in part reversible. If mixtures of virus and normal serum, or mixtures of virus and antiserum, are diluted largely with saline they regain some of their infectivity. Rather less dilution is needed with the normal serum than with the antiserum. This recovery of infectivity is shown in TABLE 8. A sample of tobacco mosaic virus was divided into three: to one portion was added an equal volume of 0.8% NaCl solution, to the second an equal volume of normal serum at a dilution of 1/10, and to the third an equal volume of tobacco mosaic virus antiserum at a

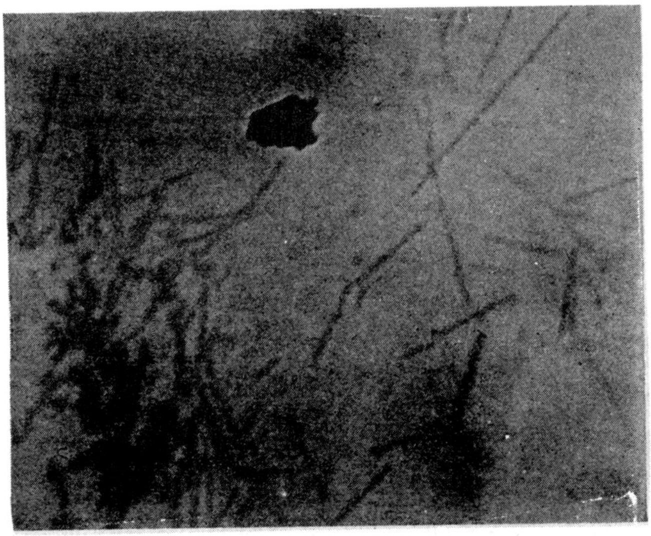

**FIG. 26.** — A mixture of tobacco mosaic and bushy stunt viruses treated with antiserum to bushy stunt virus and photographed by the electron microscope. The particles of bushy stunt virus are agglutinated but the rod-shaped particles of tobacco mosaic virus are unaffected. × 45,000. (ANDERSON, T. F. and STANLEY, W. M., 1941, J. Biol. Chem. *139*, 335).

dilution of 1/10. After thorough mixing, the fluids were allowed to stand for one hour, when they were diluted with saline and their infectivities tested at different dilutions. It will be seen that the addition of both sera reduced infectivity, but when the mixtures were diluted some was regained. This can probably be attributed to dilution causing a dissociation of non-infective complexes formed when virus and serum are present in greater concentrations.

This recovery of infectivity can be demonstrated even more strikingly in other ways. CHESTER (1936*c*) found that non-infective and serologically-inactive precipitates, produced by mixing tobacco mosaic sap and its antiserum, regained some of their infectivity if incubated with pepsin at *p*H values at which the enzyme is proteolytically active. After bringing to *p*H 7, the treated mixtures also precipitated specifically with fresh antiserum, but not with fresh virus. Hence, the enzyme had hydrolysed some of the antibodies in the precipitate, so liberating active virus.

BAWDEN and PIRIE (1937*a*) obtained similar results with purified tobacco mosaic virus. The amorphous precipitate of virus and antibody was incubated with pepsin at *p*H 3, when it soon developed a sheen. After sufficient incubation, the virus was regained with all its characteristic liquid crystalline properties, in addition to its normal infectivity and serological activity. With the less stable potato virus X, CHESTER (1936*c*) has recovered active antibody from the virus-antiserum precipitates. When the precipitates were acidified the antibody was liberated and the virus denatured. After being centrifuged and neutralised, such treated mixtures precipitated with fresh virus. From these results it is obvious that, although union between viruses and their antibodies results in the loss of the characteristic properties of both, neither is destroyed. Appropriate treatments can recover both with their original properties.

**Specificity of Serological Reactions:**—An antiserum prepared against one virus will react with the sap of any plant infected with that virus, provided the virus-content is adequate, but not with the sap of any plant infected with a different virus. This specific reaction is illustrated in Figures 26 and 27. Two samples of a solution containing both tobacco mosaic and bushy stunt viruses were mixed respectively with antiserum to each virus and the mixtures were then photographed in the electron microscope. The spherical bushy stunt particles are flocculated by the homologous antiserum but are unaffected by tobacco mosaic virus antiserum whereas the rod-shaped tobacco mosaic virus particles are flocculated by the latter and not by antiserum to bushy stunt virus. Virus-antibody precipitates have also been studied in the electron microscope by BLACK, PRICE and WYCKOFF (1946) who found that combination with antibodies upsets the normal regularity with which particles of bushy stunt and southern bean mosaic viruses arrange themselves and that it separates the particles by about twice the normal spacing. They were able to detect specific flocculation of particles with as little as $10^{-12}$ g. of virus.

To identify viruses by studying symptoms, host range and properties *in vitro*, is both time-consuming and uncertain, whereas serological methods can give unequivocal results within a few minutes. The methods have mostly been used to show that viruses which were previously thought to be unrelated, because they cause different symptoms or have different host ranges, are related strains, but they have also been valuable in demonstrating that some clinically similar diseases may be etiologically different. For example, top-necrosis of potato can be caused by several unrelated viruses, and tobacco necrosis was assumed to have only one cause until it was demonstrated that infected plants contained serologically unrelated viruses (BAWDEN, 1941).

Serological methods do not identify viruses precisely, for the reactions are only group specific. If sap precipitates specifically with antiserum to tobacco mosaic virus, the presence of a strain of this virus is established, but not the identity of the particular strain. In general, different strains of a virus react in essentially the same manner with both their own and each other's antisera (*see* TABLE 13, Chapter 8), and customary tests for precipitation, complement fixation, or virus-neutralisation, fail to differentiate between strains. By refining the methods of testing, however, differences between strains are demonstrable. The fact that one strain reacts with antiserum to another shows that the two are related and share antigens, but it does not prove that they are antigenically identical. Viruses are not single antigens, and each has several determinant groups, so that antisera also contain several antibodies, each reacting specifically with its particular determinant group. Two viruses which have one or more determinant groups in common will precipitate with each other's antiserum. But if, in addition to the common groups, each strain contains specific ones, then the strains

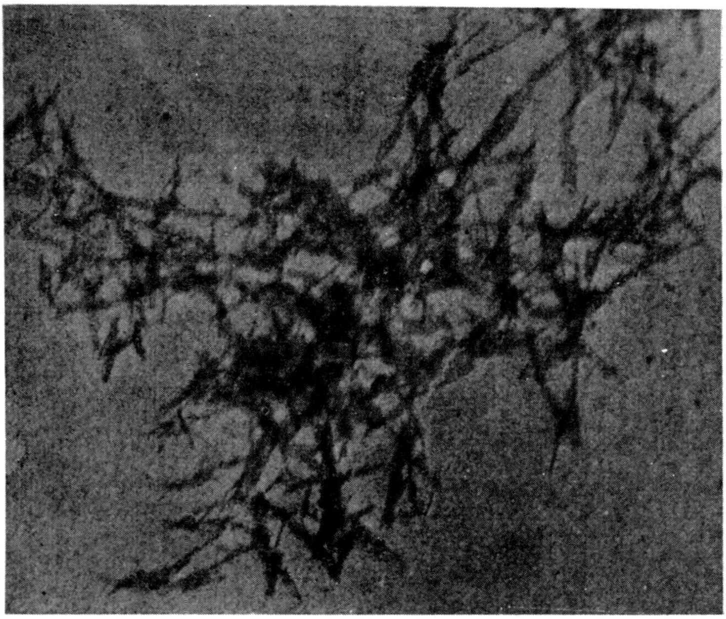

**FIG. 27.** — A mixture of tobacco mosaic and bushy stunt viruses treated with antiserum to tobacco mosaic virus and photographed by the electron microscope. The rod-shaped particles of tobacco mosaic virus are agglutinated but the smaller particles of bushy stunt virus are unaffected. (ANDERSON, T. F. and STANLEY, W. M., 1941, J. Biol. Chem. *139*, 339).

will affect their homologous and heterologous antisera differently. Each strain will react fully with its homologous antiserum, removing all the antibodies from it. On the other hand, it will remove from a heterologous antiserum only those antibodies for which it has determinant groups; the others will be unaffected and they will still precipitate the homologous strain. Using small letters to represent determinant antigenic groups and capital letters for the corresponding antibodies, three strains of the same virus and their antisera might be represented by the following formulae.

| *Antigenic groups:—* | | *Antibodies:—* | |
|---|---|---|---|
| Strain 1. | a, b, c, d, e. | Antiserum 1. | A, B, C, D, E |
| Strain 2. | c, d, e, f, g, h. | Antiserum 2. | C, D, E, F, G, H. |
| Strain 3. | a, b, g, h, j, k. | Antiserum 3. | A, B, G, H, J, K. |

**TABLE 9.** — *Summarised results of cross-absorption experiments:—*

| *Antiserum absorbed* | *Antigen used for absorption* | Precipitation tests with absorbed sera and antigens — *Antigen*: Tobacco mosaic virus | Aucuba mosaic virus | Enation mosaic virus | Cucumber virus 3 | Cucumber virus 4 |
|---|---|---|---|---|---|---|
| Tobacco mosaic virus | T.M.V. | — | — | — | — | — |
| | A.M.V. | — | — | — | — | — |
| | E.M.V. | ++ | ++ | — | — | — |
| | C.V. 3 | ++++ | ++++ | ++++ | — | — |
| | C.V. 4 | ++++ | ++++ | ++++ | — | — |
| Aucuba mosaic virus | T.M.V. | — | ++ | ++ | — | — |
| | A.M.V. | — | — | — | — | — |
| | E.M.V. | ++ | ++ | — | — | — |
| | C.V. 3 | ++++ | ++++ | ++++ | — | — |
| | C.V. 4 | ++++ | ++++ | ++++ | — | — |
| Enation mosaic virus | T.M.V. | — | + | ++ | — | — |
| | A.M.V. | — | — | + | — | — |
| | E.M.V. | — | — | — | — | — |
| | C.V. 3 | ++++ | ++++ | ++++ | — | — |
| | C.V. 4 | ++++ | ++++ | ++++ | — | — |
| Cucumber virus 3 | T.M.V. | — | — | — | ++++ | ++++ |
| | A.M.V. | — | — | — | ++++ | ++++ |
| | E.M.V. | — | — | — | ++++ | ++++ |
| | C.V. 3 | — | — | — | — | — |

— indicates that there is no precipitation.

\+ signs indicate the degree of precipitation at the optimum.

As each antiserum contains some antibodies that combine with antigens of each strain, mixing any strain with any antiserum will produce a precipitate. But the effect on the antiserum will depend upon the particular strain mixed with it; if antiserum 1 reacts fully with strain 1 all five antibodies will be bound (absorbed) and removed in the precipitate, so that the supernatant fluid will not precipitate any additional virus. On the other hand, if antiserum 1 is fully absorbed with strain 2, the supernatant will still contain antibodies A and B; although it will no longer precipitate when more strain 2 is added, it will precipitate with both strains 1 and 3. Similarly, if it is fully absorbed with strain 3, antibodies A, B, will be removed, but C, D, E, will remain, and a further precipitate will be obtained when either strain 1 or 2 is added. In this way tobacco mosaic virus and potato virus X have been shown to be complex antigens and antigenic differences found between the individual strains.

CHESTER (1936*b*), using clarified infective sap as his antigen preparations, found that antiserum to tobacco mosaic virus after being absorbed fully with tomato aucuba mosaic virus still precipitated with tobacco mosaic virus. He also found that some of the strains isolated by JENSEN (1933) were serologically similar to aucuba mosaic virus, and others to the parent tobacco mosaic virus. By serum absorption experiments CHESTER also distinguished between three strains of potato virus X, and indicated the serological differences detected by the following antigenic formulae:

| | |
|---|---|
| Mottle strain | a, b, c. |
| Masked mottle strain | a, b, d. |
| Ringspot strain | a, d, e. |

Using purified virus preparations as antigen preparations, BAWDEN and PIRIE (1937*b*) made cross-absorption experiments with various strains of tobacco mosaic virus and their antisera and their results are summarized in TABLE 9. In addition to the antigens present in all three strains, tobacco mosaic and aucuba mosaic viruses contain an antigen absent from enation mosaic virus, aucuba mosaic and enation mosaic viruses contain one absent from tobacco mosaic virus, and enation mosaic virus contains one absent from both the others. If each antigenic difference is represented by a letter, the differences are most simply expressed by the formulae:

| | |
|---|---|
| Tobacco mosaic virus | w, x. |
| Aucuba mosaic virus | w, x, y. |
| Enation mosaic virus | w, y, z. |

There is, however, no reason to believe that each antigenic difference detected is a single antigen, and it is probable that by using more strains in such experiments more antigens would be detected. Cucumber viruses 3 and 4, which are serologically related to tobacco mosaic virus but have a quite different host range, were also included in these tests. Serologically they differ from the other strains of tobacco mosaic virus much more than these differ from one another; the cucumber viruses have only few antigens in common with tobacco mosaic virus, whereas aucuba and enation mosaic viruses have many. This is shown by the varying ability of sera to precipitate their homologous strains when they have been absorbed with others. When antiserum prepared against any of the three strains of tobacco mosaic virus is absorbed with cucumber virus 3, it still precipitates its homologous antigen almost as well as unabsorbed serum, the optional precipitation point shifting only slightly. Similarly, the reaction of cucumber virus 3 antiserum with cucumber 3 is only slightly altered by absorption with tobacco mosaic virus, showing that only a few antibodies are removed. On the other hand, when antiserum to tobacco mosaic virus is fully absorbed with enation mosaic virus, its ability to precipitate tobacco mosaic virus is much reduced and the optimal precipitation point is shifted considerably showing that many antibodies have been removed.

The tests with cucumber virus 3 also suggest that the antigenic fraction w, which is common to aucuba, enation and tobacco mosaic viruses, is not a single antigen. After absorption with either enation mosaic or aucuba mosaic viruses, tobacco mosaic virus antiserum loses its power of precipitating cucumber virus 3, which means that the antigens shared by cucumber virus 3 and tobacco mosaic virus are in the fraction w. Removing all the antibodies to w, by absorption with enation or aucuba mosaic viruses, greatly reduces the serum's power to precipitate. As removal of the antibodies that combine with cucumber virus 3 affects antiserum to tobacco mosaic virus only slightly, it is apparent that these can correspond to only a minor part of the whole of w. Further evidence for the antigenic complexity of virus particles is provided by the fact that different antisera made against the same virus sometimes contain different antibodies, strong sera usually differing qualitatively as well as quantitatively from weak sera (KLECZKOWSKI, 1941*a*). This, of course, makes the results of absorption tests equivocal, for minor antigenic differences between virus strains might show in tests with some antisera but not others, and possibly accounts for CHESTER (1936*b*) finding an antigen in tobacco mosaic virus that was not present in aucuba mosaic while BAWDEN and PIRIE (1937*b*) and KLECZKOWSKI (1941*a*) did not. It is, however, equally possible that the different workers were using different strains under the same names.

It has been stated that straight-forward precipitin tests with virus strains and their homologous and heterologous antisera usually indicate no differences in the serological reactions of the strains. The relationships between cucumber virus 3 and tobacco mosaic virus are such that the two can be distinguished by their different behaviour when titrated against both antisera. The serological titre given by preparations of either virus, *i. e.*, the greatest dilution at which a visible precipitate is obtained, is nearly independent of the serum used, but the range of antigen dilution over which precipitation occurs varies widely with the two sera. When cucumber virus 3 is titrated against a constant amount of antiserum to any of the three strains of tobacco mosaic virus, there is a large zone in the region of antigen excess where no precipitation occurs. Similarly, when any of the three strains of tobacco mosaic virus used is titrated against a constant amount of cucumber virus 3 antiserum, there is a large zone of non-precipitation in the antigen excess region. By contrast, when cucumber 3 is titrated against a constant amount of its homologous antiserum, or when any of the three strains of tobacco mosaic virus is titrated against their own or each other's antiserum, such zones of non-precipitation are much smaller, occurring only where the antigen is much more concentrated. The results in which tobacco mosaic virus and cucumber virus 3 preparations were titrated against the four antisera are shown in TABLE 10. It will be seen that there are differences in the range of precipitation obtained with the antisera to the three strains of tobacco mosaic virus, but they are insignificant compared with those between these sera and the cucumber virus 3 antiserum.

These results can be best explained by postulating quantitative as well as qualitative differences in the antigenic constitution of cucumber virus 3 and tobacco mosaic virus. Most of the antigenic groups in cucumber virus 3 are absent from tobacco mosaic virus, and *vice versa.* But it seems probable that they have two common antigens, the sum total of the two in each virus being of the same order, although tobacco mosaic virus contains a preponderance of one and cucumber virus 3 a preponderance of the other. If the common antigens are called p and q, then tobacco mosaic virus will contain many p antigenic groups and only a few q, whereas cucumber virus 3 will contain only a few p and many q. With such different quantitative compositions it is apparent that the antiserum prepared against one of these viruses will contain only a few antibodies to the major antigen of the other, which will therefore have to be diluted greatly before the antibody/antigen ratio is optimal for precipitation. KNIGHT and STANLEY (1941) have found that a virus affecting plantain reacts with tobacco mosaic virus antiserum in much the same way as do cucumber viruses 3 and 4. This virus from plantain and the cucumber viruses were found to differ widely from tobacco mosaic virus in their content of some amino-acids, whereas other strains of tobacco mosaic virus, such as aucuba mosaic virus, which react strongly with tobacco mosaic virus antiserum, did not. These differences in the quantities of individual amino-acids may represent the quantitative differences betweent the common antigens postulated in the previous paragraph, but it is unlikely that they could also explain the other large qualitative antigenic differences between tobacco mosaic virus and cucumber viruses 3 and 4.

In using cross-absorption tests for identifying virus strains so-called "mirror-tests" must be made before two strains can be shown to be antigenically identical. Antisera must be prepared separately against each, and it must be shown that neither reacts with its homologous antiserum after absorption with the other strain. The necessity for the mirror tests is indicated in the results given in TABLE 9 with tobacco mosaic and aucuba viruses. After absorption with aucuba mosaic virus, antiserum against tobacco mosaic virus gives no further reaction with tobacco mosaic virus. It might be assumed, therefore, that the two are identical. However, after absorption with tobacco mosaic virus, antiserum to aucuba mosaic virus still reacts with aucuba mosaic virus. The two are not identical, but aucuba mosaic virus contains all the antigens of tobacco mosaic virus and at least one in addition.

It has been stated above that if the sap of a plant infected with an unknown virus reacts with a serum prepared against a known virus then the unknown can be identified as a strain of the known. The converse is not necessarily true, for the absence of a reaction does not prove the unknown virus to be serologically distinct from the known. This can be proved only by preparing antisera against both the known and the unknown viruses, and showing that neither reacts with the other's antiserum in the same conditions as it reacts with its own. Such reciprocal

TABLE 10.—*Precipitation of tobacco mosaic virus and cucumber virus 3 with different antisera:—*

1. Tobacco Mosaic Virus as Antigen

| ANTISERUM | TIME | DILUTION OF ANTIGEN (1/1 = 1 mg. per ml.) | | | | | |
|---|---|---|---|---|---|---|---|
| | | **1/1** | **1/4** | **1/16** | 1/64 | 1/256 | 1/1024 |
| Tobacco mosaic virus | 2 min. | ++ | ++ | + | — | — | — |
| | 2 hr. | ++++ | ++++ | ++++ | +++ | ++ | + |
| | 24 " | ++++ | ++++ | ++++ | ++++ | +++ | ++ |
| Aucuba mosaic virus | 2 min. | ++ | ++ | + | — | — | — |
| | 2 hr. | ++++ | ++++ | ++++ | +++ | ++ | + |
| | 24 " | ++++ | ++++ | ++++ | ++++ | +++ | ++ |
| Enation mosaic virus | 2 min. | — | ++ | + | — | — | — |
| | 2 hr. | — | ++++ | ++++ | +++ | ++ | + |
| | 24 " | — | ++++ | ++++ | ++++ | +++ | ++ |
| Cucumber virus 3 | 24 " | — | — | — | — | + | + |
| | 2 " | — | — | — | + | ++ | ++ |

2. Cucumber Virus 3 as Antigen

| ANTISERUM | TIME | DILUTION OF ANTIGEN (1/1 = 1 mg. per ml.) | | | | | |
|---|---|---|---|---|---|---|---|
| | | 1/1 | 1/4 | 1/16 | 1/64 | 1/256 | 1/1024 |
| Tobacco mosaic virus | 1 hr. | — | — | — | + | ++ | — |
| | 24 " | — | — | — | ++ | +++ | ++ |
| Aucuba mosaic virus | 2 " | — | — | — | — | + | — |
| | 24 " | — | — | — | — | ++ | ++ |
| Enation mosaic virus | 2 " | — | — | — | — | + | — |
| | 24 " | — | — | — | — | ++ | ++ |
| Cucumber virus 3 | 2 min. | +++ | ++ | + | — | — | — |
| | 2 hr. | ++++ | ++++ | ++++ | +++ | ++ | — |
| | 24 " | ++++ | ++++ | ++++ | ++++ | +++ | ++ |

**In all tests the antiserum was used at a dilution of 1/50. 1 ml. of antiserum was added to tubes containing 1 ml. of antigen at given dilutions, and the tubes were immediately placed in a water-bath at 50° C. + signs indicate the degree of precipitation. — signs indicate that there is no precipitation.**

tests are necessary because the virus content of sap infected with different strains can vary widely, and the absence of a reaction with one strain may merely mean that the sap contains insufficient virus to produce a visible precipitate. Therefore, it is necessary to show that expressed sap contains sufficient antigen to produce a visible precipitate, by obtaining a reaction with homologous antiserum, before a negative result can be accepted as proving no serological relationships.

In general, viruses that immunise plants against one another also react with each others' antisera. However, PRICE (1935, 1937) has shown that plants infected with celery mosaic and lily mosaic viruses are protected against further infection with cucumber mosaic virus, whereas CHESTER (1937*c*) obtained no precipitin reaction when sap from plants infected with these two was mixed with cucumber mosaic virus antiserum. This may be an example of viruses immunising plants against one another without being serologically related. On the other hand, as no antisera were prepared against celery mosaic or lily mosaic viruses, it is equally possible that they are serologically related to cucumber mosaic virus, but that the sap of plants infected with them contains too little virus to give a visible reaction. This quantitative effect is clearly shown with strains of henbane mosaic virus. During work on insect-transmission, WATSON (1938) isolated a variant producing only mild symptoms in tobacco, and the sap of infected plants produced fewer lesions than that of plants infected with the virulent strain. When tested against antiserum to the original strain, either using crude or clarified sap, extracts of plants infected with the attenuated strain do not precipitate, but the mild strain is precipitated if it is concentrated sufficiently. With purified preparations of different viruses, from 0.005 to 0.0005 mg is needed to produce a visible precipitate, and unless infective sap contains more than these minimal amounts of virus per ml., it will be unlikely to give a precipitin reaction, though the simultaneous flocculation of unstable substances from the sap may sometimes give results with smaller quantities of virus than would alone produce a visible precipitate.

Serological techniques have now been successfully applied to about 15 different plant viruses, but they have failed with many others. CHESTER (1937*c*) using infective sap for immunisation and testing, got no precipitin reaction with a number of different viruses, and since then other workers have failed with additional ones. It may be that some of these differ fundamentally from those that are demonstrably antigenic, but it is more likely that quantitative differences in the concentrations at which they occur in sap account for success with some and failure with others. There are wide differences between the serological activity of sap from plants infected with the viruses to which serological methods have been successfully applied, and the differences are correlated with virus content. Sap that is infective at high dilutions, for example from plants with tobacco mosaic, is also highly antigenic and precipitates at high dilutions with antiserum. Less infective saps produce good antisera less readily and give smaller precipitin titres; for example, sap from plants infected with potato virus Y rarely precipitates at dilutions greater than 1 in 20. Clearly viruses that occur at much less than one-twentieth of this concentration, and there is abundant evidence that many do, are unlikely to give any specific precipitation. The application of the more sensitive techniques of complement fixation and neutralisation of infectivity might provide evidence of the antigenicity of some of these viruses, or, as with the attenuated strains of henbane mosaic virus, precipitation may be demonstrable after they have been concentrated though it is not with sap. So far serological methods have been mainly successful with viruses that are readily transmitted by sap-inoculation, but KLECZKOWSKI and WATSON (1944) prepared antisera that reacted specifically with sap from sugar beet infected with yellows virus, which had not then been transmitted mechanically, although it now has with the aid of an abrasive (KASSANIS, 1949).

**Effect of Particle-Shape on Serological Reactions:**—Bacteria with flagella give a different type of precipitate with antisera from non-motile individuals of the same species. When mixed with their antisera, the flagellate, or H-type, bacterial antigens are agglutinated rapidly, and form large clumps with a fluffy, open structure. By contrast, the somatic, or O-type, bacterial antigens are agglutinated more slowly and form smaller clumps which are dense and granular. Essentially similar differences are found between the form of precipitate produced by different viruses with their antisera. Strains of tobacco mosaic virus, cucumber viruses 3 and 4, potato viruses X and Y, and henbane mosaic virus, all give bulky flocculent precipitates which form rapidly. Tomato bushy stunt, tobacco necrosis and tobacco ringspot viruses, on the other hand, give granular precipitates which settle into small compact masses. These differences are illustrated in Fig. 28. The three left-hand tubes contain 1 ml. of bushy stunt virus antiserum at 1/50 and 0.1, 0.05 and 0.025 mg. respectively of bushy stunt virus in 1 ml. The three right-hand tubes contain 1 ml. of tobacco mosaic virus antiserum at 1/50 and 0.1, 0.05 and 0.025 mg. respectively of tobacco mosaic virus in 1 ml. After adding the serum to the virus solutions, the tubes were immediately placed in a water-bath at 50°C. A precipitate formed at once in the most concentrated solution of tobacco mosaic virus, and within two minutes there was a large precipitate in this and precipitation was also obvious in the more dilute solutions. No visible floccules were formed in the most concentrated solution of bushy stunt virus until the tubes had been in the bath for six minutes, and it was half an hour before a precipitate was obvious in the most dilute solution. After twelve hours in the water-bath, the tubes were left undisturbed for a further twelve hours at room temperature when they were photographed.

The character of the specific precipitate is determined by the shape of the virus particle. Tobacco mosaic virus normally has rod-shaped particles, and all the other viruses that give nebulous precipitates also have particles that are readily orientated by streaming, whereas bushy stunt, tobacco necrosis and tobacco ringspot viruses have particles that are spherical or nearly so. Structurally, therefore, the viruses giving H-type precipitates resemble flagella and those giving O-type precipitates resemble the bodies of bacteria, and the differences in character between the types presumably occur because spheres pack more tightly than rods and so give a more compact precipitate.

Antisera to somatic type bacterial antigens are known to differ in several ways from those to flagellar antigens; they lose their ability to cause agglutination more easily on heating or ageing and they also lose this ability when treated with formalin. In all these properties, the antisera to bushy stunt and tobacco mosaic viruses also differ. After heating for 10 minutes at 75°C., bushy stunt virus antisera cease to precipitate the virus, whereas tobacco mosaic virus antisera still cause precipitation when heated to over 80°C. Similarly, after treatment with formalin, bushy stunt virus antisera do not cause precipitation, although tobacco mosaic virus antisera do, and antisera to bushy stunt virus may cease to cause precipitation after 3 or 4 years at 1°C. whereas those to tobacco mosaic virus still precipitate after 10 years.

The effects of heating on antisera to the two kinds of antigen have been studied in detail by KLECZKOWSKI (1941*b*). It was widely believed that the antibodies to O- and H-type antigens differed in their resistance to heating, but KLECZKOWSKI has disproved this by isolating globulin fractions of the two antisera and heating them separately, when he found that both behave in the same way and continue to cause precipitation until heated for 10 minutes at 90°C. When whole sera are heated at temperatures between 75°C. and 90°C., antibodies apparently combine

FIG. 28. — Precipitates of tomato bushy stunt and tobacco mosaic viruses with their homologous antisera. Three left-hand tubes, 0.1, 0.05 and 0.025 mg. of bushy stunt virus; three right-hand tubes, same weights of tobacco mosaic virus; central tube saline control. Note dense, granular precipitate of bushy stunt and bulky, flocculent precipitate of tobacco mosaic virus. (BAWDEN, F. C. and PIRIE, N. W., 1938, Brit. J. exp. Path *19*, 251).

with other proteins that are undergoing heat denaturation. The behaviour of such complexes in precipitation tests depends on the protein with which the antibody has combined and on the antigen. Complexes with euglobulin behave much like unchanged antibody, whereas complexes with albumin, which predominate when whole sera are heated, can combine with their antigens but not precipitate them. Antisera to H-type antigens appear to be more heat-stable because these antigens are more readily flocculated by mixtures of normal antibody and antibody-albumin complexes than are O-type antigens.

The difference between the serological behaviour of the two viruses probably lies in the fact that the elongated tobacco mosaic virus becomes insoluble much more easily by combination with antibody than does bushy stunt virus; its floccules not only separate sooner, but they are produced with less antibody and over a wider range of antigen/antibody ratios. It is probable that incorporating relatively small amounts of soluble protein in the particles formed by the union of antigens and their antibodies is sufficient to keep them in solution if the antigens are O-type but not if they are H-type.

Some viruses also combine with other proteins to form complexes whose serological behaviour differs from that of the viruses (BAWDEN and KLECZKOWSKI, 1941, 1942). When tomato bushy stunt virus and albumin are heated together in suitable conditions the two combine, and the virus then ceases to be precipitable by its antiserum. The complex is still antigenic and when injected into rabbits it produces antisera apparently identical with ordinary virus antiserum. The complex still combines with antibodies to the virus, for it fixes complement with virus antiserum and it also inhibits the precipitation of unchanged bushy stunt virus. No such effects are obtained when tobacco mosaic virus is heated with albumin.

The dependence of serological reactions on the particle shape of the antigen has been most clearly demonstrated by work on preparations of tobacco mosaic virus in different physical states. By using special methods for extracting tobacco mosaic virus from leaves, BAWDEN and PIRIE (1945) were able to avoid much of the aggregation that normally accompanies extraction and to obtain preparations consisting of shorter particles. The serological behaviour of such preparations approximates closely to that of somatic antigens, but the particles readily aggregate linearly to form long rods, when, with the same antiserum, the preparations take on the character of flagellar antigens. The effect of this change in producing precipitation more rapidly and over a much wider range of antigen/antibody ratios is shown in TABLE 11, which compares the

TABLE 11.—*Effect of aggregation of the precipitin behaviour of tobacco mosaic virus:—*

| *Antigen* | Precipitation after | Dilution of antigens 1/1 | 2 | 4 | 8 | 16 | 32 | 64 | 128 |
|---|---|---|---|---|---|---|---|---|---|
| *Before aggregation* | 1 hour | — | — | — | — | — | — | + | — |
| | 2 hours | — | — | — | — | — | + | ++ | + |
| *After aggregation* | 3 minutes | ++++ | ++++ | +++ | ++ | + | — | — | — |
| | 30 minutes | ++++ | ++++ | ++++ | ++++ | ++++ | +++ | ++ | + |
| | 1 hour | ++++ | ++++ | ++++ | ++++ | ++++ | ++++ | +++ | ++ |

Antiserum used at a constant dilution of 1:400.

behaviour of the same preparation before and after it was aggregated by incubation with trypsin. There seems to be no difference in the antigenicity of virus in the two physical states, but to precipitate equal weights of antigen calls for more antibody when the particles are small than when large. Similarly, when the rods are disrupted by ultrasonic radiation, there is no qualitative change in their antigenic content, but the small particles produced need to be combined with a greater proportion of antibody for precipitation to occur (MALKIEL, 1947*a*; 1947*b*). In the equivalence zone, *i. e.*, when antigen and antibody are fully combined, tobacco mosaic and tomato bushy stunt viruses both combine with rather more than one-fifth of their weight of antibody, but both can be precipitated by much less than this, and aggregated tobacco mosaic virus can be precipitated by combination with less than one-three-hundredth of its weight of antibody (KLECZKOWSKI, 1941*a*; MALKIEL and STANLEY, 1947).

It is not only the precipitin behaviour of tobacco mosaic virus that alters with the change in size and shape of particles, but in different states of aggregation this antigen shows all the various characters that have previously been thought to be specific to flagellar and somatic antigens. For instance, antiserum treated with formalin or heated suitably, will cease to precipitate virus preparations containing predominantly small particles while it still remains able to precipitate them if they are aggregated.

## *References:-*


ANDERSON, T. F. and STANLEY, W. M. (1941) : J. Biol. Chem. *139*, 339.
BAWDEN, F. C. (1935) : Brit. J. exp. Path. *16*, 435.
—— (1941) : *ibid. 22*, 59.
——, and KLECZKOWSKI, A. (1941) : Brit. J. Exp. Path. *22*, 208.
—— and —— (1942) : *ibid. 23*, 178.
—— and PIRIE, N. W. (1936) : Brit. J. exp. Path. *16*, 64.
—— and —— (1937*a*) : Proc. Roy. Soc. B. *123*, 274.
—— and —— (1937*b*) : Brit. J. exp. Path. *18*, 275.
—— and —— (1938*a*) : *ibid. 19*, 66.
—— and —— (1938*b*) : *ibid. 19*, 251.
—— and —— (1945) : *ibid. 26*, 294.
—— PIRIE, N. W. and SPOONER, E. T. C. (1936) : Brit. J. exp. Path. *17*, 204.
BEALE, H. PURDY (1928) : Proc. Soc. Exp. Biol. and Med. *25*, 702.
—— (1929) : Jour. Exp. Med. *49*, 919.
—— (1931) : Contrib. Boyce Thompson Inst. *3*, 529.
—— (1934) : *ibid. 6*, 407.
—— and SEEGAL, B. C. (1941) : Contrib. Boyce Thompson Inst. *11*, 441.
BIRKELAND, J. M. (1934) : Bot. Gazette *95*, 419.
BLACK, L. M., PRICE, W. C., and WYCKOFF, R. W. G. (1946) : Proc. Soc. Exp. Biol. Med. *61*, 9.
CALDWELL, J. (1936) : Proc. Roy. Soc. B. *119*, 493.
CHESTER, K. S. (1934) : Phytopath. *24*, 1180.
—— (1935) : *ibid. 25*, 702.
—— (1936*a*) : *ibid. 26*, 715.
—— (1936*b*) : *ibid. 26*, 778.
—— (1936*c*) : *ibid. 26*, 949.
—— (1937*a*) : Quar. Rev. Biol. *12*, 19, 165, and 294.
—— (1937*b*) : Phytopath. *27*, 903.
—— (1937*c*) : *ibid. 27*, 722.
DALE, H. H. (1913) : J. Pharm. exp. Ther. *4*, 167.
DOUNIN, M. S. and POPOVA, N. N. (1938) : Rev. Appl. Mycol. *17*, 762.
DVORAK, M. (1927) : Journ. Infect. Dis. *41*, 215.
GRATIA, A. (1933*a*) : Compt. Rend. Soc. Biol. *114*, 923.
—— (1933*b*) : *ibid. 114*, 1382.
JENSEN, J. H. (1933) : Phytopath. *23*, 964.
KASSANIS, B. (1943) : Brit. J. exp. Path. *24*, 152.
—— (1949) : Ann. appl. Biol. *36*, 270.
KLECZKOWSKI, A. (1941*a*) : Brit. J. exp. Path. *22*, 44.
—— (1941*b*) : *ibid. 22*, 192.
—— and WATSON, M. A. (1944) : Ann. appl. Biol. *31*, 116.
KNIGHT, C. A. and STANLEY, W. M. (1941) : J. Biol. Chem. *141*, 39.
MALKIEL, S. (1947*a*) : J. Immunology *57*, 51.
—— (1947*b*) : *ibid. 57*, 55.
—— and STANLEY, W. M. (1947) : J. Immunology *57*, 31.
MARRACK, J. R. (1938) : Special Rep. Ser. Med. Res. Council, London, No. 230.
MATSUMOTO, T. (1930) : Jour. Soc. Trop. Agric. *1*, 291.
—— and SOMAZAWA, K. (1932) : Jour. Soc. Trop. Agric. *4*, 161.
—— and —— (1933) : *ibid. 5*, 37.
MULVANIA, M. (1926) : Phytopath. *16*, 853.
NEWTON, W. and EDWARDS, H. I. (1936) : Canadian J. of Res. *14*, 412.
PRICE, W. C. (1935) : Phytopath. *25*, 947.
—— (1937) : *ibid. 27*, 561.
SEASTONE, C. W., LORING, H. S. and CHESTER, K. S. (1937) : Journ. Immunology *33*, 407.
SILBERSCHMIDT, K. (1933) : Beiträge Biol. Pfl. *20*, 105.
VAN SLOGTEREN, E. (1945) : Meded. N. A. K. No. 1.
SPOONER, E. T. C. and BAWDEN, F. C. (1935) : Brit. J. exp. Path. *16*, 218.
STANLEY, W. M. (1936) : Science *83*, 626.
TOPLEY, W. W. C. and WILSON, G. S. (1937) : The Principles of Bacteriology and Immunity. Edward Arnold and Co., London.
WATSON, M. A. (1938) : Proc. Roy. Soc. B. *125*, 305.
WELLS, H. G. and OSBORN, T. B. (1911) : J. Infect. Dis. *8*, 66.
YOUDEN, W. J., BEALE, H. P. and GUTHRIE, J. D. (1935) : Contrib. Boyce Thompson Inst. *7*, 37.

*Chapter 8*

# QUANTITATIVE METHODS OF ASSAYING FOR VIRUSES

In studying such problems as multiplication or inactivation of viruses, as well as many others, the need for quantitative measurements is too obvious to warrant stressing. The large amount of information that exists on the behaviour and properties of viruses which can be assayed quantitatively and our almost complete ignorance about the many which cannot, in itself adequately indicates their value. Quantitative assays can be made in three ways; the virus may be isolated and weighed, or the amount present may be estimated from either infectivity or serological tests. The first method seems the most precise for it gives an absolute answer whereas the others give only relative values. However, it is too limited to be generally used. Only a few viruses have yet been purified and even with these the method can be applied only to preparations containing weighable amounts. Purification procedures are too laborious and time-consuming for routine use, and the precision obtained by weighing an end-product may be largely illusory because of uncertainties about its exact nature. The uncertainties are dealt with in other chapters; here we need note only the opportunities for losses of virus during isolation and that the absence of critical tests for homogeneity makes it impossible to translate weights of end-products into numbers of active virus particles. Indeed no method has yet been devised by which the number of active particles can be estimated even approximately, but the relative infectivities and serological activities of different preparations can be measured with reasonable accuracy. Infectivity and serological tests can be made relatively easily and can be used with preparations containing too little virus for isolation to be possible. Each method has advantages the other lacks; in some conditions the two give results that agree closely, in others there is wide disagreement. This chapter is largely concerned with discussing these variations and describing the application and limitations of the two methods.

**The Use of Local Lesions:** — The production of systemic infection has little value for quantitative work, because it merely shows that there was enough virus to give infection; the inoculum might almost equally well have contained 10, 100 or 1,000 times this amount and the result would have been the same. By diluting the preparations suitably so that only some inoculated plants become infected, the infectivity of different preparations can be compared, but for even a small degree of accuracy large numbers of plants must be used. Had this method remained the only one, advances in the study of viruses would have been slow, but quantitative methods were revolutionised by HOLMES (1929) when he observed that some species of *Nicotiana*, especially *N. glutinosa,* produced countable necrotic lesions when their leaves were rubbed with tobacco mosaic virus.

Analogous effects have since been described with other viruses and hosts and the production of such local lesions now provides the basis of most quantitative infectivity tests.

HOLMES showed that the number of lesions varied with the virus content of the inoculum, for leaves rubbed with undiluted sap produced some hundreds per leaf whereas those rubbed with sap diluted 1 in 1,000 with water produced an average of 10 or so (Fig. 29). He compared the method with Koch's plate method for counting bacteria and many subsequent workers have also drawn analogies between the production of bacterial colonies on agar plates and of local lesions by viruses. Although the two may be similar in some respects, there are important differences. Identical agar plates can be obtained in unlimited numbers, but identical plants are as yet only an ideal, and individual differences between plants greatly influence the number of lesions produced by a given inoculum.

FIG. 29. — The effect of dilution on the number of local lesions caused by tobacco mosaic virus in leaves of *Nicotiana glutinosa*. (Undiluted sap, 1:3.1, 1:10, 1:100, 1:1000). (HOLMES, F. O., 1929, Bot. Gaz. *87*, 39).

Also, any part of an agar plate is as good as any other for the establishment of a bacterial colony, but a virus particle can cause infection only if it alights on a suitable entry point. HOLMES noted the individual plant variations and that the number of lesions produced by the same inoculum varied with the method of inoculation and with different leaves on the same plant. These observations have been repeatedly confirmed, and subsequent work has been largely devoted to finding experimental arrangements of test plants that reduce the errors produced by such variations, and, by statistical analysis, to evaluate the significance of lesion counts in terms of virus content. SAMUEL and BALD (1933) showed that there was little difference between the reaction of opposite halves of the same leaves, and that by comparing preparations on opposite half-leaves fewer plants could be used and more accurate results obtained. Since then the half-leaf method has come into general use, with modifications to allow for the comparison of more than two samples. The simplest is to select one preparation as a standard and apply it to one-half of every leaf, while the other half-leaves are apportioned between the other preparations. Each preparation can then be compared directly with the standard and indirectly through the standard with any other. With the half-leaf method, because it is difficult to rub equally on both halves, the preparations should be systematically varied, so that each occurs on an equal number of left-hand and right-hand halves, or they should be distributed over the opposite halves at random.

To compare a number of preparations without the need of referring each to a standard, YOUDEN and BEALE (1934) suggested designs by which results could be subjected to analysis of variance and differences arising from individual plant variations or from leaf position could be assessed and eliminated. Of these the chief is the Latin square, in which the treatments are arranged so that each appears the same number of times on each plant at each leaf position. An example of comparing six preparations is given below, in which columns represent plants and rows represent leaves (or half-leaves) occupying the same relative position.

| | | | | | |
|---|---|---|---|---|---|
| A | F | B | E | D | C |
| D | C | F | A | B | E |
| E | A | D | B | C | F |
| B | D | C | F | E | A |
| C | E | A | D | F | B |
| F | B | E | C | A | D |

The Latin square or some similar design is convenient with plants such as *N. glutinosa* which produce six or more inoculable leaves, but is obviously inapplicable when the number of treatments exceeds the number of leaves. However, various modifications of the half-leaf method can be used, such as the incomplete block (YOUDEN, 1937*a*), or any other arrangement that distributes the different preparations equally over the test plants. The need for such experimental designs is set by individual-plant variations and much could no doubt be done to increase the accuracy of the lesion-count method were more attention paid to the raising of test plants. Much variation probably comes from genetic differences, and the breeding of pure lines of test plants might overcome this. However, cultural conditions are also important. SAMUEL and BALD (1933) noted that the number of lesions produced by *N. glutinosa* plants of the same age varied and that the variation could be correlated with their appearance, "small hard" plants giving only one-tenth as many lesions as "larger and better" plants, and that the poor plants produced more lesions if they were supplied with ammonium phosphate. They also found that if half-leaves were covered with black paper for some days before inoculation they produced more lesions than the corresponding uncovered halves, and SAMUEL, BEST and BALD (1935) found that plants reacted more uniformly, and produced more lesions, if they were kept in a dimly-lit room for a day before inoculation than if they were kept in the glasshouse. Greater uniformity was also obtained if the apices of *N. glutinosa* were removed a few days before the plants were to be inoculated.

In recent years the French bean *(Phaseolus vulgaris)* has been increasingly used for local-lesion work with tobacco mosaic and other viruses. It grows more rapidly than *N. glutinosa,* the primary, heart-shaped leaves being well developed about ten days after sowing. A disadvantage of beans is that only four treatments can be compared on one plant, but they react more uniformly than *N. glutinosa,* possibly because they are used much younger and so have fewer opportunities of being affected by fluctuations in watering, nutrient supply and illumination. For work with tobacco mosaic virus, the variety Early Golden Cluster is mainly used, but there are others that also react by producing necrotic lesions (PRICE, 1930; SILBERSCHMIDT and KRAMER, 1941). At Rothamsted we have found beans less suitable than *N. glutinosa* for tobacco mosaic virus, because the same inoculum not only produces many fewer lesions but beans behave erratically and concentrated inocula sometimes produce no lesions at all. SPENCER and PRICE (1943), at Princeton, find that more reliable results are obtained during the winter if the bean plants are illuminated with 1000 watt lights and are kept at 30-35°C. for a day or so after they are inoculated. Early Golden Cluster beans are also suitable for local-lesion work with tobacco necrosis, alfalfa mosaic, tobacco ringspot and southern bean mosaic viruses (PRICE and SPENCER, 1943*a*; PRICE, 1945); Canadian Wonder, Prince, Bountiful, or many other varieties, are equally suitable for work with tobacco necrosis viruses; the primary leaves become less susceptible as they age, and plants must be raised under equal illumination if they are to react uniformly, because shading before inoculation greatly increases the numbers of lesions produced (BAWDEN and ROBERTS, 1948). The variety Bountiful also gives countable necrotic local lesions with cucumber mosaic virus during the winter in the Rothamsted glasshouses, but appears to be immune during the summer.

In making local-lesion tests, plants all of one age should be used, and they should be raised under as similar conditions of manuring, watering and illumination as possible. In any one test all the plants should be the same size and uniform in colour; any looking different from the majority should be rejected. An experimental design should be used that distributes the treatments at random over the different half-leaves or that distributes them systematically in some pre-arranged manner, and the inocula should be rubbed as evenly as possible over them. Each treatment should be applied to the same number of left-hand and right-hand half-leaves. Provided there is enough fluid to wet the whole surface, the volume of inoculum applied seems to be unimportant (SAMUEL, BEST and BALD, 1935; YOUDEN, 1937*b*), but the degree of rubbing affects the number of lesions. SAMUEL (1931) suggested inoculating with ground-glass spatulas having flat blades long enough to reach across a half-leaf, but most workers have rubbed either with a piece of cheese-cloth or with their fore-fingers wet with the inoculum. With sap from plants infected with some viruses, *e. g.* cucumber mosaic (COSTA, 1944), these methods of inoculation produce too few lesions for accurate quantitative work. More lesions can be obtained by incorporating a suitable abrasive in the inocula, but this, of course, also introduces an additional variable and the amounts used need controlling carefully. When working with sap or with virus preparations containing salts, the leaves should be rinsed with water as soon as possible after inoculation. The washing does not interfere with the establishment of infection and removes materials that might injure the leaves if allowed to dry.

The accuracy of comparisons made by the local-lesion method is, of course, increased by increasing the number of replications, and the number needed depends on the accuracy sought. Unless the differences between the preparations to be compared are large, less than six replications are undesirable. In each test it is usually necessary to compromise between what is desirable and what is expedient. Increasing the random distribution of treatments and the number of replications also increases both the chances of errors in the distribution of treatments and the time needed for inoculation. If the plants are exposed to fluctuating conditions during the period of inoculation, what is gained by elaborating the experimental design or increasing replication may be lost by changes in susceptibility. This source of error can be reduced by keeping the test plants in low light intensities for 24 hours before they are inoculated and avoiding bright light or any conditions that encourage wilting during the period of inoculation.

For any accurate comparisons of infectivity it is also desirable that the different inocula should produce approximately the same numbers of lesions, because there is no direct way of relating the infectivities of preparations that produce few lesions with those that produce many. The comparison is most sensitive when preparations are producing an average number of lesions per half-leaf of between 15 and 40, and to ensure comparisons over this range different preparations should be tested at a number of different dilutions. The need for this is shown by considering the curve obtained when the numbers of lesions produced by one preparation at various dilutions is plotted against the logarithm of the dilution. The dilution curve for tobacco mosaic virus divides into three sections: *(1)* a flat portion at high virus concentrations, where a large change in concentration produces only a small change in the number of lesions, *(2)* a steeply-sloping portion, where a change in concentration produces an almost equivalent change in the number of lesions, and *(3)* another flat portion at high dilutions, where the fall in number of lesions is also proportionally smaller than the dilution (BEST, 1935). For translating lesion counts into relative infectivities, tests should be made so that numbers fall in the middle portion of the curve where there is an approximately linear relationship between lesions and dilution. With tobacco mosaic virus this relationship usually holds at dilutions of infective sap around 1/1,000 or with solution of purified virus containing about $10^{-6}$ g. per ml., but the range of dilutions over which it holds varies considerably with different batches of test plants. This type of curve is given by many viruses in addition to tobacco mosaic, but probably not by all. For example, BEST (1935) finds that tomato spotted wilt virus in tobacco gives a third section differing from that with tobacco mosaic virus, for at high dilutions the fall in number of lesions is greater than the fall in concentration.

Various workers have attempted to assess the accuracy of the local-lesion method for measuring infectivity by comparing different dilutions of one preparation by some experimental design and then analysing the results statistically to see what differences in concentration of virus give significantly different lesion counts. BEALE (1934) concluded that, by the half-leaf method, differences in virus-content of from 25 to 50% could be detected by inoculating 50 or more leaves of *N. glutinosa.* Similarly, LORING (1937) found that, provided the inoculum contained around $10^{-6}$ g. per ml., differences of 10% in concentration of purified tobacco mosaic virus were detectable by the half-leaf method using from 40 to 50 beans. At other virus concentrations, or using *N. glutinosa* as a test plant, the smallest difference that could be distinguished was 20%. PRICE and SPENCER (1934*a* and *b*) have pointed out that the slope of the dilution curve is not constant and cannot be predicted. To determine the relative infectivities of two preparations they consider that it is necessary to measure, not only the numbers of lesions produced by the two, but also the slope of the dilution curve. They consider that the best and simplest arrangement for comparing two preparations on beans is to inoculate them at two dilutions, so that each dilution of each preparation is inoculated to one-half of every plant. The error in estimating relative infectivities increases with increasing differences between the preparations being compared, but they find that when dilutions are adjusted suitably and 36 plants inoculated, differences of 10% in activity are detectable with tobacco mosaic virus (SPENCER and PRICE, 1943), of 11, 14 and 18% respectively with tobacco necrosis, tobacco ringspot and alfalfa mosaic viruses (PRICE and SPENCER, 1943*a*) and 10 to 15% with southern bean mosaic virus (PRICE, 1945).

In analysing the results of lesion-count experiments, some workers have applied statistical methods directly to the numbers of lesions whereas others have applied them to logarithms of the numbers. Statistical methods, however, are based on the assumption that the variate is normally distributed, and methods that are applicable to lesion numbers clearly cannot also be applicable to the logarithms of those numbers. KLECZKOWSKI (1949) has studied the distribution of local lesions caused by tobacco mosaic and tomato bushy stunt viruses in *N. glutinosa,* and a tobacco necrosis virus in beans, and has found that with none of them are the numbers of local lesions normally distributed around their means. The distribution is skew and, what is more important, the standard error increases with the increasing mean. Conversion to logarithms is not a suitable transformation, because the distribution around the common mean is again skew, though this time the standard error decreases with increasing mean. A satisfactory transformation is obtained by the formula $y = \log (x + c)$, in which x is the number of lesions and c is a constant which can be calculated. No great accuracy is needed in calculating c, for in the example quoted by KLECZKOWSKI a satisfactory distribution was obtained with values ranging from 15 to 80. Thus, if the results of local-lesion tests are to be subjected to statistical analysis for assessing significant differences, it seems that the numbers cannot legitimately be used as such, but they should first receive an appropriate subsidy and then be converted into logarithms.

There are two main reasons for stressing the need for suitable experimental designs, many replications, and appropriate analysis of results, before differences in lesion-numbers can be interpreted in terms of active virus present. The first is to prevent workers from attaching undue importance to differences in numbers of local lesions when they have no idea what the probable error of their estimates may be. The second reason is to counteract the tendency of some workers to translate differences in numbers of lesions directly to differences of virus content. The last is a common practice, but is indefensible. The local-lesion method can give valuable information without an impractical number of replications or tiresome analysis of results, as long as large effects are being sought and no attempt is made to calculate precise quantitative effects on the virus from the results. For instance, if a virus preparation is subjected to a certain treatment for five different periods, and the five treated samples are compared with a control on a Latin square design with six replications, the lesions produced might average 100 per half-leaf for the control, and 90, 75, 35, 25 and 5 for the samples treated for increasing lengths of time. From such figures it is quite plain that the treatment is inactivating the virus and that the amount inactivated is increasing with length of treatment, but one cannot assume that 10% of the virus was inactivated by the shortest treatment and 95% by the longest. Before any such quantitative assumptions become legitimate, detailed tests and analysis of the kind outlined are essential.

Even when every precaution has been taken to ensure both the adequacy of the experimental design and the statistical significance of the results, differences may still not be interpretable in terms of relative virus contents. The local-lesion method measures the relative infectivities of preparations and it is important to realise that this is not necessarily correlated with their virus contents. Provided samples under test are identical in every respect other than their concentration of infective virus, and are inoculated at suitable dilutions, the differences in infectivity probably indicate their relative virus contents. On the other hand, if the samples differ in other respects, it is by no means certain that they will. Differences in the salt content, or *p*H value, the presence or absence of different substances, or the presence of different amounts of the same substance, may all play important parts in determining the number of lesions produced by a given amount of virus when inoculated to the same plants. To cite an extreme example; a preparation containing $10^{-6}$ g. of tobacco mosaic virus per ml. might produce an average of 10 lesions per half-leaf, but the same preparation with some abrasive present might give 200 and if ribonuclease were present instead of abrasive, it would give no lesions at all, although the addition of neither substance alters virus content. The effect of abrasives in increasing infectivity, and of ribonuclease and many other substances in decreasing it, are described in other chapters. They are mentioned here only to illustrate the uncertainties attached to interpreting lesion-count differences in terms of virus content and the need before doing so of ensuring that preparations are equivalent in all other respects. The fact that virus preparations containing inhibiting materials such as ribonuclease often produce more lesions when diluted than when concentrated, further emphasises the danger of assessing virus concentration from infectivity. Such increases in infectivity with dilution are not restricted to mixtures of virus and inhibitor for YOUDEN (1937*a*) found that purified preparations of tobacco mosaic containing about 0.1 mg/ml also caused more lesions if slightly diluted.

Many attempts have been made to interpret the dilution curve, not only to increase the accuracy of local lesion work, but also in attempts to differentiate between treatments that alter infectivity by acting on the virus and on the host, and to gain information about whether infection is initiated by single particles or by many acting together. It cannot be said that these have produced any very precise results. For distinguishing between substances that reduce infectivity by acting on the virus and those that act on the plant, STANLEY (1934) and CHESTER (1934) suggested that infection tests should be made with varying concentrations of virus in the presence of constant amounts of the substance. If infectivity is reduced proportionally more at low than at high virus concentrations, the substance is acting on the virus, but if it is reduced equally at all concentrations or proportionally more at high virus concentrations, it is acting on the plant. In practice, however, the method rarely gives unequivocal results, and different workers have disagreed about whether some substances are having one action or the other. More often than not, the only conclusion is that they are affecting both virus and host; this perhaps is only reasonable, for in the establishment of infection both are concerned, and with our lack of knowledge of the processes involved it may be pointless to attempt to distinguish between those that are peculiar to the viruses or to the host cells. Some of the inhibiting substances, which were first thought to act by increasing host resistance, are now known to combine with the virus particles. From this it might be assumed that their action is on the viruses, but as there is no proportionality between the concentration of inhibitor and virus that leads to complete loss of infectivity (KASSANIS and KLECZKOWSKI, 1948), such as occurs with the neutralisation of toxins by antitoxins, the relevance of this combination to loss of infectivity is uncertain. Most likely the entry into cells of such inhibitors along with virus particles interferes with some essential step in the complex reaction between virus and host that precedes virus multiplication.

Such uncertainties about virus-host interactions and the mechanism of infection complicate interpretations of the dilution curve. YOUDEN, BEALE and GUTHRIE (1935) pointed out that there was no reason why essential relationships between virus concentration and infectivity should be obvious from the appearance of curves obtained simply by plotting the two variables against one another, or by plotting one against the logarithm of the other. By analogy with bacterial counts and on the assumption that lesions are caused by single virus particles, they derived an equation from the Poisson series relating virus concentrations to numbers of lesions. In their suggested formula

$$y = N\ (1 - e^{-ax}),$$

$y$ is the number of lesions produced by any given relative concentration $x$ of the virus; $N$ is a constant, being the theoretical maximum number of lesions obtainable; $a$ is a constant, representing for undiluted inoculum the average number of particles per entry point, and $e$ is the base for natural logarithms. They considered that the dilution curves obtained by different workers fitted this equation satisfactorily, but YOUDEN (1937*a*) later concluded that, although it was in accord with the general form of the dilution curve, the agreement was not entirely satisfactory. BALD (1935, 1937*a*) independently arrived at essentially the same equation, but concluded that only purified virus preparations could be fitted. However, even with these, the fit was often far from satisfactory, and BALD (1937*b*) suggested that some deviations could be accounted for by postulating that dilution disrupts virus aggregates into smaller particles. This formula is based on the assumption that single virus particles are responsible for infection and LAUFFER and PRICE (1945) have derived other equations, also from the Poisson series, on the assumption that two and three particles are needed. When these equations are plotted it is apparent that they fit the dilution curves less satisfactorily than the formula for one particle, but there are many examples in which the fit given by this one is also far from good.* From the formula $y = N\ (1 - e^{-ax})$ it should be possible, by fitting data from

*Statistical tests made recently (KLECZKOWSKI, A. (1950): J. gen. Microbiol. *4*, 53) indicate that, unless auxiliary assumptions are made, the effect of dilution on the numbers of local lesions is incompatible with the hypothesis that lesions result from chance encounters between single virus particles and susceptible regions of a uniform type, but is compatible with the hypothesis that susceptible regions vary in susceptibility and that different numbers of particles are needed to produce a lesion in different regions.

tests at different concentrations, to determine whether substances that reduce infectivity are acting on the virus particles or on the host, for it should be seen whether results were best fitted by postulating a change in *N*, entry points, or *x*, virus concentration. However, at the present stage of our knowledge the method can hardly be recommended with confidence.

The production of lesions depends on so many unknown factors that it may be unreasonable to expect any simple relationship between virus concentration and lesion counts, and the derivation of formulae involves many assumptions that may be unwarranted. Analogies between the production of local lesions and of bacterial colonies should not be pressed too far, for in spite of their superficial similarities, the two may differ fundamentally. Also, the application of the Poisson series assumes a uniformity of both virus particles and of host cells for which there is no experimental evidence. For all that is known to the contrary individual particles of one virus may have different infective powers, and individual cells may vary in susceptibility, so that particles able to infect some cells may be unable to infect others. It seems likely that single lesions can be and often are caused by single virus particles, but it would be rash to assume that all lesions are. Furthermore, little is known of what constitutes an entry point for viruses and of the virus-host interactions that are concerned in the establishment of infection. It was shown in Chapter 4 that there is every reason to think that wounding is necessary, but from the behaviour of tobacco mosaic and tomato bushy stunt viruses in different leaves of *N. glutinosa*, it is also obvious that a wound *per se* does not necessarily constitute an entry point; the physiological state of the injured cells seems to determine whether infection occurs or not, and conditions that favour establishment of one virus may be inimical to another. It is rather odd that there has been so much written on the dilution curve and so few experiments made, successive interpreters continuing to use the same data that have been in the literature for more than 10 years. Much more experimental work is needed on these problems before any definite interpretations are possible, but that does not prevent the local-lesion method from being a practical asset of great value, without which virus research would have made much slower progress.

**Starch-Iodine Lesions:**—For local-lesion tests hypersensitive hosts are the best, for discrete necrotic spots are the most convenient to count. Nevertheless the method can be used with virus-host combinations that give other reactions. When necrotic, easily-visible, local lesions are produced by viruses that cause systemic infections, for example, strains of potato virus X that produce ringspot in tobacco, or tomato spotted wilt in tobacco, the method is similar to using a hypersensitive host, the only difference being the need to count the lesions before they coalesce. Hosts that react with mosaic symptoms usually produce either no local lesions or diffuse chlorotic spots that are too indistinct for accurate counting. At first sight it would seem that these are not amenable to quantitative study by the local-lesion method, but by varying the technique it can be applied.

The modifications required are treatments of the inoculated leaves to make the sites of local infection obvious and for this purpose the

FIG. 30.—Different types of local lesion produced in tobacco by severe etch virus at different times. *Left:* Necrotic rings and spots. *Centre:* Diffuse chlorotic spots. *Right:* Starch-iodine lesions on a leaf which showed no local lesions before it was decolourised and stained.

effects of the viruses on the synthesis and translocation of carbohydrates can be utilised. Like the whole local-lesion technique this method derives from an original observation of HOLMES (1931). He noticed that tobacco leaves rubbed with tobacco mosaic virus sometimes showed vague yellowish spots and he found that these could be made more definite if the leaves were decolourised with alcohol and then stained with iodine. HOLMES used two methods, one depending on the fact that infection decreases the rate of synthesis of starch when leaves are exposed to light and the other on the fact that synthesized starch is translocated more slowly from infected than from uninfected cells. In the first method, the leaves are taken from the plants in the evening, decolourised and treated with iodine, when the local lesions stain less deeply than the rest of the leaf, and show as light spots on a dark background. In the second method, whole plants are placed in the dark for several hours before the leaves are picked, decolourised and stained, when the lesions then show as dark spots against a light background.

Similar results have since been obtained in tobacco with a number of viruses other than tobacco mosaic, for example, with potato X and Y, henbane mosaic, cucumber mosaic and tobacco etch, and the method may be generally applicable to hosts that react with mosaic symptoms. The second method is usually preferable, for by it the local lesions are more clearly differentiated. Plants are placed in the dark for an appropriate length of time, when the leaves are picked, plunged into boiling water, decolourised with ethyl alcohol, washed with water and then stained in a solution of iodine in potassium iodide. Under favourable conditions, the method works quite satisfactorily and the lesions are well defined and easy to count, but to ensure success it needs careful controlling. Plants must be kept in the dark long enough for most of the starch to leave the uninfected cells, but not long enough for it to leave the lesions. The length of time needed for this differential action will depend on the temperature and on the amount of starch in the leaves, which in turn will depend on the conditions to which the plants were exposed before they were placed in the dark. The test plants must all be equally illuminated after inoculation, for treatments that will show starch-iodine lesions in plants grown under bright light will fail with those in shade. The leaves must also be picked at an appropriate time after inoculation. If they have been infected too long, the virus will have spread from the original lesions to nearby cells, so that the lesions coalesce and cease to be countable. If the interval between inoculation and picking is too short, the difference in starch content between the lesions and uninfected cells is too small to show clearly.

Under the fluctuating conditions experienced in most glasshouses, the starch-iodine method cannot be relied on and total failures are not uncommon, but it would probably not be difficult so to standardise conditions that countable lesions could regularly be obtained. With a thorough study of the effect of environment on symptoms, however, it is possible that the local lesion method could be used with some of the mosaic-producing viruses, without the need for iodine-staining. Potato Y, henbane mosaic and tobacco etch viruses all occasionally produce local lesions, usually only diffuse chlorotic areas but sometimes fairly well defined and isolated lesions. The various local effects produced by severe etch viruses in tobacco are illustrated in Fig. 30; of these, only the starch-iodine lesions regularly occur, and no reasons are known for the occasional production of visible necrotic or chlorotic reactions. By determining the environmental conditions which lead to the production of countable necrotic lesions, plants could probably be raised so that they would be suitable for quantitative infectivity tests.

**The Use of Precipitin Tests:**—The precipitin test provides a quantitative method that is more rapid than infectivity tests, giving results in minutes or hours instead of in days, and one that has fewer limitations and uncertainties. Probably its greatest value is that, unless preparations have been subjected to unusual treatments, the results can usually be translated directly into relative antigen concentrations, regardless of how widely the concentrations of the samples differ. Differences in numbers of local lesions are not proportional to virus concentration unless inocula differ only little in concentration and are strictly comparable in all other respects, whereas differences in precipitin reactions are proportional to concentration over the whole range to which the method is applicable. For example, if one inoculum produces half as many lesions as another, it may contain half as much virus, but if it produces a fiftieth the number, it probably contained much less than a fiftieth of the virus, but no accurate comparison of the two is possible. In precipitin tests, on the other hand, large differences are, if anything, more accurately determinable than small ones, and factors of two or fifty are all proportional to antigen concentration. Another great merit of the precipitin test is that, because antisera remain relatively constant for long periods and tests can be made under standardised conditions, results obtained at different times are as comparable as those obtained at one time. Also, even if little is known of the precipitin behaviour of a virus, reasonable guesses can be made about its concentration, and once its behaviour has been established from tests on purified preparations, results can be interpreted as absolute weights with considerable precision.

It is rather surprising, considering the many advantages of the precipitin test, that most workers have neglected it and have made quantitative estimations, exclusively by the more laborious and less accurate local-lesion method. Partly this is attributable to a natural reluctance to trust results obtained by an unfamiliar technique, and there is no doubt that the precipitin tests could with advantage be used much more than it is. Nevertheless, there are good reasons why, in spite of its many advantages, it should not wholly replace infectivity tests. Of these, probably the most important reason is that the two kinds of test do not necessarily measure the same things. Local-lesion counts indicate, however inaccurately, the relative infectivities of different preparations, or the relative numbers of infective virus particles, whereas precipitin tests measure the relative concentrations of serologically active material. The two activities may be correlated, so that the two methods of measurement give similar results, but often they are not and then precipitin and infectivity tests may give results that conflict widely. The presence of many different substances in inocula affects lesion counts without influencing precipitability with antiserum, and changes in virus particles may make them non-infective without destroying serological activity or may alter precipitin behaviour without greatly affecting infectivity. Hence the two methods of measurement are not simply convenient alternatives, they are complementary and are most valuable when used together rather than separately.

Quantitative comparisons can be made from precipitin tests in two ways, by determining either the precipitin titre, *i. e.*, the highest dilution at which a visible precipitate is obtained after a given period of incubation with antiserum, or the optimal precipitation point, *i. e.*, the ratio of antigen to antiserum at which precipitation first occurs. The former is more accurate, but the latter gives results more quickly. Provided concentrations are suitably chosen, readings of both points can be made with one set of titrations; for example, this can be done with the method of making precipitin tests described in Chapter 7, when a constant amount of antiserum at a suitable dilution is added to each of a series of tubes containing decreasing amounts of antigen. Measurements by the two methods usually agree closely; if the precipitin titre of one sample is four times that of another, then it will also precipitate optimally at four times the dilution of the other. Provided that the antiserum is not used so con-

TABLE 12.—*Tobacco mosaic virus. Comparison of infectivity and antigen content as indicated by precipitation end points:—*

| *Precipitation end point* | *Dilution* of inoculum* | *Mean number of lesions per half leaf* |
|---|---|---|
| 1/88 | 1/22 | 66 |
| 1/88 | 1/22 | 74 |
| | | |
| 1/14 | 1/7 | 72 |
| 1/32 | 1/16 | 65 |
| | | |
| 1/520 | 1/260 | 48 |
| 1/65 | 1/32.5 | 53 |
| | | |
| 1/65 | 1/32.5 | 104 |
| 1/130 | 1/65 | 92 |

*Inoculum diluted so that each pair has equal antigen content.

centrated that inhibition occurs because of antibody excess, the precipitin titres of virus solutions are largely independent of the concentration of antiserum. On the other hand, optimal precipitation occurs at a given ratio of antigen to antibody and, to ensure covering this ratio, the relative concentrations of antiserum and antigen must be appropriate; this is sometimes more conveniently done by keeping antigen concentration constant and varying the antiserum. When using the optimal precipitation point for comparing antigen concentrations, it is rarely worth making serial dilutions with a factor of less than two, for precipitation occurs too rapidly over a zone of antigen/antibody ratios for differences of less than 50% to be accurately assessed. When using the precipitin titre, eye-estimates of the bulk of precipitate produced can indicate differences smaller than this, even when serial dilutions with a factor of two are employed, and by making dilutions with a smaller factor the method can be refined to show differences down to 20%. The accuracy with which small differences can be measured and taken as proportional to antigen concentration, varies with different viruses. With those that have particles of a uniform size, *e. g.*, tomato bushy stunt and tobacco necrosis viruses, reproducible results with a high degree of accuracy are normally attainable, but with those that occur in particles of variable size and shape, results are less reproducible and less certainly proportional to antigen content.

The precipitin test was first used quantitatively in work on plant viruses by BEALE (1934), to gain information on the relation between tobacco mosaic virus and the specific antigen present in sap from infected plants. She estimated the relative antigen contents of various lots of sap by finding their precipitin titres, diluted them appropriately so that paired lots contained equal amounts of antigen, and then found that they were also equally infective. The results of a few comparisons are given in TABLE 12. Similarly, BAWDEN (1935) found that the infectivities of partially purified preparations of potato virus X (strain S) were proportional to their relative antigen contents as estimated from measurements of their optimal precipitation points. Although these correlations between quantitative results obtained by infectivity and precipitin tests suggested that the two methods were measuring the same thing, and that the viruses were the specific antigens, there have been many examples since in which the two methods have given widely divergent results. Provided preparations that are to be compared by the two methods have the same antecedents, then their relative infectivities are usually proportional to their relative contents of antigen, but no such agreement is to be expected if the samples have been treated differently. It is widely recognised that some treatments destroy infectivity without changing serological specificity, and these must obviously alter the ratio of infectivity to serological activity. But the ratio is also affected by treatments not normally considered to inactivate, such as varying the method of extraction from infected plants, exposure to salts and pH changes, centrifugation, ageing, and many others whose actions are ill-understood. Also, different viruses behave differently, and treatments that alter the ratio of infectivity to serological activity with some may not with others. The precipitin test measures the total quantity of material in a preparation that reacts with the antiserum, without distinguishing between what is infective and what is not. Although this obviously limits the value of the test when it is used alone, it is not altogether disadvantageous. Infectivity tests indicate the relative numbers of infective particles in different preparations, but give no information about the relative amounts of non-infective material with the same general constitution. This material can be detected and measured only by serological methods, without which its existence might still go unrecognised. For all we know, the bulk of all virus preparations may consist of such material, and it is worth stressing that we have no method at present of measuring the absolute number of infective particles present in any one preparation. Local-lesion counts can tell us if preparations differ from one another in their relative infectivities, and precipitin tests if they differ in their total content of material that resembles virus particles. By combining the two methods, it is easy to show that different preparations contain varying proportions of infective to non-infective particles, as, for example, in the experiments made by BEALE and LOJKIN (1947) on different purified preparations of tobacco mosaic virus, but the relative proportions of the two kinds of particles present in any one preparation remains undeterminable.

The precipitin test not only fails to distinguish between infective and non-infective virus particles, but also fails to distinguish between related strains of the same virus. Thus, if mixtures of virus strains that differ in their pathogenicity are compared in infectivity and precipitin tests, the two methods may give quantitatively different results. For example, some strains of tobacco mosaic virus produce local lesions in tobacco and *Nicotiana sylvestris* and others do not; mixtures containing different quantities of the two types might give similar precipitin titres, but widely different numbers of local lesions. Similarly, only some strains of potato virus X produce local lesions in tobacco, and the effect of mixing one that does with one that does not is shown in TABLE 13 (BAWDEN, 1935). All the samples tested contained approximately the same total amount of antigen, which is indicated by the constancy of the optimal precipitation point (serum dilution 1/32), but they produced widely different numbers of lesions. Also, the numbers of lesions are not proportional to the relative concentrations of strain S, which produces them, for the

TABLE 13. — *Potato virus X. Comparison of mottle (G) and ringspot (S) strains, and mixtures of the two, for precipitating optima and production of local lesions*:—

| Antigen | Time | Serum dilution | | | | | | | | Mean number of lesions |
|---|---|---|---|---|---|---|---|---|---|---|
| | | 1/8 | 1/16 | 1/32 | 1/64 | 1/128 | 1/256 | 1/512 | 0 | |
| G | 6 min. | — | — | + | — | — | — | — | — | 0 |
| | 11 — | + | ++ | +++ | ++ | + | + | — | — | |
| | 20 — | +++ | ++++ | ++++ | ++++ | +++ | +++ | ++ | — | |
| 3/4 G 1/4 S | 5 — | — | — | + | — | — | — | — | — | 13 |
| | 9 — | + | ++ | +++ | ++ | + | + | — | — | |
| | 20 — | +++ | ++++ | ++++ | ++++ | +++ | +++ | + | — | |
| 1/2 G 1/2 S | 7 — | — | — | + | — | — | — | — | — | 20 |
| | 11 — | — | + | +++ | + | + | — | — | — | |
| | 20 — | ++ | ++++ | ++++ | ++++ | +++ | ++ | + | — | |
| 1/4 G 3/4 S | 6 — | — | — | + | — | — | — | — | — | 63 |
| | 10 — | — | + | ++ | + | — | — | — | — | |
| | 18 — | ++ | ++++ | ++++ | ++++ | +++ | ++ | + | — | |
| S | 8 — | — | — | + | — | — | — | — | — | 120 |
| | 12 — | — | + | ++ | + | + | — | — | — | |
| | 20 — | ++ | +++ | ++++ | ++++ | +++ | +++ | + | — | |

+ Signs indicate the degree of precipitation

strains interfere with one another and diluting strain S in fluids containing strain G reduces lesion numbers much more than would equal dilution with fluid free from strain G.

When working with infective extracts as usually made, precipitin results on different preparations can usually be taken as proportional to their relative contents of total antigen. However, there are circumstances that can upset this straight-forward relation and that may destroy the customary agreement between estimations based on optimum precipitation and precipitin titre. Three such complicating circumstances warrant some comment, of which the last is the most important. They are, first, the possibility of substances other than the virus precipitating from the samples during tests, secondly, the possibility of the virus not precipitating because it is combined with other proteins to form complexes, and, thirdly, the possibility of the virus particles occurring in different sizes and shapes.

The precipitate that forms as a result of union between virus and antibody may carry with it other materials, which were previously dissolved or were forming stable suspensions, and this may alter the bulk or visibility of the resulting floccules. Hence, if samples contain different amounts of such removable constituents, errors may arise in interpreting precipitin results quantitatively. Complications from the occurrence of virus in complex particles can arise because such particles may not only fail to precipitate with virus antiserum themselves, but their presence may also interfere with the precipitation of free virus (BAWDEN and KLECZKOWSKI, 1941). Such complexes are commonly formed when some viruses are heated in the presence of other proteins, but are probably rare in most preparations, though what are thought to be examples of this phenomenon have been encountered in extracts from plants infected with tomato bushy stunt virus (BAWDEN and PIRIE, 1944) and sugar beet yellows virus (KLECZKOWSKI and WATSON, 1944).

The greatest uncertainties in interpreting the results of precipitin tests as quantities of antigen arise with viruses that can occur in particles of various sizes and shapes. BAWDEN and PIRIE (1945) separated preparations of tobacco mosaic virus into fractions containing particles of different average lengths and found that particle length greatly influenced the precipitin behaviour. They found that, unless particles of approximately the same average length are being compared, not only is there no correlation between infectivity and serological activity, but there is also none between weight of antigen and precipitin titre or optimal precipitation point. Short particles readily increase in length by aggregating linearly, and this change alters precipitin behaviour strikingly. An example of this effect is recorded in TABLE 11 (Chapter 7), which shows that increasing the average length of the particles affected both the precipitating optimum and the titre. Without any change in virus concentration, the titre increased by a factor of about three, while the optimal precipitation point changed by a factor of more than sixty, and whereas the change in titre suggests an increase, the change in optimum suggests a reduction in virus concentration. In such conditions it is clear that precipitin measurements give conflicting and misleading results, and for results to be reliable the samples being compared should be in similar physical conditions. An adequate degree of uniformity is easily achieved with tobacco mosaic virus if the particles are aggregated, either by heating the samples to 70°C. at *p*H 5.5 or by incubating them with trypsin.

Such uncertainties are not restricted to work with tobacco mosaic virus but are probably general to all those that occur in rod-shaped particles. Most have not yet been studied, but BAWDEN and CROOK (1947) have made preparations of potato virus X containing particles of different average sizes and have found that the serological behaviour is closely related to particle length. Virus X particles can aggregate to such an extent that they become entangled and tend to come out of solution. A similar effect also occurs when the virus particles combine with some substances such as ribonuclease (BAWDEN and KLECZKOWSKI, 1948). This introduces a further complication into the quantitative interpretations of precipitin tests, because the same weight of virus gives a higher precipitin titre when freely soluble than when forming an unstable suspension.

The special value of precipitin tests for quantitative work with viruses that are difficult to transmit mechanically, or that form no countable local lesions, is too obvious to stress, but with others they are no less useful. The rapidity with which tests can be made would alone make them worth while even if the results were only approximations, but, provided a few elementary precautions are taken, measurements can be relied upon to show differences with considerable precision. For reasons already stressed, the precipitin test should not be used instead of, but in addition to, infectivity tests; the two combined not only give more accurate determinations, but often provide more and different information than could come from either method alone.

*References:-*


BALD, J. G. (1935) : Nature *135*, 996.
—— (1937*a*) : Ann appl. Biol. *24*, 33.
—— (1937*b*) : Austral. J. Exp. Biol. Med. *15*, 211.
BAWDEN, F C. (1935) : Brit. J. exp. Path. *16*, 435.
—— and CROOK, E. M. (1947) : Brit. J. exp. Path. *28*, 403.
—— and KLECZKOWSKI, A. (1941) : Brit. J. exp. Path. *22*, 208.
—— and —— (1948) : J. gen. Microbiol. *2*, 173.
—— and PIRIE, N. W. (1944) : Brit. J. exp. Path. *25*, 68.
—— and —— (1945) : *ibid. 26*, 294.
—— and ROBERTS, F. M. (1948) : Ann. appl. Biol. *35*, 418.
BEALE, H. P. (1934) : Contrib. Boyce Thompson Inst. *6*, 407.
—— and LOJKIN, M. E. (1947) : Contrib. Boyce Thompson Inst. *14*, 457.
BEST, R. J. (1935) : J. Austral. Inst. Agric. Sic. *1*, 159.
CHESTER, K. S. (1924) : Phytopath. *24*, 1180.
COSTA, A. S. (1944) : Phytopath. *34*, 288.
HOLMES, F. O. (1929) : Bot. Gaz. *87*, 39.
—— (1931) : Contrib. Boyce Thompson Inst. *3*, 163.
KASSANIS, B. and KLECZKOWSKI, A. (1948) : J. gen. Microbiol. *2*, 143.
KLECZKOWSKI, A. (1949) : Ann. appl. Biol. *36*, 139.
—— and WATSON, M. A. (1944) : Ann. appl. Biol. *31*, 116.
LAUFFER, M. A. and PRICE, W. C. (1945) : Arck. Biochem. *8*, 449.
LORING, H. S. (1937) : J. Biol. Chem. *121*, 637.
PRICE, W. C. (1930) : Amer. J. Bot. *17*, 694.
—— (1945) : *ibid. 32*, 613.
—— and SPENCER, E. L. (1943*a*) : Amer. J. Bot. *30*, 340.
—— and —— (1943*b*) : *ibid. 30*, 720.
SAMUEL, G. (1931) : Ann. appl. Biol. *18*, 494.
——, BEST, R. J. and BALD, J. G. (1935) : Ann. appl. Biol. *22*, 508.
—— and BALD, J. G. (1933) : Ann. appl. Biol. *20*, 70.
SILBERSCHMIDT, K. and KRAMER, M. (1941) : Phytopath. *31*, 430.
SPENCER, E. L. and PRICE, W. C. (1943) : Amer. J. Bot. *30*, 280.
STANLEY, W. M. (1934) : Phytopath. *24*, 1055.
YOUDEN, W. J. (1937*a*) : Contrib. Boyce Thompson Inst. *9*, 41.
—— (1937*b*) : *ibid. 9*, 49.
——, BEALE, H. P. and GUTHRIE, J. D. (1935) : Contrib. Boyce Thompson Inst. *7*, 37.
—— and BEALE, H. P. (1934) : Contrib. Boyce Thompson Inst. *6*, 437.

*Chapter 9*

# THE PURIFICATION OF VIRUSES

The discovery that tobacco mosaic virus can be precipitated by certain protein precipitants without losing infectivity (MULVANIA, 1926; VINSON and PETRE, 1929, 1931) aroused little interest until STANLEY (1935*a*, 1936*a*) reported its isolation in the form of a crystalline protein (Fig. 31). Since then additional methods have been described whereby tobacco mosaic and several other viruses can be purified, and the study

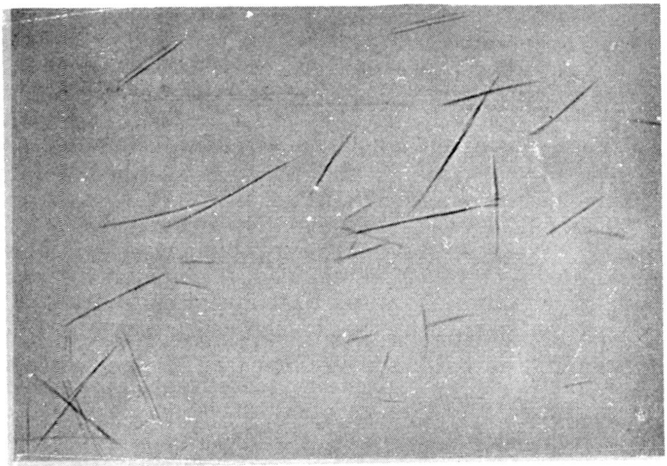

FIG. 31. — Needle-shaped paracrystals of tobacco mosaic virus produced by precipitation with acid and ammonium sulphate. × 675. (STANLEY, W. M., 1937, Amer. J. Bot. *24*, 59).

of purified preparations has become a major part of virus research. The methods give infective preparations of some viruses that satisfy all the customary tests for homogeneity. There is much other evidence that also closely relates the isolated proteins with the viruses themselves, but it should be stressed that the precise relationship between the two remains uncertain. Some workers, impressed by the apparent uniformity of their products, have unreservedly identified the proteins with the viruses and called the virus particles molecules, but this seems to prejudge what is still an open question. The word molecule implies a uniformity of structure that cannot be demonstrated with particles as large as viruses, and it is unjustified to assume that all the particles in purified preparations resemble one another in every respect, or that they are identical with the particles as they are formed in infected cells. Even if the particles are all identical when the original plant extracts are made, and this is by no means certain, ample opportunities occur for changes during the course of purification. Purified preparations that appear identical by chemical, physical and serological tests often differ considerably in infectivity, showing that customary tests for homogeneity often fail to distinguish between infective and non-infective particles. Because of this limitation, it is probably safest to assume heterogeneity even when none is detectable and even the most infective preparations remain suspect as mixtures.

Our knowledge of the chemical nature of viruses derives from the study of purified preparations and the need for caution when applying this knowledge to viruses in infected cells will be clear from considering what are the principles and aims of purification. Extracts from infected cells are subjected to various treatments that separate virus particles from other components, with the object of obtaining a product that has the maximum biological activity in a system containing the smallest weight of material and the fewest components. The activity most often considered is infectivity, and any materials that are not directly concerned with this are regarded as inessential impurities. Many of the substances discarded during fractionations are undoubtedly normal host constituents, but others may well be products specific to infected cells and as such it would be equally reasonable to regard them as parts of the virus as to regard them as parts of the host. It is worth reflecting on the fact that, if we applied the same type of reasoning to animals as we do to viruses and considered ability to multiply as the only essential activity, we might conclude that we had purified a rabbit by cutting of its ears and tail. The other property of virus preparations that can be assayed quantitatively is serological activity; when this increases proportionally with infectivity and the end-products also appear to be homogeneous when subjected to different tests, it is tempting to assume a uniform product. However, serological activity often continues to increase when infectivity either does not, or even decreases. These discrepancies pose problems that are not yet resolved; the relationship between serologically active material that is infective and non-infective is uncertain, as are also many changes that lead to loss of infectivity.

An ideally pure virus preparation would consist of particles all identical in size, chemical composition and biological activity (PIRIE, 1940), and there is no evidence that such preparations have yet been achieved. Deviations from this ideal may arise in several different ways. An obvious one is that preparations may contain normal host constituents, but their detection should present no insuperable difficulties and it is probable that preparations of some viruses have been made free from significant quantities of such impurities. A more serious cause of inhomogeneity is that the fractionations needed to get rid of contaminants alter the virus, so that although the end-products may consist exclusively of virus particles, these may differ from one another and from the original particles in size, shape, activity or composition. All preparations of tobacco mosaic virus probably fall into such a category, for most treatments cause changes in the sizes of the virus particles. The first preparations of tomato bushy stunt virus, in spite of their high degree of physical homogeneity, were also mixtures, for the sap was heated and this is now known to cause considerable loss of infectivity (BAWDEN and PIRIE, 1938*b*; STANLEY, 1940). Similarly, the fall in infectivity when the Rothamsted culture of tobacco mosaic virus is purified shows that the final crystalline products must be mixtures of infective and non-infective material (Bawden and Pirie, 1945*a*). With treatments that produce large changes in activity, such as those mentioned in the two preceding sentences, detecting heterogeneity is relatively simple, but with treatments that modify the particles only slightly a mixed end-product may well pass unnoticed. Apart from changes that may be produced during purification, perfectly homogeneous preparations may be unattainable because virus particles may not all be identical as they occur in extracts of infected plants. For example, with mutable viruses such as tobacco mosaic, most extracts contain mixtures of virus strains; also as the extracts can be separated by ultra-centrifugation into fractions containing particles of different sizes and activities, it is obvious that purified preparations as customarily made must be heterogeneous (Bawden and Pirie, 1945*b*).

Until tests for infectivity have been refined sufficiently to work with individual particles, unequivocal evidence about the homogeneity and identity of purified preparations is unlikely to be obtained. Nevertheless, there is little doubt that the bulk of material in the purified preparations which will be described below is closely related to or derived from virus particles and it is reasonably assumed that the properties of these preparations have a bearing on those of the virus particles.

**Treatment of Plants:** — The ease with which different viruses can be purified varies greatly and methods that work reasonably well with some fail with others. To be isolated a virus must be stable enough to withstand fractioning treatments and have properties sufficiently different from normal host constituents to be separable from them by such treatments, and purification will be facilitated by increasing virus concentration and decreasing the quantities of host constituents. Hence it is a valuable preliminary to choose the most suitable host plants and to pay attention to the conditions under which they are raised and infected. The concentration both of viruses and of materials that interfere with purification may vary with different host species, and with their age, nutrition and conditions under which they are kept, particularly with the intensity of illumination. In general, young seedlings that have received balanced dressings of nitrogen, phosphate and potash are most suitable, for they are easier to infect, produce a higher concentration of virus and contain fewer complicating impurities than old plants or those that have been inadequately manured. Some viruses, for example, tobacco ringspot (Stanley, 1939*a*) and alfalfa mosaic (Ross, 1941*a*), reach a maximum concentration a fortnight or so after plants are infected and then decrease. This phenomenon may be general with viruses that first produce severe symptoms from which plants later recover, and the highest virus content is obtained by harvesting leaves when they are reacting most severely. It does not seem to occur with viruses that continue to cause symptoms, though with most seedlings it is usually preferable to pick leaves within a month of infection. To get a high yield of tobacco necrosis viruses, which produce local lesions only, a concentrated inoculum is needed and it should be applied with a suitable abrasive. The use of an abrasive also increases the concentration reached by tomato bushy stunt virus, for it produces a more rapid and thorough systemic invasion of inoculated tomato plants. The yield of tobacco necrosis and tomato bushy stunt viruses is also higher if the test plants have been raised under low light intensity than if they have been brightly illuminated, and shading infected plants also facilitates purification of these and some other viruses (BAWDEN and ROBERTS, 1947). Placing infected plants in the dark for 24 to 48 hours immediately before the leaves are picked is another useful practice which removes some troublesome host constituents (RAWLINS, ROBERTS and UTECH, 1946).

Stems and petioles contain less of most viruses than do leaf laminae, and purification is often easier if they are not included in the tissues to be macerated. To prepare infective extracts most workers have macerated leaves, either fresh or when frozen, with a pestle and mortar or a meat mincer. FRAMPTON (1942*a*) has described an alternative method which has the advantage of producing extracts containing fewer coloured components; the leaves are first cytolysed with ether, the vacuolar sap is expressed and discarded, and the leaves then minced and extracted with phosphate buffer. The sap obtained by squeezing the macerated leaves in a cloth bag was generally assumed to contain all the virus present in the leaves, for washing the fibrous residues with water produces no more virus than would be expected from their remaining content of sap. The fibrous residues, however, still contain as much virus as occurs in the sap, but special treatments are needed to release it. The simplest treatment is to grind the fibrous residues finely in a power-driven roller mill; water extracts of the finely-ground fibre from plants infected with tomato bushy stunt, tobacco necrosis and tobacco mosaic viruses contain as much virus as is present in the sap (BAWDEN and PIRIE, 1944, 1945*a*, 1945*b*). The fine grinding does not liberate all the virus from the fibrous residues and seems to destroy some. Approximately twice as much tobacco mosaic virus can be obtained by incubating the residues with the enzyme mixture obtained from snails' alimentary tracts as by grinding, though this is not a practical method for obtaining virus for purification (BAWDEN and PIRIE, 1946). Grinding in a roller mill is ineffective in releasing potato virus X from leaf fibre, possibly because it destroys most of the virus, and it produces less than one-twentieth the amount that is obtained by incubation with snail enzymes (BAWDEN and CROOK, 1947).

The relationship between virus obtained in the sap and that remaining in the fibrous residues is uncertain. The two may differ qualitatively, for the virus in the fibre may be insoluble because it is combined with some host constituent which is destroyed by snail enzymes and fine grinding. On the other hand, the two may be essentially similar but merely come from different cells. Mincing seems to damage all the leaf cells enough for their sap to be released, for no more fluid is obtained if the minced leaves are finely ground. However, mincing may not disrupt all the cells sufficiently for the virus in them to escape with the sap. If this is so, virus in the sap will come only from the most damaged cells, whereas in the others it will be retained and become increasingly concentrated as the fluid is expressed. Snail enzymes and fine grinding may merely disrupt the walls of these relatively undamaged cells enough to make virus extractable. Over a wide range of virus contents, the ratio of virus in the sap to that retained in the fibre is reasonably constant, which does perhaps suggest that mincing opens a constant proportion of cells rather than that the two lots of virus are produced differently or differ in any other qualitative manner. No differences have been found between tomato bushy stunt virus obtained in the sap and from the fibre, but tobacco mosaic virus from the fibre contains a greater proportion of smaller particles and is less infective than that in the sap, and the small amount of potato virus X obtained by grinding the fibre also contains an unusually high proportion of small particles. These differences may, of course, be artefacts, produced by the effects of fine grinding.

The first step in purification is to clarify the sap, and the ease with which this can be done varies with different plant species. With sugar beet no satisfactory method has yet been devised, and this possibly accounts for the failure to isolate sugar beet yellows virus, which seems to precipitate in much the same conditions as most of the normal leaf proteins (KLECZKOWSKI and WATSON, 1944). With tobacco and many

TABLE 14. — *Approximate weights of different viruses in a litre of infective sap*:—

| *Virus* | *Host plant* | *Weight of virus* |
|---|---|---|
| Tobacco mosaic | Tobacco | 2.0 g. |
| ——— | Tomato | 1.3 |
| ——— | Spinach | 0.15 |
| Tomato enation mosaic | Tomato | 1.5 |
| Masked tobacco mosaic | Tobacco | 1.0 |
| Potato X Strain S | Tomato | 0.7 |
| ——— | Tobacco | 0.3 |
| Potato X Strain Y | Tomato | 1.5 |
| ——— | Tobacco | 0.7 |
| Cucumber 3 | Cucumber | 0.3 |
| Tomato bushy stunt | Tomato (winter) | 0.15 |
| ——— | Tomato (summer) | 0.01 |
| Tobacco necrosis (Rothamsted culture) | Tobacco (winter) | 0.20 |
| | Tobacco (summer) | 0.01 |
| Tobacco necrosis (Tobacco culture) | Tobacco (winter) | 0.004 |
| | Bean (winter) | 0.001 |
| Alfalfa mosaic | Tobacco | 0.2 |
| Turnip yellow mosaic | Chinese cabbage | 0.17 |
| Southern bean mosaic | French bean | 0.4 |
| Tobacco ringspot | Tobacco with symptoms | 0.012 |
| ——— | Tobacco recovered | 0.002 |
| Severe etch | *Nicotiana glutinosa* | 0.004 |
| ——— | Tobacco | 0.007 |
| Potato Y | Tobacco | 0.002 |
| Henbane mosaic | Tobacco | 0.002 |

other plants, low-speed centrifugation is adequate to produce a clear fluid, though more host constituents are removed if the leaves are frozen before mincing, or the sap itself is first frozen and thawed, or if some sodium phosphate is added. Heating to 60°C also coagulates many of the sap constituents and provides a useful fractionation, but it cannot be recommended for general use because it affects the activity and properties of many viruses. Some components of sap denature fairly rapidly at room temperature and clarification is easier and more complete if the sap is kept over-night before centrifuging than if it is centrifuged when fresh.

The various amounts of different viruses that have been isolated from clarified sap are shown in TABLE 14. The figures given are only approximations, for, as has already been stated, the concentration of any one virus in the same host may vary considerably depending on the conditions under which the plants are grown, the length of time they have been infected, and the method of inoculation. However, the table does show the great differences in concentration reached by different viruses, and that the concentration reached with one virus may vary considerably with the species of host plant and the particular virus strain. With present techniques it is unlikely that viruses occurring at less than about 1 mg. per l. of sap could be isolated, unless they were unusually stable or had widely different properties from all normal host constituents. Hence until new techniques are developed, there will be no information on the chemical nature of the many viruses that seem to occur in smaller concentrations than this.

For details of the methods used to purify individual viruses, original papers should be consulted. Here only the types of fractionating treatments can be indicated, but methods that require only ordinary laboratory centrifuges are described for purifying tobacco mosaic and tomato bushy stunt viruses because these may be of general interest (BAWDEN and PIRIE, 1943).

**Tobacco Mosaic and Other Anisometric Viruses:** — To each litre of clarified infective sap, 250 g. of ammonium sulphate is added and mixed until dissolved, when the resulting precipitate is separated by centrifugation. The supernatant fluid is discarded and the precipitate is extracted with water until the extracts no longer have an obvious shimmer when stirred. Four or five extractions are usually needed, giving a final volume equal to about one-quarter of the original sap. Dilute NaOH is added in a fine stream with stirring to bring the *p*H to about 6.5, and the process of precipitation with ammonium sulphate and resolution is repeated until the supernatant fluid obtained by centrifuging is only faintly coloured. The precipitate is suspended in 10 times its volume of water and centrifuged to free it from insoluble material. The supernatant fluid is brought to *p*H 3.3 with dilute HCl and the precipitate is removed by centrifugation, washed by mixing with water and centrifuging again, and then transferred with the minimum amount of water to a dialysis sac. After 24 hours' dialysis against running tap water, the *p*H will have risen enough for the virus to have dissolved, but in these conditions most of the remaining normal plant proteins are insoluble. The contents of the sac are diluted to reduce viscosity and then centrifuged. If the precipitate is bulky, it may contain some virus and should be extracted again with water. The further treatments needed can be gauged from the appearance of the fluids. If the preparations are from old plants grown under bright light, they will probably still be brown, when precipitation with ammonium sulphate at *p*H 6.5 and with HCl at *p*H 3.3 should be repeated. Pale or colourless preparations are dialysed against distilled water and then centrifuged to free from insoluble material. When stood undisturbed for a few days at 1°C. such preparations, if they are more concentrated than about 2-3%, should separate into two layers, the lower of which is liquid crystalline. This method recovers most of the virus present in the original sap, causes little inactivation and gives products with reproducible analytical figures. However, the purified preparations are inhomogeneous for they contain particles differing greatly from one another in length. Also, they are probably still slightly contaminated with normal plant constituents, for of the preparations tested by the sensitive anaphylactic test for such contaminants only those that have been incubated with trypsin have failed to react.

When observed by reflected light solutions of tobacco mosaic virus show a satin-like sheen when shaken and observations on this phenomenon provide a convenient check on the presence or absence of virus in different fractions. A more sensitive test is provided by observations on the extent to which fluids show the phenomenon of anisotropy of flow when they are tipped from end to end of half-filled tubes about 0.5 cm. in diameter between crossed polarisers. Examination of anisotropy

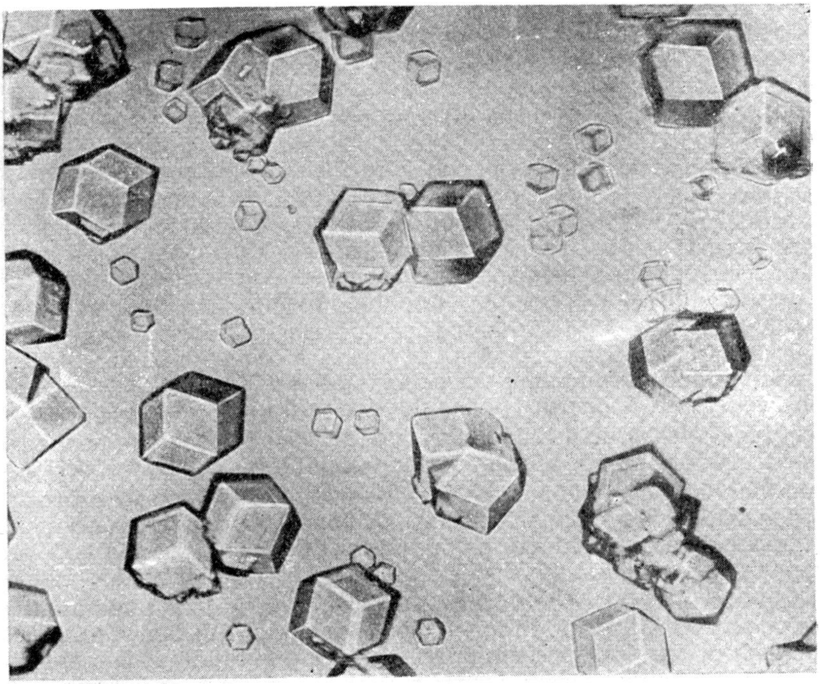

FIG. 32. — Dodecahedral crystals of tomato bushy stunt virus produced by slow precipitation with ammonium sulphate in the cold. The crystals were photographed on the wall of a vessel in which the same preparation was twice crystallised. The larger crystals were produced by the first crystallisation and the smaller by the second. × 250. (BAWDEN, F. C. and PIRIE, N. W., 1938, Brit. J. exp. Path., *19*, 251).

of flow is almost essential when purifying potato virus X, for otherwise large losses of virus are likely to occur on discarded precipitates. This virus can be purified by precipitation methods of the same type as that described for tobacco mosaic virus, but much less reliably. After precipitation with ammonium sulphate, or with acid at *p*H 4.5, the virus tends to become insoluble and the chocolate-coloured precipitates that fail to dissolve in water at *p*H 7 may contain most of the virus. The virus in such precipitates can be brought into solution by incubation with trypsin but this must be done cautiously because prolonged treatment will hydrolyse all the virus. Also, only part of the virus present precipitates at *p*H 4.5, and tests for anisotropy of flow or serological activity are needed on all fractions to check the occurrence of the virus (BAWDEN and PIRIE, 1938*a*; BAWDEN and KLECZKOWSKI, 1948). Sedimentation in the ultracentrifuge provides a safer method of purifying virus X (LORING, 1938*b*), but even with this method much of the sedimented virus may fail to re-dissolve and virus that has remained soluble during the course of several successive precipitations or sedimentations may become insoluble on dialysis.

Potato virus Y and henbane mosaic virus, which occur at much lower concentrations than potato virus X, are even more difficult to purify. They can be precipitated with ammonium sulphate, but they are too labile to withstand *p*H values below 5 and they are easily adsorbed on to precipitates of normal host constituents. Infected tobacco plants also contain a lipoid-like material which has been separated from potato virus Y only by repeated differential ultra-centrifugation (BAWDEN and PIRIE, 1939).

Using a method similar to that described above for tobacco mosaic virus, involving precipitation with acid and salt, RAYCHAUDHURI (1947) claims to have isolated a virus that causes a mosaic of sann-hemp *(Crotalaria juncea)* in the form of "acicular crystals". Insufficient detail is given for any positive identification, but it seems that these may resemble the paracrystalline needles produced by precipitating tobacco mosaic virus.

**Tomato Bushy Stunt and Other Crystalline Viruses:** — Ammonium sulphate at the rate of 280 g. per l. is dissolved in clarified sap from infected tomato leaves and the mixture centrifuged after standing at room temperature for an hour or more. The precipitate is suspended in water, using one-tenth the volume of the original sap, centrifuged and the precipitate extracted again with a smaller volume of water. The supernatant fluids are mixed, brought to *p*H 4.5 with acetic acid and the bulky precipitate is centrifuged off, when it is washed twice with water and the washings are added to the original supernatant fluid. Ammonium sulphate is added to the fluid until there is a slight turbidity. The mixture is now left to stand and within a few hours a precipitate separates that contains most of the virus. After centrifuging, the precipitate is extracted two or three times with acetate buffer at *p*H 4.5, using a total of about 10 ml. per l. of sap. If the preparation is from young plants, it will now be slightly coloured and faintly opalescent, but if it is from old plants it may still be dark brown; it should then be precipitated again with ammonium sulphate and the precipitate taken up in a minimum amount of water and dialysed. As dialysis proceeds a brown precipitate separates, which is removed by centrifugation, and the supernatant fluid is now ready for crystallization. Saturated ammonium sulphate solution is added drop by drop with constant stirring until the fluid becomes permanently turbid at temperatures between 20 and 25°C., which occurs at salt concentrations between 10 and 15% depending on the virus concentration. If the fluid is kept at room temperature a precipitate soon separates, but it largely dissolves again on cooling to 0°C. After 2 or 3 hours at 0°C., the fluids are centrifuged at 0°C., when the supernatant fluid should be clear and colourless. If the fluid is warmed, the virus will separate rapidly as an amorphous precipitate, whereas at

0°C. it separates slowly as rhombic dodecahedral crystals (Fig. 32). Provided the concentration of ammonium sulphate is accurately gauged, crystallization should begin within a few hours and be complete within a week. The crystals dissolve slowly in water, and after decanting the mother liquor some residual contaminants can sometimes be removed by washing the crystals quickly with water. When the crystals have dissolved, the solutions are dialysed and centrifuged to free from any insoluble material.

Similar methods can also be used to obtain crystalline preparations of some of the viruses that cause tobacco necrosis, several of which crystallize from ammonium sulphate solutions at 0°C. (PIRIE *et al.*, 1938; BAWDEN and PIRIE, 1942). The different tobacco necrosis viruses crys-

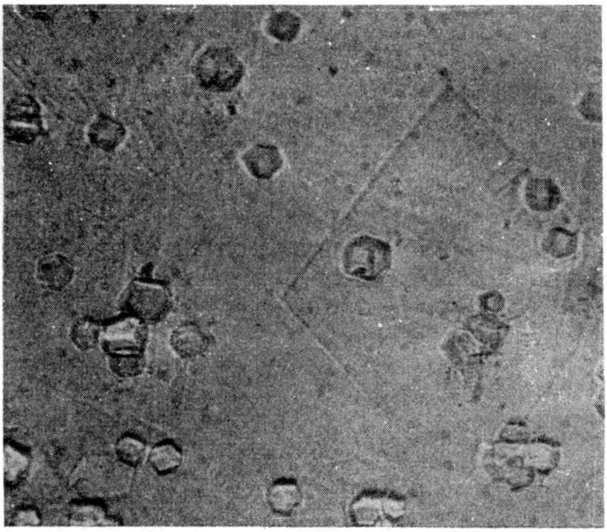

FIG. 33. — Crystals from a mixed culture of tobacco necrosis viruses obtained from the roots of a naturally infected tobacco plant. Flat plates and dodecahedra are found together. × 70.

tallize in distinct forms and when plants are infected simultaneously with more than one, the purified preparations give mixtures of crystals (Figs. 33, 34, 35). One of the tobacco necrosis viruses, the Rothamsted culture (BAWDEN and PIRIE, 1945*a*), resembles the others in being more soluble in ammonium sulphate solution at 0°C. than at 20°C., but it does not crystallize from the cold salt solution. It is also largely inactivated by exposure to salts and it is best isolated by ultra-centrifugation, which causes less loss of infectivity than other fractionating treatments.

Precipitation with ammonium sulphate after preliminary fractionations with alcohol or high-speed centrifugation have also given crystalline preparations of southern bean mosaic virus (PRICE, 1945, 1946), turnip yellow mosaic virus (MARKHAM and SMITH, 1946, 1949) and squash mosaic virus (TAKAHASHI, 1948). Turnip yellow mosaic virus seems to be the simplest to obtain in crystalline form. After precipitating the virus from clarified sap of infected Chinese cabbage with 30% alcohol and redissolving in 0.5% salt solution, the virus crystallizes on the addition of ammonium sulphate. Alternatively, it can be crystallized by adjusting solutions containing 20% alcohol to *p*H 3.7.

**High Speed Centrifugation: —** Virus particles readily sediment from neutral solution when centrifuged at high speed, and from 1 to 3 hours in centrifugal fields of from 40,000 to 80,000 g. is adequate to sediment completely any of the plant viruses yet studied. High-speed centrifugation provides a valuable method of fractionating and concentrating viruses and with the development of air-driven machines capable of holding 100 ml. of fluid it has been increasingly used for the isolation of viruses directly from clarified sap. The sediment obtained by high-speed centrifugation is suspended in water or phosphate buffer and centrifuged at low speed to get rid of insoluble matter, and the sequence of alternate high- and low-speed centrifugation is repeated two or three times, the pellets from the high-speed centrifugation being suspended in decreasing volumes each time. Because of the infrequency with which stable substances with large particles occur in sap from many plant species, the method alone gives good preparations of some viruses, but better ones are usually obtained by combining it with other fractionating treatments. Also, except for viruses that occur in sap at unusually high concentrations, the method is laborious for preparing large quantities and a preliminary method of concentration is desirable. Precipitation with ammonium sulphate provides one such method but if exposure to strong salt solutions is undesirable, as with the Rothamsted culture of tobacco necrosis, the virus can be considerably concentrated by removing most of the water as ice. This is done conveniently by freezing the sap until about 95% has changed to ice and then centrifuging in a basket centrifuge, or in tubes with perforations, so that ice and fluid are separated. The virus is concentrated in the first-formed fluid and about four-fifths of the total is obtained in the first 100 ml. that runs from 1 l. of frozen sap (PIRIE, 1948). LAUFFER and PRICE (1947) have concentrated southern bean mosaic virus by dialysing infective sap against egg white, but this method is not likely to be practical in most countries.

Using only differential centrifugation, preparations of tobacco mosaic virus (LAUFFER and STANLEY, 1938) and potato virus X (LORING, 1938*b*) have been made that form liquid crystalline layers, which suggests that they contain little material other than virus. With tobacco mosaic virus the analytical composition of preparations made by ultracentrifugation also closely resembles that of preparations made by precipitation methods. With potato virus X and tomato bushy stunt virus, on the other hand, preparations made by centrifugation alone usually contain more carbohydrate than those subjected to other fractionations. Similarly, chromoprotein and brown pigments are often difficult to separate from the viruses by sedimentation, whereas colourless preparations can readily be made by precipitation with acid or salts. Preparations of southern bean mosaic virus, especially those from plants grown under high light intensities, often contain a coloured contaminant, which is not separated either by sedimentation, precipitation or crystallization, but is removed by electrophoresis (LAUFFER and PRICE, 1947).

BAWDEN and PIRIE (1937*a*, 1938*a*) found that the filterability and infectivity of tobacco mosaic virus and potato virus X were reduced by purification and suggested that the treatments used were causing the virus particles to aggregate linearly. LORING, LAUFFER and STANLEY (1938) confirmed this with tobacco mosaic virus, but they claimed that aggregation could be avoided by using ultra-centrifugation as the sole method of purification. Although sedimentation produces less striking changes than precipitation with acid and salts, and much less than incubation with trypsin or heating, it does have some effect on the physical state of the virus particles. WYCKOFF (1937) found that preparations of tobacco mosaic virus that previously seemed homogeneous in the ultracentrifuge became inhomogeneous when exposed to phosphate buffer at *p*H 7 and LORING, LAUFFER and STANLEY (1938) allowed that there were detectable changes when the virus was sedimented four times. BAWDEN and PIRIE (1945*b*) have shown that sap contains substances

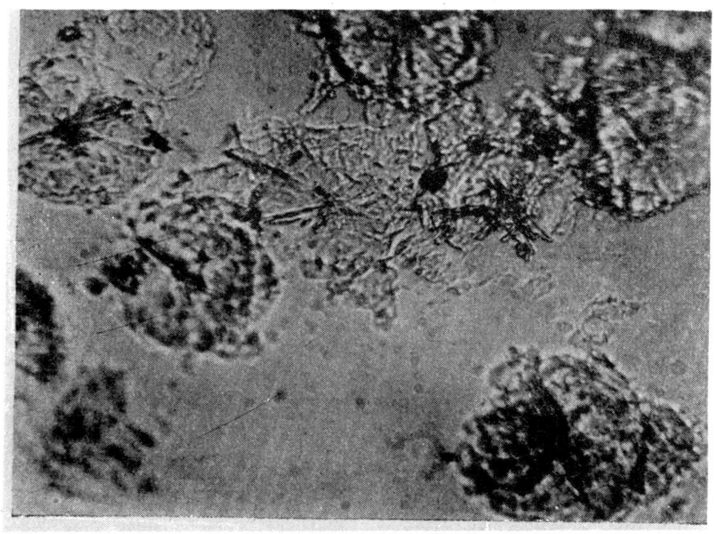

**FIG. 34. — Crystals of a tobacco necrosis virus in the form of round laminae. An alternative crystal form of this virus is shown in FIG. 35). × 200.**

that aggregate the virus particles and that infective extracts as usually obtained already contain particles of widely different sizes. By centrifuging sap at various speeds they obtained fractions containing particles that differed widely in their average lengths and infectivities. As all or most of these different particles would be sedimented by the speeds normally used to prepare the virus by high-speed centrifugation, it is obvious that the apparent centrifugal homogeneity of such products is no evidence that they contain only one kind of particle and that aggregation has been avoided. Electron micrograms of centrifugally isolated preparations of tobacco mosaic virus also provide abundant evidence for the presence of particles of widely different sizes. Similarly, with potato virus X, although sedimentation may cause less change than precipitation with acid or salts, there is no reason to think that it gives homogeneous preparations; indeed those made solely by ultra-centrifugation gave diffuse boundaries when examined in a qualitative centrifuge and contained particles giving different sedimentation constants (LORING and WYCKOFF, 1937). LORING (1938*b*) attributed the greater infectivity of centrifugally prepared virus X to the fact that precipitation with acid and salts was causing slight chemical changes in the virus particles. An alternative explanation, however, is that precipitation methods cause more aggregation, perhaps simply because they produce a chemically purer product. The available evidence on tobacco mosaic virus and virus X suggests that any process that concentrates them or that removes contaminants from them is likely to cause aggregation (BERNAL and FANKUCHEN, 1941; PIRIE, 1945). Hence aggregation may be inseparable from a high degree of impurity, and claims to have isolated homogeneous, unaggregated particles may be equivalent to claiming incomplete purification.

In the early use of ultra-centrifuges for isolating viruses there was a tendency to regard sedimentation as a specific property of the viruses, but this uncritical attitude is fortunately now less common. It may have arisen from the statement of WYCKOFF, BISCOE and STANLEY (1937) that healthy plants contain no demonstrable amounts of molecules with weights greater than 30,000 and that there was no sedimentation from a solution of normal plant protein maintained in a field of 180,000g. for 90 minutes. However, this is far from generally true, and the absence of large proteins in these preparations can probably be explained by the fact that the leaves were frozen over-night at —14°C., a treatment that would denature many of the normal plant proteins. Since then large proteins have been found in many plants, and the difficulty of removing coloured contaminants by centrifugation has also shown the limitations of the method. The ultra-centrifuge provides a method of concentrating and fractionating that is likely to be applicable to most viruses and it is particularly valuable with those that cannot withstand treatment with customary protein precipitants, but too much faith should not be placed on its ability to produce a pure product. When the final preparation is coloured, as with tobacco ringspot (STANLEY, 1939), tomato bushy stunt (STANLEY, 1940) and alfalfa mosaic (ROSS, 1941), there are obvious grounds for suspecting the purity. But when products are colourless, evidence of more than centrifugal homogeneity is needed before purity is assumed, because components that have sedimented together two or three times in the quantitative centrifuge would also be expected to behave alike when examined in the qualitative centrifuge.

**Activity of Purified Preparations:** — After removing salts by dialysis, the solid content of purified preparations can be found by drying a sample, or the protein content can be deduced from micro-Kjeldahl, and the activity of the product can then be determined on a weight basis. Most workers now do this in two ways, by local-lesion counts at various dilutions and by finding the smallest weight that will give a visible precipitate with antiserum. Some representative results with preparations of different viruses are recorded in TABLE 15. It will be seen that the weight required to produce infection with different viruses ranges from 0.01 to 0.0000001 mg. but how far this represents differences in the actual infectivities of the viruses is uncertain. Purification causes much loss of infectivity with some viruses and relatively little with others; also the minimum weight required to cause infection, and the numbers of lesions formed by a given weight, vary widely with different batches of host plants and with the method of inoculation. Using highly susceptible plants and rubbing with an abrasive in the inoculum, infections can sometimes be produced with as little as $10^{-11}$ or $10^{-12}$ g. of tobacco mosaic virus. Even these small weights are equivalent to a million or more particles and it is a moot point whether this amount is necessary because most particles are wasted and never encounter suitable entry points when rubbed over the leaves or because most of the particles are not infective.

Viruses with elongated particles, such as potato X and strains of tobacco mosaic virus, give higher precipitin titres than those with spherical particles. There is also more variation between the precipitin titres of different preparations of viruses with elongated particles; this in part

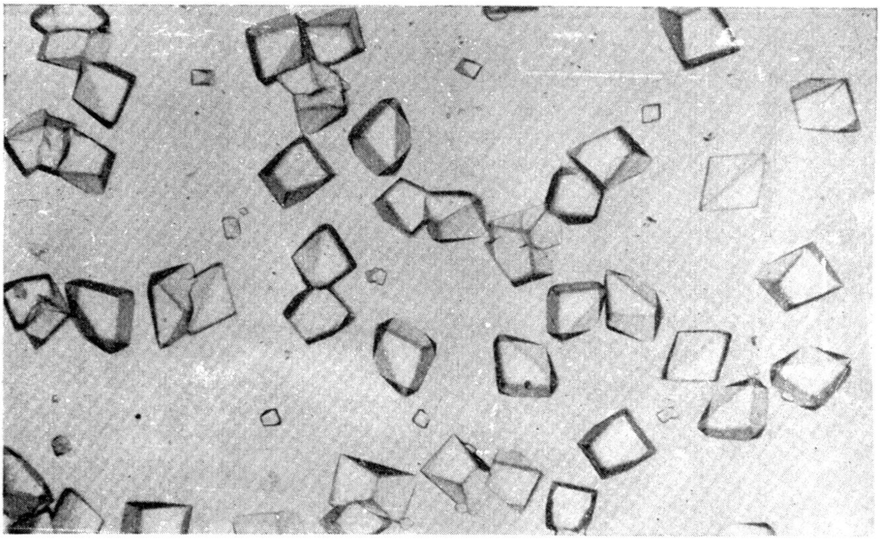

FIG. 35. — Crystals in the form of bi-pyramids produced by the same virus as shown in FIG. 34. A further form of crystallisation, dodecahedra, is shown in FIG. 33. The bi-pyramids are birefringent. × 70.

reflects the varying degrees to which the purification procedures have caused the virus particles to aggregate linearly, but as titres obtained with aucuba mosaic are consistently higher than those obtained with tobacco mosaic virus prepared in the same manner, there also seem to be differences between virus strains that affect their precipitin titres.

**Proteins from Healthy Plants:** — No proteins similar to the purified viruses have yet been isolated from uninfected plants and the methods of purification described above for tobacco mosaic and tomato bushy stunt viruses isolate nothing from the sap of healthy tobacco and tomato plants. This is, of course, presumptive evidence that the proteins from infected plants are the viruses and that these differ qualitatively from normal plant constituents, but it is far from conclusive. Only a few species of plants have yet received attention, so that others may contain

**TABLE 15. — *Activity of purified virus preparations:—***

| *Virus* | *Serological titre* | *Infectivity at various dilutions. Average number of lesions per leaf.* $10^{-4}$ | $10^{-5}$ | $10^{-6}$ | $10^{-7}$ | $10^{-8}$ | $10^{-9}$ | *Host plant* |
|---|---|---|---|---|---|---|---|---|
| Tobacco mosaic | 1/8,000,000 | 416 | 240 | 56 | 6 | 2 | 0.5 | *N. glutinosa* |
| | 1/2,000,000 | 242 | 98 | 29 | 5 | 2.1 | 0.8 | *P. vulgaris* |
| Aucuba mosaic | 1/10,000,000 | 210 | 140 | 34 | 6 | 0.4 | 0 | *N. glutinosa* |
| | 1/8,000,000 | 290 | 82 | 39 | 11 | 3 | 1 | *N. glutinosa* |
| Enation mosaic | 1/6,000,000 | 352 | 152 | 26 | 2.6 | 1 | 0.3 | *N. glutinosa* |
| Potato X | 1/6,000,000 | 660 | 220 | 4 | 7 | 1 | 0 | *N. Tabacum* |
| | 1/6,000,000 | 196 | 38 | 9 | 2 | 2 | 0.3 | *N. Tabacum* |
| Tomato bushy stunt | 1/600,000 | 105 | 45 | 10 | 3 | 0.5 | 0 | *N. glutinosa* |
| | 1/600,000 | 86 | 30 | 15 | 1.5 | 0.2 | — | *N. glutinosa* |
| Tobacco necrosis, Potato culture | 1/400,000 | — | 350 | 115 | 23 | 3 | — | *P. vulgaris* |
| ———, Rothamsted culture | 1/300,000 | — | 119 | 36 | 9 | 1 | — | *P. vulgaris* |
| Turnip yellow mosaic | 1/500,000 | — | — | — | *8/8 | 3/5 | 2/8 | *B. chinensis* |
| Southern bean mosaic | 1/500,000 | 170 | 57 | 6.2 | 1.4 | 1.1 | — | *P. vulgaris* |
| Tobacco ringspot | 1/1,000,000 | — | 29.8 | 13.8 | 0.2 | — | — | *V. sinensis* |
| Alfalfa mosaic | — | 30 | 2 | — | — | — | — | *P. vulgaris* |

*No. of plants infected/No. inoculated.
The serological titres are expressed at the highest dilution, in grams per ml., at which there was a visible precipitate when 1 ml. of the virus solutions was incubated with 1 ml. of diluted antiserum. The dilutions in the infectivity tests are given as grams of protein per ml.

such proteins. Also, with the species that have been examined, complete absence of such proteins cannot be proved, for negative results are never unequivocal; the most that can be concluded is that, if they occur, they do so in much smaller quantities than in infected plants. If on present evidence we conclude that healthy plants contain no proteins similar to the viruses, we must also conclude that no such proteins occur in plants infected with many different viruses, for the purification procedures so far tried also isolate nothing from them. It is worth stressing here that, for purification to be possible, a virus, or a normal plant protein, must be reasonably stable, occur at more than a minimum concentration, and differ in some specific manner from other proteins present. While the failure to purify most viruses is attributed to the fact that they are too unstable or occur in too small quantities, it is obviously premature to attribute any significance to the fact that healthy plants apparently contain nothing similar to virus particles. The failure of the purification procedures to isolate anything from healthy plants is often advanced as evidence that the purified virus preparations are free from normal host constituents, but again this is of doubtful value. Normal host constituents, which alone are unaffected by precipitation or sedimentation, might well be carried along with the virus particles when these are removed from solution.

Apparently pure preparations of stable proteins have been made from seeds and grains, but none has yet been made from healthy leaves. However, only a few species have been examined and most of the work done on them has been for other purposes than the isolation of single proteins. It is clear that "normal leaf-protein" is a mixture of many different proteins, but different species behave so differently when treated in the same ways that no general statements about the properties of healthy leaves are possible. This variation is shown strikingly by comparing the behaviour of tobacco and strawberry leaves when minced. From 100 g. of minced tobacco leaves, about 70 ml. of sap can be expressed which will contain about 2 mg. of nitrogen per ml., although the precise amount of sap and nitrogen will vary with the age of the plants and the conditions under which they are raised. From minced strawberry leaves, on the other hand, little or no sap can be expressed, and aqueous extracts contain no protein and little except sugars and tannin (BAWDEN and KLECZKOWSKI, 1945).

CHIBNALL's (1923, 1924) work showed that leaf-proteins need gentle handling if denaturation is to be avoided and this has been emphasised by later results. Indeed, the most striking differences between the viruses that have been purified and the normal leaf-proteins are the ability of the viruses to withstand many treatments which render the other proteins insoluble and the much greater antigenicity of the viruses. Failure to appreciate the greater instability of normal leaf-proteins is probably responsible for some of the earlier claims that tobacco plants suffering from mosaic contained more protein than healthy plants (STANLEY, 1937*c*, 1937*e*) and that healthy plants contained no protein that sedimented in the high-speed centrifuge (WYCKOFF, BISCOE and STANLEY, 1937). These claims were made largely from a consideration of the behaviour of clarified sap and much of the normal protein had been rendered insoluble by the treatments used to facilitate clarification. In leaves fully infected with tobacco mosaic virus, the virus itself may amount to as much as 10% of the total dry matter and to one-third of the nitrogen (BAWDEN and PIRIE, 1946). There is no comparable increase in the total nitrogen, for the slight increase in percentage of nitrogen produced by infection (MARTIN, BALLS and MCKINNEY, 1938, 1939; TAKAHASHI, 1941) is more likely to reflect a reduction in other leaf components than an increase in nitrogen. Hence it seems that the virus is synthesized mainly or exclusively at the expense of some normal leaf proteins. There is no conclusive evidence on the type of normal protein that is replaced by the virus, but MARTIN, BALLS and MCKINNEY (1939) and TAKAHASHI (1941) found that increase of virus was accompanied by a decrease in normal protein and WOODS and DU BUY (1941) state that in leaves deficient in nitrogen virus is produced instead of chromo-protein. Although the total leaf protein is unaltered by infection, the protein content of clarified sap is increased, the increase depending on the method employed for clarification. If the leaves or sap are frozen, if phosphate is added or the sap is heated to 60°C., tobacco mosaic virus may account for as much as four-fifths of the total protein in clarified sap. Some strains of potato virus X also amount almost to this proportion, but other viruses form only a minor part of the protein content.

Like the viruses, less than half the total normal proteins of the leaf occur in the sap expressed from minced leaves, and further treatments such as fine milling (CROOK, 1946) or incubating the residues with snail enzymes are needed to extract the remainder. When sap from healthy tobacco or tomato leaves is centrifuged at 3,000 r.p.m., a green precipitate separates leaving an opalescent and darkly-coloured supernatant. If this is heated to 60°C., acidified below *p*H 5, frozen and thawed, or is allowed to stand at room temperature overnight, bulky precipitates are produced, which are readily removed by centrifugation at 3,000 r.p.m. and leave a clear brown supernatant. The addition of sodium phosphate produces a smaller precipitate. All these precipitates consist of material the greater part of which has been irreversibly coagulated.

When centrifuged sap is quarter saturated with ammonium sulphate a precipitate separates which dissolves readily in water at *p*H 7 to give a dark brown viscous solution. These proteins can be separated into fractions that differ in their precipitability with ammonium sulphate, but which otherwise have similar properties. They are moderately stable at neutrality and when kept cold, but they are denatured in a few days at room temperature, when the *p*H is reduced below 5, when they are frozen and thawed, heated to 55°C. or higher, and when they are exposed to 35% alcohol in the presence of salts. They are also readily hydrolysed by incubation with trypsin. All the preparations obtained by fractional precipitation with ammonium sulphate contain proteins with large particles, and when neutral solutions are ultra-centrifuged they deposit isotropic sediments which re-dissolve when taken up in water. Except for their large sizes, these normal proteins have little in common with the viruses yet isolated. They are much more unstable, they contain little or no nucleoprotein, and they are almost inactive as producers of precipitating antibodies when injected into animals (BAWDEN and PIRIE, 1938*c*). The unstable substances that sediment from cucumber, *Vigna sinensis* and *Nicotiana glutinosa* (PRICE and WYCKOFF, 1938, 1939) and from peas and beans (LORING, OSBORN and WYCKOFF, 1938) are presumably proteins similar to those that occur in tobacco, though those from peas and beans are stated to contain a pentose and so may be nucleoproteins. The proteins from *V. sinensis* gave a sedimentation constant of 51 $\times$ $10^{-13}$ and those from all the other species 75 $\times$ $10^{-13}$. French beans contain a protein that accompanies tobacco necrosis viruses when these are ultra-centrifuged or precipitated with ammonium sulphate, but it is insoluble at *p*H 4.5 and can be separated from the viruses by centrifuging preparations when acid (BAWDEN and PIRIE, 1945*a*).

Unstable proteins form the greater part of the protein obtained in sap from healthy tobacco and tomato plants. When they have been removed by appropriate treatments, little remains that is precipitable with ammonium sulphate. A small precipitate is obtained at one-quarter saturation and more material separates at two-thirds saturation. The second precipitate contains both phosphorus and carbohydrate and seems to be largely nucleoprotein, which denatures and releases a nucleic acid when heated to 90°C. Although resistant to heat, it does not withstand repeated treatment with acid or strong salt solutions and usually denatures after two or three precipitations. It is susceptible to proteolytic enzymes that do not hydrolyse tobacco mosaic virus.

TAKAHASHI and RAWLINS (1933*a*) first noticed that clarified sap from frozen leaves of tobacco mosaic plants showed the phenomenon of anisotropy of flow (stream double refraction), and as sap from similarly treated healthy leaves did not, they suggested that the virus particles were rod-shaped. That the virus particles are responsible for this phenomenon has been repeatedly confirmed since and none of the proteins that has been isolated from healthy plants shows it (BAWDEN and PIRIE, 1937*a*; LAUFFER and STANLEY, 1938). Nevertheless, some normal leaf components may be anisometric, for TAKAHASHI and RAWLINS (1933*b*) detected anisotropy of flow in clarified sap from unfrozen healthy leaves and in uncentrifuged sap from frozen leaves, but the materials responsible are so unstable that they are destroyed by customary methods of clarification. If there are any stable ones, they must be dilute. STANLEY (1937*c*) recovered tobacco mosaic virus from mixtures containing one part of purified virus to 100,000 parts of normal leaf extract, so that if any similar materials occur in healthy sap they can amount only to a very minor component. However, the great stability of tobacco mosaic virus, and of those that have been isolated in crystalline forms, is not characteristic of all viruses, and the properties of normal leaf proteins may well approximate more closely to those of the many viruses for which as yet no methods of purification have been devised.

*Chapter 10*

# CHEMICAL AND PHYSICAL PROPERTIES OF PURIFIED VIRUS PREPARATIONS

**Elementary Composition:** — Elementary analyses of purified preparations of five strains of tobacco mosaic virus and of eight other viruses are given in TABLE 16. Different figures for some viruses have been reported from time to time, the main disagreements being in the contents of phosphorus and carbohydrate. Tobacco mosaic virus was first stated to contain neither of these constituents and was described as a globulin (STANLEY, 1935*a*, 1936*a*) as was tomato aucuba mosaic virus (STANLEY, 1937*d*). Another strain of tobacco mosaic virus, green mosaic, was also considered to be free from phosphorus (THORNBERRY and MCKINNEY, 1939), and southern bean mosaic virus was first described as containing no carbohydrate (PRICE, 1946). However, none of these claims has been confirmed, and phosphorus and carbohydrate, which were first shown to occur in the form of a nucleic acid in strains of tobacco mosaic virus (BAWDEN, PIRIE, BERNAL and FANKUCHEN, 1936), are now accepted as essential components of all the plant viruses that have been critically studied. In spite of the wide differences between their general properties, these have now all been found to consist largely, and probably exclusively, of nucleoprotein. Too few viruses have yet been purified for any generalisation to be safe, but it can reasonably be concluded from the chemical uniformity of these that many other viruses are also nucleoproteins.

Different viruses contain similar amounts of carbon, hydrogen and nitrogen, and in their content of these elements they also resemble other proteins so that analytical figures for them are of little value in assessing the purity of a preparation. The figures for phosphorus and carbohydrate vary with different viruses, and the ratio between the two is a useful indication of purity. The most common contaminants of incompletely purified preparations are rich in carbohydrates but contain little phosphorus, and if preparations contain more than five times as much carbohydrate as phosphorus, they can usually be further fractionated. A high carbohydrate content seems to be common with preparations of some viruses made solely by differential centrifugation, and repeated fractionations are needed to obtain values as small as those shown in TABLE 16 for potato virus X and tomato bushy stunt virus. Whether this extra carbohydrate was originally attached to the virus particles, from which it is removed by some slight chemical changes, as suggested by LORING (1938*b*) for virus X, or is a contaminant adsorbed onto the particles, is unknown. The same uncertainty applies to many other materials that are discarded during the course of purification, and it must be stressed here that the products whose compositions are being described may be less complex than the original virus particles and be merely the chemical minima that are capable of initiating infection.

From denatured tobacco mosaic virus enough nucleic acid can be obtained to account for so much of the phosphorus and carbohydrate that there is no need to assume the presence of any other component containing these two substances (BAWDEN and PIRIE, 1937*a*; LORING, 1938*a*, 1939*a*, 1939*b*). This is probably also true of the other viruses, but with them the evidence is less conclusive. However, it is clear that different viruses contain different amounts of nucleic acid and the ultraviolet absorption spectra show that these amounts are roughly proportional to the phosphorus contents. Solutions of the purified viruses are colourless and there is little absorption of visible light, but in the ultra-violet range absorption increases to a maximum at 260 mμ, followed by a minimum at 250 mμ and then general absorption at shorter wavelengths. Absorption in the region of 260 mμ is characteristic of the purines contained in the nucleic acid and of other substances that contain heterocyclic compounds, and the absorption by different viruses in this region is correlated with their phosphorus and carbohydrate contents (Figs. 36*a* and *b*). Viruses such as tobacco mosaic and potato X, which contain about 0.5% phosphorus, absorb less strongly at 260 mμ than tomato bushy stunt and tobacco necrosis viruses, which contain about 1.5% phosphorus, and these in turn absorb less strongly than tobacco ringspot virus which has 4% phosphorus. Although quantitative isolations of nucleic acid have not been made from most viruses, sufficient has been done to show that whereas about 6% of the weight of tobacco mosaic virus is nucleic acid, the figures for tomato bushy stunt (BAWDEN and PIRIE, 1938*b*; STANLEY, 1940) and tobacco ringspot (STANLEY, 1939) are about 15% and 40% respectively. If these virus preparations contain any constituents other than nucleic acid and protein, they must be minor ones, and it is their chemical simplicity that separates them most sharply from organisms. Protein and nucleic acid are, of course, important constituents of all organisms, but even the simplest bacteria contain a wide range of other substances, both diffusible and indiffusible. By contrast, the virus particles in their simplest forms seem to consist exclusively of nucleoprotein, and as such resemble cellular constituents rather than the cells themselves. It is, perhaps, in this manner that the purified viruses should be regarded, because their chemical simplicity may be partly illusory, a transient stage produced by purification and lost again when the particles infect susceptible cells, from which they may need to borrow other constituents to equip themselves as functioning systems.

TABLE 16. — *Analyses of purified virus preparations:—*

| *Virus* | *Carbon* | *Hydrogen* | *Nitrogen* | *Sulphur* | *Phosphorus* | *Ash* | *Carbohydrate* |
|---|---|---|---|---|---|---|---|
| Tobacco mosaic | 50% | 7% | 16.6% | 0.2% | 0.53% | 1.5% | 2.5% |
| Tomato aucuba mosaic | 50% | 7% | 16.7% | 0.2% | 0.52% | 1.5% | 2.5% |
| Tomato enation mosaic | 50% | 7% | 16.7% | 0.3% | 0.53% | 1.0% | 2.5% |
| Rib-grass mosaic | 50% | 7% | 15.7% | 0.6% | 0.54% | 2.27% | 2.35% |
| Cucumber 3 and 4 | 50% | 7% | 15.4% | 0.0% | 0.54% | 2.28% | 2.3% |
| Potato X | 49% | 7% | 16.4% | — | 0.45% | 2.2% | 2.7% |
| Potato Y | 50% | — | 16.0% | — | 0.4% | — | 3.0% |
| Alfalfa mosaic | 53% | 6.7% | 16.2% | 0.6% | 1.4% | — | 9.0% |
| Tomato bushy stunt | 49% | 7.5% | 16.1% | 0.6% | 1.5% | 3.0% | 6.0% |
| Tobacco necrosis | | | | | | | |
| Potato virus | 45% | 6.5% | 16.3% | 1.6% | 1.65% | 7.0% | 6.5% |
| Rothamsted virus | 48% | — | 16.2% | — | 1.7% | — | 7.0% |
| Tobacco ringspot | 51% | 7.6% | 14.6% | 0.4% | 4.1% | — | 18.0% |
| Southern bean mosaic | 46% | 6.5% | 17.0% | 1.3% | 1.9% | 5.7% | — |

TABLE 16 shows that the different strains of tobacco mosaic virus differ in composition very little, the strains that resemble one another closely giving identical analytical figures and the more remotely related strains, cucumber 4 and rib-grass virus, differing from the others only in their content of sulphur (KNIGHT and STANLEY, 1941; KNIGHT, 1943). Undenatured tobacco mosaic virus contains no free sulphydryl groups, but after denaturation these are produced in sufficient amount to account for the total sulphur (LAUFFER and STANLEY, 1939). However, there is some uncertainty about the exact nature of the sulphur, for ROSS and

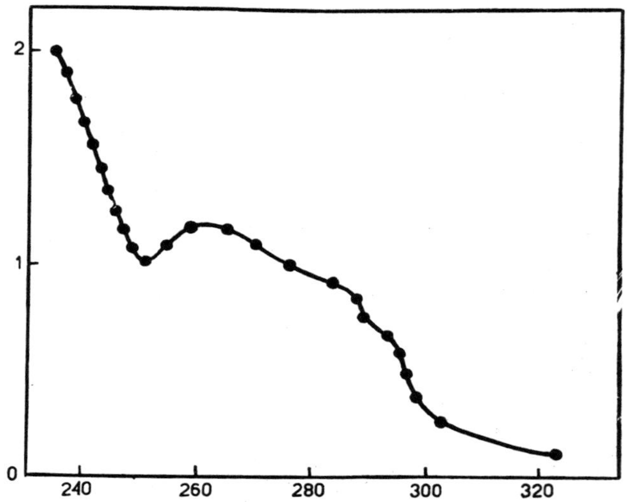


FIG. 36*a* and *b*. — Ultra-violet absorption spectra of purified virus preparations. In each of the curves the abscissa is the wavelength of light in m$\mu$ and the ordinate is the $\log_{10}$ of the amount of light of that wavelength that is incident on a 2 cm. deep layer of solution to the amount of light that is transmitted through the solution. — *a:* 0.02% solution of tobacco mosaic virus (BAWDEN, F. C. and PIRIE, N. W., 1937, Proc. Roy. Soc. B., *123*, 274). — *b:* 0.012% solution of tomato bushy stunt virus (BAWDEN, F. C. and PIRIE, N. W., 1938, Brit. J. exp. Path., *19*, 2151).

STANLEY (1939) stated that the virus contained both methionine and sulphate sulphur, whereas later workers have found no methionine and ROSS (1940) considered that all the sulphur was accounted for by cysteine. The extra sulphur in the rib-grass strain of tobacco mosaic virus seems to be accounted for by the presence of methionine (KNIGHT, 1943). Various values up to 0.8% have been reported for the sulphur content of purified preparations of cucumber viruses 3 and 4 (BAWDEN and PIRIE, 1937*b*; KNIGHT and STANLEY, 1941), but KNIGHT (1949) states that the viruses contain no sulphur and that the variable results are caused by an unidentified constituent that renders sulphur analyses unreliable. Except for these two and tobacco ringspot virus (STANLEY, 1939), all the purified viruses that have been tested give a positive nitroprusside test for sulphydryl groups after denaturation.

The ash remaining when virus preparations are burnt is at first sight unusually large. The figures given in TABLE 16 are for residues obtained from neutral preparations, and as the viruses all have acid isoelectric points part will be derived from the NaOH used for neutralisation. But even after prolonged dialysis or electro-dialysis at the isoelectric points, the viruses still leave a high percentage of ash. Only a small part of this is probably the actual metal content of the protein, for most of it will be derived from the phosphorus of the nucleic acid which will be converted to metaphosphate by burning (PIRIE and others, 1938). The ash content of most viruses has not been examined, but that from tomato bushy stunt has been found to contain mainly calcium and sodium with traces of twelve other metals (PIRIE, 1945).

Turnip yellow mosaic virus has not been included in TABLE 16, because purified preparations prepared by crystallization contain two

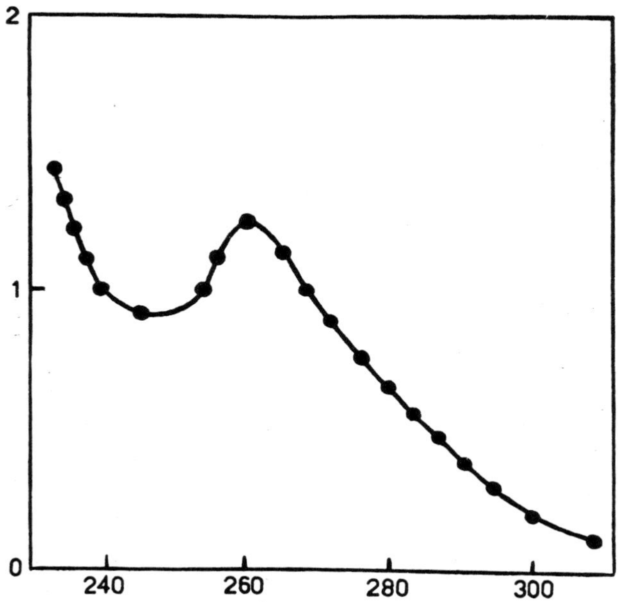


FIG. 36*b* (*see* p. 186).

kinds of particles with widely different chemical compositions. These particles can be partially separated by differential centrifugation, and 80% of preparations as normally made seem to consist of nucleoprotein, with a nucleic acid content of about 22%, whereas the remainder are composed of protein only and are free from nucleic acid. Only the particles containing nucleic acid seem to be infective (MARKHAM and SMITH, 1949).

**Isolation and Properties of Nucleic Acid:** — The nucleic acid in all the plant viruses so far examined is of the same type, but insufficient work has yet been done to know whether it is merely similar or identical. It resembles yeast nucleic acid and differs from thymus nucleic acid, which is characteristic of nuclei, in that the carbohydrate present is a pentose and not a desoxypentose. It is generally assumed to be ribose, but this has been demonstrated only with nucleic acid from tobacco mosaic virus (SCHWERDT and LORING, 1947). The ease with which nucleic acid can be separated from the protein varies considerably with different viruses. From tobacco mosaic virus, nucleic acid is readily prepared by heat denaturation. When neutral solutions of the virus are heated for a few minutes at 90°C. or higher, they become slightly acid and most of the protein coagulates. The addition of a little salt precipitates more protein and the supernatant fluid now contains little protein and most of the nucleic acid, which precipitates in characteristic resinous masses when acid is added. Nucleic acid can also be obtained by treating tobacco mosaic virus with NaOH at 0°C., or with glacial acetic acid or 30% pyridine. When virus solutions are mixed with five times their volume of glacial acetic acid, the nucleic acid precipitates while the protein moiety remains in solution. With pyridine, the virus solutions become clear, and after the protein has been precipitated with ammonium sulphate, the addition of acid precipitates the nucleic acid (BAWDEN and PIRIE, 1937*a*). About 6% of the weight of the virus can be isolated as nucleic acid containing about 33% carbon, 14% nitrogen, 8% phosphorus, and 30% carbohydrate, figures similar to those given by yeast and tritico nucleic acid, which also have the same solubility as the virus nucleic acid. LORING (1938*a*, 1939*a*, 1939*b*) isolated guanine, adenine and cytosine from the virus nucleic acid and found them to resemble those from yeast nucleic acid, but preliminary work suggested that the uridylic acid isolated from the virus was isomeric rather than identical with that in yeast. However, as a result of a more thorough examination of the pyrimidine nucleotides, uridylic and cytidylic acids, and the purine nucleotide, guanylic acid, SCHWERDT and LORING (1947) have concluded that the three from the virus are identical with their analogues from yeast. This establishes the fact that the virus contains a ribonucleic acid, but, as SCHWERDT and LORING emphasise, it is insufficient evidence to conclude that yeast and the virus nucleic acids are identical as they occur in their respective nucleoproteins.

The nucleic acid when freshly prepared from the virus has larger particles than that from yeast, but this may merely reflect the more gentle methods that can be used with the former, because nucleic acid readily depolymerises. COHEN and STANLEY (1942*b*) found that nucleic acid from heat-denatured tobacco mosaic virus is birefringent and has highly anisometric particles with an average weight equivalent to a molecular weight of 300,000, but these decompose spontaneously to form particles about one-fifth this weight and cold alkali reduces them still further to a molecular weight of about 15,000. They point out that the large polymerised particles are too long to lie across the width of the virus particles, and they suggest that the nucleic acid is bound along the length of the virus at or near the surface, but an alternative interpretation is that the nucleic acid particles are coiled instead of extended when in the virus. Also, the possibility has not been excluded that the treatment used to obtain the birefringent nucleic acid may itself have produced polymers, especially as heating at *p*H 5 to 6 would cause the virus particles to aggregate linearly before they were denatured (BAWDEN and PIRIE, 1945*b*). However, there is evidence from other sources that the nucleic acid may be attached at the surface. The complete and ready detachment of the nucleic acid from the protein by most denaturing treatments would be explained by such a union, and the manner in which solutions of tobacco mosaic virus, orientated by flowing in quartz capillary tubes, absorb polarised ultra-violet light is also stated to suggest that the nucleic acid lies on the outside of the rod-shaped particles (BUTENANDT *et al.*, 1942).

The enzyme ribonuclease hydrolyses the isolated nucleic acid from tobacco mosaic virus, but does not attack the undenatured virus, though it combines with it (LORING, 1942; KLECZKOWSKI, 1946). SCHRAMM (1941) reported that the nucleic acid could be separated from the virus by a nucleophosphatase derived from the intestinal mucosa of calves, but this was not confirmed by COHEN and STANLEY (1942*a*). PFANKUCH (1940) suggested that the differences between strains of tobacco mosaic virus might lie in their nucleic acids, for he found that the protein fractions from two strains gave a single electrophoretic boundary whereas mixtures of the two intact strains gave a double boundary. The nucleic acids of different strains may differ, but further work is needed to establish the fact. PFANKUCH's reason for suggesting differences rests on the assumption that electrophoresis necessarily distinguishes between different proteins. KNIGHT and LAUFFER (1942) showed this to be untrue, by demonstrating that mixtures of the proteins from ordinary tobacco mosaic and the rib-grass strain, which are known to differ in constitution, migrate as a single component.

Nucleic acid can also be obtained from potato virus X after denaturation with 80% acetic acid or heating, but it separates from the protein less completely than with tobacco mosaic virus (BAWDEN and PIRIE, 1938*a*; LORING, 1938*b*). Some of the denatured protein produced by heat does not precipitate until the *p*H is brought to 4, and this carries some nucleic acid with it. A higher yield is obtained by adding 4 volumes of acetic acid to the precipitate obtained by centrifuging virus preparations at *p*H 4.5, when most of the proteins are dissolved leaving a residue of crude nucleic acid. By incubation with 5% NaOH at 0°C. for two hours, about three-quarters of the original phosphorus can be isolated in the form of a protein-free nucleic acid (BAWDEN and KLECZKOWSKI, 1948). When the incubated preparations are brought to *p*H 6 with acetic acid, the denatured protein can be separated and this contains little phosphorus but will contain some carbohydrate if the original virus preparation had a carbohydrate content exceeding 2.5%. Nucleic acid can now be precipitated from the clear supernatant fluid by adjusting it to *p*H 3 with HCl and adding an equal volume of alcohol. The dried material contains about 7% phosphorus, 14% nitrogen and 33% carbohydrate. It dissolves in water at *p*H values above 5, but in alkaline solutions it depolymerises rapidly at room temperature and soon ceases to be precipitable by 0.5M HCl. It gives qualitative tests for pentose and is depolymerised by both ribonuclease and alkali at the same speed as nucleic acid isolated from tobacco mosaic virus by comparable methods.

The nucleic acid in viruses with spherical particles is also of the yeast type, containing a pentose and not a desoxypentose, but it seems to be bound differently from that in tobacco mosaic and potato virus X. When tomato bushy stunt, tobacco necrosis and southern bean mosaic viruses are denatured by heat, the protein coagulum carries with it all the phosphorus and carbohydrate; from preparations disrupted by spreading agents, acetic acid or alkali, the nucleic acid can be isolated in amounts roughly proportional to the phosphorus contents of the viruses (BAWDEN and PIRIE, 1938*b*, 1945*a*; STANLEY, 1940; MILLER and PRICE, 1946*b*). Tobacco ringspot virus contains the unusually large amount of 40% nucleic acid and some of this seems to be separable by treatment with ammonium sulphate (STANLEY, 1939). It seems likely that other treatments than strong salt may also separate nucleic acid from this virus, for the simplest explanation that can be offered for the variable phosphorus contents attributed to preparations of the virus made solely by ultra-centrifugation (STANLEY and LORING, 1938, 1939) is that the preparations had suffered losses of varying amounts of nucleic acid.

Turnip yellow mosaic virus is denatured at room temperature and *p*H 7 by exposure to alcohol more concentrated than 30%, when nucleic acid is liberated as a large polymerised particle (MARKHAM and SMITH, 1948). Because of this, it is difficult to assess the significance of the particles free from nucleic acid that occur in purified preparations made by alcohol precipitation; obviously they may be artefacts, but the fact that they consist of undenatured protein suggests that they may occur as such in infective extracts and not be produced by the treatment applied. These particles are apparently non-infective, but they share many properties of the infective ones; they have the same electrophoretic mobility as the nucleoprotein and form mixed crystals with it; although much less effective than the nucleoprotein in stimulating the production of antibodies, they appear to contain the same antigens and to react fully with virus antisera (MARKHAM, MATTHEWS and SMITH, 1948). As the properties common to the two kinds of particle are largely those usually attributed to external groups of the particles, MARKHAM and SMITH (1949) suggest that the nucleic acid in the virus is held internally surrounded by a hollow shell of protein. X-ray measurements on crystals of the two kinds of particle suggest that these containing nucleic acid are slightly smaller (BERNAL and CARLISLE, 1948). Thus, if the two have the same protein structure, it seems that the nucleic acid holds the particles in a tighter mass while leaving the superficial attractive surfaces unaffected.

**The Denatured Protein:** — After separating nucleic acid from virus particles, protein preparations can be obtained with various properties depending on the method of denaturation employed. When inactivated by alkali, solutions of tobacco mosaic virus lose their opalescence and anisotropy of flow, showing that the virus has been split into smaller and less anisometric particles. Centrifugal data on such products suggest that the particles are disrupted in a series of steps to produce progressively smaller fragments. After exposure to alkali around *p*H 9.8, for example, the split products sediment only a little more slowly than the untreated virus, but above *p*H 11 the products sediment at about the same speed as small proteins such as egg albumin (ERIKSSON-QUENSEL and SVEDBERG, 1936; WYCKOFF, 1937). All these products are more unstable than the intact virus, but some of them are stable enough to be precipitated and re-dissolved. Soluble proteins that can be separated into fractions showing different properties are also obtained when tobacco mosaic virus is disrupted by surface active substances (SREENIVASAYA and PIRIE, 1938). When incubated in the presence of 1% sodium dedecyl sulphate at *p*H 8, solutions of the virus lose all their characteristic properties and cease to be sedimented by centrifugation at 16,000 g. After removing the dodecyl sulphate by dialysis, the proteins can be precipitated from solutions more alkaline than *p*H 8 by one-tenth saturation with ammonium sulphate, while the nucleic acid remains in solution.

When centrifuged from ammonium sulphate solution, the proteins form clear gelatinous pellets, which dissolve again in water. If precipitation is repeated many times, however, they change partly into an insoluble fraction and partly into one that is precipitable with as little as 2% saturated ammonium sulphate. A similar change is also produced by exposure to acid or by ageing for a few days at room temperature. The insoluble fraction resembles the protein coagulum produced by heat-denaturation of the virus in that it is readily hydrolysed by trypsin, whereas the soluble fraction resembles the intact virus in resisting attack by trypsin. The soluble proteins seem reasonably homogeneous, but need careful handling as they are much more unstable than the intact virus and readibly become insoluble. Preliminary studies with tomato bushy stunt virus showed that this also breaks into smaller components when denatured with acid or alkali, but fewer were noted than from tobacco mosaic virus and the smallest still sedimented at 20,000 g. (McFarlane and Kekwick, 1938)

When boiled with acid, the viruses, like other proteins, are hydrolysed to produce amino acids. Only tobacco mosaic viruses have yet been analysed for their content of various amino acids, and, as is not unusual with subjects in which new techniques are constantly being developed, the contents recorded for some have varied widely from time to time. In general, values obtained by successive workers have tended to increase, but tryptophan is exceptional. For example, the values given by Ross (1940, 1941*b*, 1942) for aspartic acid, glutamic acid, leucine, phenylalanine, proline, threonine and valine are all much below those subsequently recorded, and glycine and lycine were stated by Ross to be absent from the type strain of tobacco mosaic virus whereas they have since been found to be present (Stokes *et al.*, 1945; Knight, 1947). On the other hand, Ross (1941*b*) and Knight and Stanley (1941*a*) using the glyoxylic acid method found 4.5% tryptophan, but, using a different colour reaction for indoles, Best and Lugg (1944) found 2.1% and the smaller value has been subsequently confirmed by microbiological assays (Stokes *et al.*, 1945; Knight, 1947).

Of the various strains of tobacco mosaic virus that have been analysed, some have been found to have similar amino acid constitutions, whereas others have been found to differ. The first differences were described in the contents of certain aromatic amino acids which were noted in comparative tests made by Knight and Stanley (1941*a*) on a type strain of tobacco mosaic virus, cucumber viruses 3 and 4, and Holmes' rib-grass strain. More detailed analyses have since been made by Knight (1942*a*, 1942*b*, 1943, 1947) and his latest figures for eight strains are given in Table 17. The figures given are the percentage value of the dry virus represented by each amino acid; when allowance is made for the water combined during hydrolysis and for the 6% nucleic acid, the values given account for about 98% of the total substance of the various strains, except for cucumber viruses 3 and 4 of which they amount to about 90%. However, too much weight should not be given to this apparent completeness of analysis, because uncertainties in the methods are ample to hide the occurrence of other amino acids, for which assays have not yet been made. Also, except for tryptophan, Knight's figures for most amino acids in tobacco mosaic virus are higher than those found by previous workers. This is to be expected when microbiological assays have replaced isolation methods, but KNIGHT (1947) also records higher values than STOKES *et al.* (1945) for phenylalinine, leucine and valine though the same assay methods were used. The configuration of only four amino acids has been determined, but these, aspartic acid, glutamic acid, leucine and tyrosine, all belong to the normal or l-series (SCHRAMM and MÜLLER, 1940).

Although the precise values for some amino acids in tobacco mosaic virus may still be uncertain, KNIGHT's results show clearly that the proportions in which the constituent amino acids occur may differ from strain to strain. Of the nineteen amino acids assayed for, sixteen have been found to occur in different relative proportions in the different strains, only cystine, leucine and proline being the same for the eight virus strains examined. These differences are striking, and they provide the first evidence for the types of chemical differences that may exist between virus strains. Nevertheless, they do not justify some of the confident conclusions that have been drawn from them. For example, STANLEY (1943) states that the presence of histidine in the rib-grass strain and its absence from tobacco mosaic virus "demonstrates conclusively that the mutation of a virus may be accompanied by the introduction of an entirely new amino acid", and further says "the conversion of tobacco mosaic virus into the rib-grass strain must have been accompanied by the introduction of several hundred histidine molecules into the virus structure". Such statements postulate as facts events about which there is no evidence whatsoever, for nothing is known of the origin of the rib-grass strain, although *a priori* it is perhaps unlikely that it arose as a mutation from tobacco mosaic virus. The rib-grass strain, like cucumber viruses 3 and 4, differ in many respects from tobacco mosaic virus, and the recognition of those as related strains rests largely on the fact that they share some antigens. There is not the slightest reason to consider that these can, or did, arise by mutation from tobacco mosaic virus. The most that can reasonably be assumed is that they all derive from a common ancestor, but for how long the different strains have been following different evolutionary courses, and how many separate mutations have led to their present conditions, cannot be guessed even approximately. The serological relations between these naturally occurring strains, which differ widely from one another in their pathological behaviour, suggest that they are but distantly related, for they have only a few of their antigens in common, whereas the strains that seem to arise from tobacco mosaic virus share most of their antigens with it and usually differ from it in relatively few pathogenic characters. Studies on the serological reactions of polypeptides (LANDSTEINER, 1936) show that the sequence of aminoacids, and particularly the terminal aminoacids, is important in determining specificity. Hence mutation and differences between virus strains would seem at least as likely to result from changes in sequence of aminoacids on a peptide chain, with no quantitative change, as they are to result from quantitative changes.

STANLEY (1943) also states that the differences in amino-acid constitution provide information "of the changes in chemical structure that must accompany the mutation of a virus". This seems to disregard completely the similarities in constitution of certain strains and to attribute to varying amino acid contents effects for which there is no evidence. The differences between strains that appear to arise from another by a process akin to mutation resemble the differences between the individuals of such pairs as tobacco mosaic and masked tobacco mosaic, green and yellow aucuba, and cucumber viruses 3 and 4, much more closely than they resemble differences between the different pairs. One of the most striking results recorded in TABLE 17 is that such pairs have apparently identical amino acid constitutions. A change in the amino acid constitution is, of course, as legitimately regarded as a mutation as a change in any other property, but no more so, and to attribute mutations recognised by changes in biological behaviour, to quantitative changes in amino acid content would be decidedly premature and, indeed, would go

TABLE 17. — *Amino acid content of purified preparations of some strains* of tobacco mosaic virus* (KNIGHT, 1947) :—

| AMINO ACID | STRAIN | | | | | | | | |
|---|---|---|---|---|---|---|---|---|---|
| | TMV | M | J14D1 | GA | YA | HR | CV3 | CV4 | M.D.† |
| Alanine | 5.1 | 5.2 | 4.8 | 5.1 | 5.1 | **6.4** | | **6.1** | 0.2 |
| Arginine | 9.8 | 9.9 | 10.0 | **11.1** | **11.2** | 9.9 | **9.3** | **9.3** | 0.2 |
| Aspartic acid | 13.5 | 13.5 | 13.4 | 13.7 | 13.8 | **12.6** | | 13.1 | 0.2 |
| Cysteine | 0.69 | 0.67 | 0.64 | 0.60 | 0.60 | 0.70 | **0** | **0** | |
| Cystine | 0 | | 0 | | 0 | 0 | | 0 | |
| Glutamic acid | 11.3 | 11.5 | **10.4** | 11.5 | 11.3 | **15.5** | **6.4** | **6.5** | 0.2 |
| Glycine | 1.9 | 1.7 | 1.9 | 1.9 | 1.8 | **1.3** | **1.2** | **1.5** | 0.1 |
| Histidine | 0 | 0 | 0 | 0 | 0 | **0.72** | 0 | 0 | 0.01 |
| Isoleucine‡ | 6.6 | 6.7 | 6.6 | **5.7** | **5.7** | **5.9** | **5.4** | **4.6** | 0.2 |
| Leucine | 9.3 | 9.3 | 9.4 | 9.2 | 9.4 | 9.0 | 9.3 | 9.4 | 0.2 |
| Lysine | 1.47 | 1.49 | **1.95** | 1.45 | 1.47 | 1.51 | **2.55** | **2.43** | 0.04 |
| Methionine | 0 | 0 | 0 | 0 | 0 | **2.2** | 0 | 0 | 0.1 |
| Phenylalanine | 8.4 | 8.4 | 8.4 | 8.3 | 8.4 | **5.4** | **9.9** | **9.8** | 0.2 |
| Proline | 5.8 | 5.9 | 5.5 | 5.8 | 5.7 | 5.5 | | 5.7 | 0.2 |
| Serine | 7.2 | 7.0 | 6.8 | 7.0 | 7.1 | **5.7** | **9.3** | **9.4** | 0.3 |
| Threonine | 9.9 | 10.1 | 10.0 | 10.4 | 10.1 | **8.2** | **6.9** | **7.0** | 0.1 |
| Tryptophan | 2.1 | 2.2 | 2.2 | 2.1 | 2.1 | **1.4** | **0.5** | **0.5** | 0.1 |
| Tyrosine | 3.8 | 3.8 | 3.9 | 3.7 | 3.7 | **6.8** | 3.8 | 3.7 | 0.1 |
| Valine | 9.2 | 9.0 | 8.9 | 8.8 | 9.1 | **6.2** | 8.8 | 8.9 | 0.2 |

The figures given are for the weight of the amino acid that would arise from the hydrolysis of 100 g. of dry virus. The values which KNIGHT considers to differ significantly from tobacco mosaic virus are in heavy type.

*STRAINS: TMV, tobacco mosaic virus; M, Holmes' (1934) masked tobacco mosaic virus; J14D1, a derivative from Jensen's (1937) J14; GA and YA, green and yellow tomato aucuba mosaic; HR, Holmes' (1941) rib-grass strain; CV3 and CV4, cucumber viruses 3 and 4.

†Mean deviation of the values of single determinations from the averages given. Three to five preparations of each strain were analysed for each amino acid, with the exception of cyteine, and the results were averaged to give the figure presented.

against the existing evidence. If proportions of amino acids determine such characters as symptom manifestation, then the changes involved are too small to be detectable. With particles containing a total of 30,000 or more aminoacids, the opportunities for many such unmeasurable differences are clearly great. Significant differences in amino acid constitution as so far determined are restricted to naturally occurring variants whose origins are obscure, and none has yet been demonstrated in the sort of clinically distinct strains that mutation normally seems to produce.

The one possible exception to the last statement is strain J14D1, which does differ slightly from tobacco mosaic virus in its contents of glutamic acid and lysine. In discussing this, KNIGHT (1947) states "Since this strain is believed to have arisen from tobacco mosaic virus by two mutations and since two differences in composition were found, it would seem reasonable to conclude that each mutation involved an analytically significant change in proportion of one amino acid". Were there any evidence that this strain arose from tobacco mosaic virus by two mutations, this type of argument would still savour of *post hoc ergo propter hoc*, but in fact there is none, and much that suggests otherwise. KNIGHT described this strain as "obtained by the spontaneous mutation of J14 strain" which "had in turn arisen by the mutation of ordinary tobacco mosaic", but this description postulates a fixity of bulk cultures, and a freedom from accidental contaminations, that conflicts with practical experience. JENSEN (1937) derived J14 from a bulk culture of tobacco mosaic virus, which, from the number of variants he described, obviously contained many strains other than "ordinary tobacco mosaic", and from which of these many variants J14 itself came is, of course, unknown. Similarly, J14D1 derived from bulk cultures of J14, but whether it came direct from the original J14 or from other mutants of it, is also unknown. Its degree of relationship to "ordinary tobacco mosaic virus" is uncertain, but its properties as a pathogen, such as its slow movement, necrotic local lesions in tobacco, a systemic necrotic spotting and chlorosis leading to death of tobacco plants, and its ability to kill large tomato plants, differ in many ways from those of tobacco mosaic virus and the differences suggest the occurrence of many changes, each of which would represent a single mutation. Some single mutations that affect pathogenic behaviour may, of course, result from alterations in contents of certain amino acids, but it would certainly be rash to assume that all do, and the existing evidence suggests that mutations influencing such characters as ability to cause a certain type of mosaic symptoms are not associated with such changes. Indeed, if they were, it is difficult to see how they could be detected, or what significance could be attached to amino acid determinations on different strains, for preparations of tobacco mosaic virus large enough to allow assays to be made, must almost invariably consist of a mixture of many different strains. No chemical explanation yet exists to explain the different symptoms caused by closely related strains or the changes that occur when one strain mutates to another; all that can be said is that remotely related strains have somewhat different amino acid constitutions and that the differences between the relative proportions of individual amino acids increase with increasing differences between host range and antigenic constitution. There is no evidence that these constitutional differences are correlated with, let alone responsible for, the kind of symptoms produced, and their origin and biological significance are highly speculative.

**Specific Volume and Hydration:**—To calculate the size of virus particles from sedimentation and diffusion data, and to convert size into particle weight, their partial specific volume must be known. The specific volume is the reciprocal of the density and it is readily obtained for the dry virus from a knowledge of the amount of dry virus contained in a solution of known density, but the value obtained by this method may not be relevant for the purpose of the calculations. There is good evidence that the virus particles combine with water, so that their effective volume and density are not those of the particles in the dry state, and it is the volume and density of the wet rather than the dry particles that is important.

ERIKSSON-QUENSEL and SVEDBERG (1936) measured the specific volume of material in one of STANLEY's early preparations of tobacco mosaic virus and found a value of 0.646, corresponding to a density of 1.55. They commented on the fact that this was unusual for proteins, and measurements made on later preparations of the virus have given more normal values. BAWDEN and PIRIE (1937*a*), using three different methods, found the density to be between 1.3 and 1.37. The lower values were obtained by flotation methods, in which dried virus was equilibrated in mixtures of nitrobenzene and dichlorbenzene or virus precipitated with ammonium sulphate was equilibrated in solutions of sucrose and ammonium sulphate. The higher values were obtained by pycnometric methods, in which the specific gravity of solutions of known concentration was measured. These values agree well with those obtained for most other proteins and the differences between the values obtained by the three methods also agree well with results on other proteins. The same results were obtained with three different strains, and STANLEY (1938*a*) confirmed the values of 0.77 for specific volume and 1.3 for density with tobacco mosaic virus itself. Subsequently, specific volumes have been obtained for other viruses, namely, tomato bushy stunt, 0.739 (MCFARLANE and KEKWICK, 1938), potato virus X, 0.73 (LORING, 1938*b*), a tobacco necrosis virus, 0.77 (PIRIE and others, 1938), tobacco ringspot, 0.636 (STANLEY, 1939), alfalfa mosaic, 0.68 (ROSS, 1941) and southern bean mosaic 0.696 (MILLER and PRICE, 1946*a*). The low values for tobacco ringspot and southern bean mosaic viruses may be attributable to their high contents of nucleic acid, and the preparations of alfalfa virus used were inhomogeneous.

The specific volumes of the viruses as they exist in solution probably differ considerably from these values. Dried preparations of the viruses are hygroscopic and absorb 10-15% of their weight of water when exposed to a moist atmosphere, so that it is likely that they hold at least this much water when they are dissolved. From X-ray measurements on preparations of different water contents, PIRIE (1945) concluded that tobacco mosaic virus is normally associated with about its own weight of water, and similar conclusions were reached by MARKHAM, SMITH and LEA (1942) from a consideration of the diffusion and sedimentation data. Direct measurements on the hydration of tobacco mosaic virus have been made by SCHACHMAN and LAUFFER (1948) who studied its sedimentation behaviour in serum albumin solutions of different densities and found its specific volume and density to be 0.885 and 1.13 respectively, suggesting that 65% of the effective volume of the dissolved particles is composed of water. How this water is held is uncertain. BERNAL and FANKUCHEN (1941) concluded that no water penetrated into the particles, for the X-ray pattern resulting from the internal structure of the particles was similar for dried virus and for virus in solution. It is argued that the penetration of water would cause the protein structure to swell and to give increased X-ray spacings, but no such increase is observed when water is added to a dry preparation. Thus it would seem that the particles are substantially anhydrous internally but may be surrounded by a layer of water several molecules thick.

Similarly, there is evidence from several different sources that tomato bushy stunt virus particles in solution are also hydrated. From X-ray measurements on wet and dry crystals, BERNAL, FANKUCHEN and RILEY

(1938) suggest that 63% of the wet crystals may be water. Markham, Smith and Lea (1942), from a consideration of the sedimentation and diffusion constants, suggest that dissolved particles are less hydrated than the crystals and conclude that 1 g. of dry virus is associated with 0.76 g. of water; this would give a specific volume of 0.853 for the hydrated virus (density 1.17) compared with 0.739 (density 1.35) for the dried virus. Southern bean mosaic virus also seems to be hydrated to the same extent as tomato bushy stunt. By studying the sedimentation behaviour in solvents of various densities, Miller and Price (1946*a*) found the partial specific volume of the hydrated particles to be 0.827 corresponding to a density of 1.21, and from consideration of measurements on various physical properties they suggest that 1 g. of dry virus is associated with 0.83 g. of water.

**Precipitation and Isoelectric Points:** — The viruses so far purified are soluble at neutrality and in the absence of salts, and their solutions have the characteristic opalescence of large particles. From such solutions they are readily salted out or precipitated by the addition of various substances. The salt most used for precipitation is ammonium sulphate, from one-eight to one-quarter saturation usually being adequate to cause complete precipitation, but the minimum amount needed varies with different viruses, the state of purity, temperature and *p*H value.

When ammonium sulphate is added to neutral solutions of tobacco mosaic virus, they become gelatinous and develop an intense satin-like sheen. At one-eight saturation a precipitate composed of paracrystalline needles can be separated by centrifuging at 3,500 r.p.m., but precipitation is more rapid and complete if the concentration of the salt is increased to one-fifth or one-sixth saturation. Similar fibrous precipitates are produced by the addition of many neutral salts (Best, 1940) and by agents as diverse as clupein (Bawden and Pirie, 1937*a*), dodecyl sulphate (Sreenivasaya and Pirie, 1938), salicylate (Best, 1940*a*), ribonuclease (Loring, 1942), guanidine, nicotine and arginine (Bawden and Pirie, 1940*a*), heparin, hyaluronic acid and gum arabic (Cohen, 1942), globin (Kleczkowski, 1946), and a glycoprotein from *Phytolacca esculenta* (Kassanis and Kleczkowski, 1948). Most of these substances are oppositely charged from the virus at *p*H 6-7, and the production of precipitates is probably the result of mutual discharge of the two components when mixed; most of the precipitates are dissolved by the addition of small amounts of salt (Kleczkowski, 1946). When serum albumin and tobacco mosaic virus are mixed at pH values between their respective isoelectric points similar precipitates occur which dissolve when electrolytes are added. Lauffer (1948) finds that the virus is also precipitated by the albumin when both are negatively charged; in these conditions the precipitates seem to consist exclusively of the virus and the amount of albumin needed to precipitate the virus varies inversely with the concentration of electrolytes, the product of ionic strength and albumin concentrations required to produce precipitates being approximately constant. Different strains of tobacco mosaic virus behave in the same general manner with the various precipitants, but no systematic comparisons of the behaviour of different strains have been made; the precipitates formed by clupein with aucuba mosaic and cucumber viruses 3 and 4, however, are more soluble in dilute salt solution than those formed by tobacco mosaic virus itself (BAWDEN and PIRIE, 1937*b*).

Potato virus X does not precipitate from neutral solutions until the concentration of ammonium sulphate is raised to over one-sixth of saturation, but the precipitated virus will not redissolve until the salt concentration is reduced to about one-twentieth saturation. The precipitates differ strikingly from those produced by tobacco mosaic virus; they are transparent and their formation is mainly obvious because of the manner in which the fluids hold bubbles. With salt concentrations too little to cause precipitation, the fluids develop a slight sheen and show increased anisotropy of flow, but the precipitates themselves are amorphous and show no anisotropy of flow. Precipitates formed by combination with substances such as ribonuclease are amorphous, and they dissolve in salt solutions much more slowly than the fibrous precipitates of tobacco mosaic virus (BAWDEN and KLECZKOWSKI, 1948). Potato virus Y precipitates with salt in much the same manner as virus X.

The solubility in ammonium sulphate solution of tomato bushy stunt and of several other viruses with spherical particles is greater at 0°C. that at 20°C., and the type of precipitate obtained depends on the salt concentration and temperature. Neutral solutions become increasingly opalescent as ammonium sulphate is added and at 20°C. precipitates will separate from fluids when one-sixth to one-fifth saturated. With a little more salt, or when stood for some time, most of the virus will settle as a dense, white, amorphous precipitate. If the fluids are cooled to 0°C., however, the precipitates will dissolve and the fluids regain their normal clarity, but after some hours at 0°C. the virus begins to crystallize. Greater solubility of the amorphous precipitates at 0°C. than at 20°C. is also a feature of the several tobacco necrosis viruses, turnip yellow mosaic and southern bean mosaic viruses. Crystalline precipitates are also produced when tomato bushy stunt virus is mixed with clupein (BAWDEN and PIRIE, 1938*b*) and heparin (COHEN, 1942), and amorphous precipitates with ribonuclease (LORING, 1942) and a glycoprotein from *Phytolacca esculenta* (KASSANIS and KLECZKOWSKI, 1948).

With the addition of acid, the behaviour of different viruses differs more than with salt, for whereas some are soluble over the whole *p*H range in which they are stable, others are insoluble at their isoelectric points. When acid is added to tobacco mosaic virus solutions containing a little salt, they develop an intense sheen and show increased anisotropy of flow. At about *p*H 4 the fluids become almost opaque and at *p*H 3.3 the whole preparation is precipitated in the form of paracrystalline needles and can be removed by centrifugation at 3,500 r.p.m. The addition of more acid causes the virus to redissolve. In the absence of salt, the fluids show less sheen when acid is added; precipitation is optimal at about *p*H 4.2, but longer centrifugation is needed to sediment the virus than at *p*H 3.3 in the presence of salts, and the virus dissolves again at about *p*H 3.5. Enation mosaic virus behaves in essentially the same manner as tobacco mosaic virus, precipitating optimally at *p*H 3.3 in the presence of salts and at *p*H 4.2 when salt-free. Tomato aucuba mosaic virus, however, precipitates optimally at *p*H 3.6 in the presence of salts and at *p*H 4.5 in their absence. Cucumber viruses 3 and 4 are serologically related to tobacco mosaic virus but infect only cucurbits, which have more alkaline sap than tobacco, and they have higher isoelectric points. In the presence of salts they precipitate optimally at about $p$H 4.8, whereas they are difficult to sediment from salt-free solutions at any $p$H value by centrifugation at 3,500 r.p.m. but they are least soluble at about $p$H 5.5.

Isoelectric point determinations made electrophoretically on preparations of tobacco mosaic virus have also given variable results. ERIKSSON-QUENSEL and SVEDBERG (1936) found the isoelectric point in acetate buffer to be $p$H 3.49, while LORING and STANLEY (1937), using suspensions of virus in solutions of NaCl and $(NH)_2SO_4$, found values between $p$H 3.2 and 3.5; the value varied with salt content and they found that the addition of acetic acid raised the isoelectric point by from 0.1 to 0.3 of a $p$H unit. It seems that the $p$H value of maximum insolubility is the same as that at which the virus does not migrate in an electric field, and that this value, as with some other proteins (ADAIR, 1937), depends on the concentration of electrolytes. The virus appears to be isoionic, that is to be combined with equal quantities of anions and cations at $p$H 4.2, the $p$H value at which it is isoelectric in the absence of salts (BAWDEN, 1939).

Potato virus X is also precipitated by acid, the floccules formed resembling those caused by salt in showing no anisotropy of flow, being transparent and amorphous. Between $p$H 5 and 6 the solutions show a definite sheen, but they lose this character below $p$H 5 when the virus begins to flocculate. The virus precipitates optimally at about $p$H 4.5, but unlike tobacco mosaic virus, there is no $p$H value at which all the virus present in a preparation can be separated by centrifuging at 3,500 r.p.m. The virus that remains in solution at $p$H 4.5 can usually be precipitated by adjusting the $p$H to 4 or 5, whereas the portion that precipitated at $p$H 4.5 will usually precipitate incompletely if it is dissolved at neutrality and then again brought to $p$H 4.5. In electrophoresis experiments using collodion particles coated with potato virus X, LORING (1938*b*) found the isoelectric point to vary with the previous treatment of the virus preparation. Samples recently made by ultracentrifugation were isoelectric at about $p$H 3.7 in buffer solutions containing phthalate, phosphate and borate. Samples which had been precipitated with ammonium sulphate or potassium citrate were isoelectric in these buffers at about $p$H 4.4, and samples that had been treated with 3% dibasic phosphate at $p$H 7 were isoelectric at about $p$H 5. These changes in electrophoretic behaviour are attributed by LORING to changes in the virus particles caused by the action of strong electrolytes, but the variability in precipitation at different $p$H values is encountered also with virus isolated by ultracentrifugation. Many treatments, the action of which is ill understood, cause the particles of virus X to aggregate so much that they become intertwined and poorly soluble even in neutral solutions. The varying isoelectric behaviour may result from different states of aggregation or from the presence of varying amounts of different impurities (BAWDEN and KLECZKOWSKI, 1948). Potato virus Y (BAWDEN and PIRIE, 1939) and alfalfa mosaic virus (ROSS, 1941) are also precipitated by acid, both having their minimum solubility in the region of $p$H 4.5.

The spherical viruses, tomato bushy stunt, the various tobacco necrosis viruses, southern bean mosaic and turnip yellow mosaic, are all soluble at their isoelectric points, and the addition of acid to neutral solutions causes no change in appearance. From centrifugal studies with tomato bushy stunt virus at different $p$H values, it seems that the virus particles remain unaggregated over the range $p$H 2.4 to 8.7, for the same sedimentation constant is obtained at all values. The extent to which the virus particles are hydrated may vary with the $p$H, for the character of the pellet obtained by ultracentrifugation depends on the

FIG. 37. — Electrophoretic migration of tomato bushy stunt virus in 0.02 M acetate buffer at $p$H 6.2. Potential gradient 9.5 V./cm. A single sharp boundary is present in all the exposures. (MCFARLANE, A. S. and KEKWICK, R. A., 1938, Biochem. J., *32*, 1607).

$p$H value. At or near $p$H 4 the pellets are compact and firm, whereas as the $p$H departs from this value the pellets become increasingly less rigid. From electrophoretic tests, in which the preparations gave a single boundary over a wide $p$H range (Fig. 37), the isoelectric point was found to be $p$H 4.1 (MCFARLANE and KEKWICK, 1938). The other viruses also migrate with a single boundary and their isoelectric points determined electrophoretically are $p$H 5.5 for southern bean mosaic virus (MILLER and PRICE, 1946*b*), $p$H 3.7 for turnip yellow mosaic (MARKHAM and SMITH, 1949), and $p$H 4.7 for tobacco ringspot virus (STANLEY, 1939).

*Chapter 11*

# THE CRYSTALLINITY OF PLANT VIRUSES

**Optical Properties of Crystals:** — In describing the optical properties of purified virus preparations a few terms are needed that may be unfamiliar to some biologists. To define these, properties of polarised light, and the manner in which it is affected by various objects, are briefly described. For greater details, and for the theory underlying these processes, textbooks on optics should be consulted.

Ordinary light is conveniently regarded as a series of electrical disturbances vibrating in all planes at right angles to the direction of transmission. In passing through a Nicol prism the light is resolved into two beams at right angles to one another and in each of which the vibrations are all in one plane. In the Nicol prism two pieces of Iceland spar are cemented together in such a manner that one beam is transmitted through the prism whereas the other is reflected off the cement layer and is absorbed on the blackened side of the prism. Polarising crystals such as tourmaline, or oriented suspensions of polarising crystals such as "Polaroid", similarly resolve light into two beams, one of which is transmitted whereas the other is absorbed in the body of the crystals. The beam of light transmitted by a Nicol prism or a polarising crystal thus consists of vibrations all in one plane, and is said to be plane polarised. A beam of plane polarised light is transmitted completely by a second polariser provided its plane of polarisation is parallel to that of the first. On the other hand, if the two polarisers are "crossed", that is, have their planes of polarisation at angles of 90° to one another, no light is transmitted, for the vibrations in the plane that are transmitted by the first polariser are now stopped by the second.

The manner in which light is transmitted through a medium depends, not only on the nature of the particles present, but also on the manner in which they are arranged. Amorphous objects have no regular arrangements of particles and have the same refractive index in all directions. So, too, do cubic system crystals, in which all three axes of symmetry are identical. With structural regularities intermediate between these extremes, the refractive indices differ in different directions. The light transmitted through objects that have different refractive indices in directions at right angles to the line of transmission, is conveniently regarded as the equivalent of two plane polarised rays at right angles to one another and travelling at different speeds. Objects that transmit light in this manner are called doubly refracting, birefringent or optically anisotropic, and the light they transmit is said to be elliptically polarised. Doubly refracting objects also change plane polarised into elliptically polarised light, unless the object is so positioned that one of its transmission planes is parallel to the polarising plane of the incident light, when the plane polarised light is transmitted unchanged. Elliptically polarised light has two rays and so cannot be extinguished by a

Nicol prism (if the position of the prism is such that one ray is stopped, the other will be transmitted, and *vice versa*). The change from plane to elliptically polarised light is therefore simply detected by observation between crossed Nicol prisms.

If a single crystal between crossed Nicol prisms is rotated through 360°, in four positions at right angles to one another, where one of the transmission planes of the crystal is parallel with the plane of polarisation of the first prism, no light will be transmitted through the second prism. The crystal will be visible at all positions except these four extinction positions, and mid-way between them the amount of light transmitted through the second prism will be at a maximum. If a preparation containing many separate crystals arranged at random is rotated

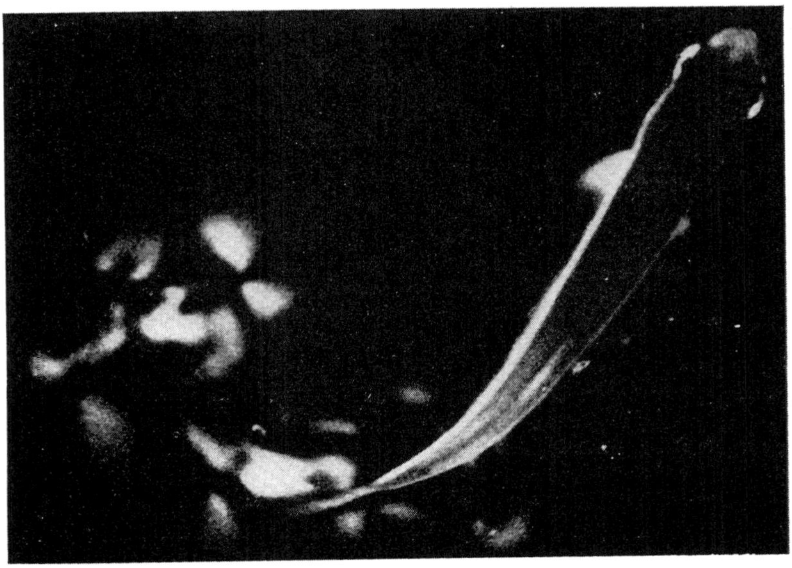

FIG. 38. — Anisotropy of flow. Wake of a goldfish swimming in a dilute solution of tobacco mosaic virus, photographed between crossed Nicol prisms. Where the fluid is stationary no light is transmitted, but where the rod-shaped particles are oriented by flow movements the fluid becomes birefringent. (BAWDEN, F. C., PIRIE, N. W., BERNAL, J. D., and FANKUCHEN, I., 1936, Nature *138*, 1051).

through 360°, each crystal will show four extinction positions, but light will be transmitted through the second prism at all positions because the transmission planes of individual crystals will be parallel with the plane of polarisation at different positions.

All crystals except these belonging to the cubic system are birefringent, but those that have two identical axes of symmetry are birefringent only when observed along one or other of these axes. The phenomenon is not confined to crystals but is shared by all systems that possess some degree of orderly arrangement. Many fibrous substances, typified by cellulose, hair and muscle, are highly birefringent, although they cannot be obtained in the form of crystals. Most liquids are amorphous and therefore optically isotropic, but in some the arrangement is sufficiently orderly for them to be anisotropic. If they are birefringent when at rest they are called liquid crystals, and it is generally accepted that liquid crystals are formed only by anisometric particles, that is particles having different dimensions in different directions, which arrange themselves parallel to one another. If the liquids are isotropic when stationary, but become birefringent when made to flow or when a shearing force is applied, they are said to show the phenomenon of anisotropy of flow; this is also known as stream double refraction or flow birefringence (Fig. 38).

Anisotropy of flow can result from one or more of a number of causes. If substances which are normally optically isotropic, such as glass and some other highly viscous gel-like fluids, become deformed when made to flow, they are subject to internal stresses and become birefringent at points of strain. This phenomenon is known as the photoelastic effect. In less viscous fluids anisotropy of flow results from the presence of small anisometric particles. When the fluid is at rest Brownian movement keeps the particles distributed at random, but when the fluid is made to flow they become orientated in directions parallel to the lines of flow, in the same manner as logs floating down a stream. The particles may be rod-shaped or plate-like and may themselves be optically isotropic or anisotropic, but the more anisometric they are, the more easily they will be orientated and show anisotropy of flow. Single anisotropic particles small relative to the wavelength of light will not produce a visible effect between crossed Nicol prisms, because they are too small to produce a detectable difference between the fast and slow rays. When arranged at random their individual effects will be compensatory instead of additive, so that a statistical optical isotropy results. If the particles are themselves optically isotropic, the double refraction obtained when they are orientated by flow results entirely from the difference between the refractive indices of the particles and the surrounding medium. On the other hand, if they are optically anisotropic, when they are orientated by flowing there will be an additional effect resulting from their intrinsic birefringence.

**Anisotropy of Flow and the Layering Phenomenon: —** For many years it was tacitly assumed that all virus particles were spherical and conclusions about particle sizes were made on this assumption. The fact that they might be some other shape was first indicated by TAKAHASHI and RAWLINS (1933*a*), who noticed that clarified sap from plants infected with tobacco mosaic virus showed the phenomenon of anisotropy of flow. They suggested that the virus particles themselves were responsible, and that, if so, these were rod-shaped. The possibility of plate-like particles being responsible was eliminated by comparing the behaviour of infective sap with that of ferric oxide sol; the entire path of flowing infective sap was birefringent, whereas with the plate-like particles of ferric oxide birefringence was restricted to the edges of the stream. The fact that the particles responsible are rod-shaped has since been confirmed by many different workers using a variety of techniques.

The identification of the rod-shaped particles in infective sap with the virus came from work on purified preparations. BAWDEN, PIRIE, BERNAL and FANKUCHEN (1936) found that purified preparations of tobacco mosaic virus showed the phenomenon of anisotropy of flow strongly and, if sufficiently concentrated, formed spontaneously birefringent liquid crystals. When allowed to stand undisturbed, neutral salt-free solutions of the virus separate into two layers, which have different solid contents and different appearances both when viewed by ordinary light and between crossed Nicol prisms. The upper layer is the more dilute and by transmitted light is opalescent, whereas the bottom layer, in spite of its greater solid content, is usually clear. In newly layered preparations, the lower layer shows an intense sheen by reflected light, probably because separation is incomplete and droplets of top-layer fluid are still present. When preparations are left undisturbed for long periods, or when they are centrifuged for short periods at 12,000 r.p.m., separation is complete and the lower layer is then usually clear whether viewed by reflected or transmitted light. The most striking difference between the two layers is evident by examination in polarised light, for the lower layer is spontaneously birefringent, whereas the upper layer is not birefringent when stationary but becomes so when gently agitated. These phenomena are illustrated in Fig. 38 and 39.

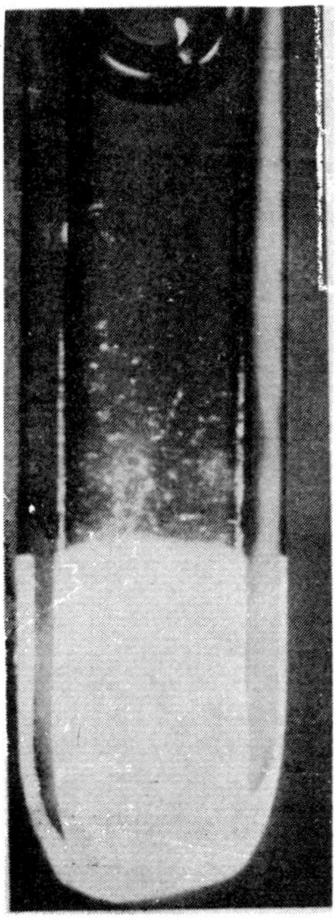

FIG. 39. — Photograph, in polarised light, of a 2% solution of purified tobacco mosaic virus which has settled into layers, showing the spontaneous birefringence of the lower layer. (BAWDEN, F. C. and PIRIE, N. W., 1937, Proc. R. Soc. B., *123*, 274).

As a preliminary to the separation of a liquid crystalline layer, birefringent, spindle-shaped droplets (tactoids) appear throughout the originally homogeneous solution (Fig. 40). Each droplet consists of a region in which the virus particles are arranged approximately parallel; they form because the particles are unable to move freely and to be accommodated in the available space must become aligned. This close packing leads to a higher virus concentration than in the surrounding fluid, and the denser tactoids thus slowly settle under gravity. The formation and settling of tactoids continue until the concentration of virus in the upper layer is reduced to a level at which the particles have sufficient space to rotate freely in at least two dimensions, when Brownian movement will keep them distributed at random. In this condition the upper layer will be isotropic but will become birefringent as soon as any shearing force is applied to orientate the particles. The bottom layer is not a single homogeneous structure, but consists of a collection of liquid crystals arranged at random in three dimensions. Each liquid crystal when rotated in polarised light shows the usual four extinction positions, but the layer as a whole gives no extinction positions because different birefringent tactoids extinguish at different positions. When examined in polarised light through a lens or microscope, the individual liquid crystals are seen as coloured regions of various hues. This indicates the random arrangement and variable size of the liquid crystals, for the colour produced depends on the path difference between the fast and slow ray in each liquid crystal, and this depends in turn on the thickness of the crystal and on the angle made by the orientated rods with the plane of polarisation. The average size of the individual liquid crystals depends on the concentration of the virus preparation, being large in dilute solutions and small in concentrated ones.

BAWDEN and PIRIE (1937*a*) noted that the purity of preparations affected both the concentration at which layering occurred and the concentration of the two layers when in equilibrium. Highly impure preparations did not layer at all, though birefringent gels could be sedimented from them by high speed centrifugation. Less impure preparations layered if more concentrated than about 5% and the virus concentration in the lower layer was then sometimes twice as much (8%) in the lower as in the upper (4%). By contrast, highly purified preparations separated into layers at concentrations as low as 1.6%, when there was also very little difference between the virus concentration in the two

TABLE 18.—*Activity of top and bottom layers separating from slightly impure preparations:—*

| *Virus* | *Layer* | *Serological titre* | *Infectivity at various dilutions** Mean No. of lesions per half-leaf | | | |
|---|---|---|---|---|---|---|
| | | | $10^{-4}$ | $10^{-5}$ | $10^{-6}$ | $10^{-7}$ |
| Tobacco mosaic | Top | 1/6,000,000 | 165 | 99 | 23 | 4 |
| | Bottom | 1/10,000,000 | 161 | 80 | 25 | 2.6 |
| Aucuba mosaic | Top | 1/6,000,000 | 144 | 51 | 24 | 5 |
| | Bottom | 1/10,000,000 | 151 | 74 | 26 | 4 |
| Enation mosaic | Top | 1/3,000,000 | 75 | 27 | 6.5 | 0.16 |
| | Bottom | 1/6,000,000 | 72 | 27 | 7.5 | 0.5 |

*Dilution given as the weight of protein in g. per ml.

layers. Impurities tend to concentrate in the upper layer and a useful fractionation can often be produced by allowing layering to proceed and then separating the lower from the upper. If the lower layer is diluted slightly and again allowed to stand undisturbed, it will in turn produce a liquid crystalline layer. This time there is less difference between the solid contents of the two layers in equilibrium, and if the process is repeated preparations are obtained which change their state from being wholly birefringent to being wholly isotropic on slight change in concentration or temperature. This suggests that the coexistence of two layers depends on some heterogeneity and would not occur with perfectly homogeneous preparations. The behaviour implies an equilibrium system in which the alteration of one variable causes a continuous transition from a distinct two phase system to a critical condition in which only one phase is possible. The variable is almost certainly the length of the particle. The possibility of maintaining the parallel alignment of particles that characterizes the liquid crystals will depend on the ratio of the particle length to their distance apart, and beyond a certain critical ratio Brownian movement will make such an arrangement unstable. As the length of the particles increases, the concentration at which solutions form liquid crystals decreases. The fractionation obtained by separating successively formed bottom layers presumably results from the selection of the longest particles, which also seem to be the most highly purified. In preparations that form layers only when more concentrated than about 2.5%, and in which there are considerable differences between the solid contents of the layers, the material in the two layers differs appreciably in certain characters. When compared on equal solid contents, the material from the bottom layer shows greater anisotropy of flow, and gives a higher serological titre than material from the top layer, but it is usually no more infective (TABLE 18).

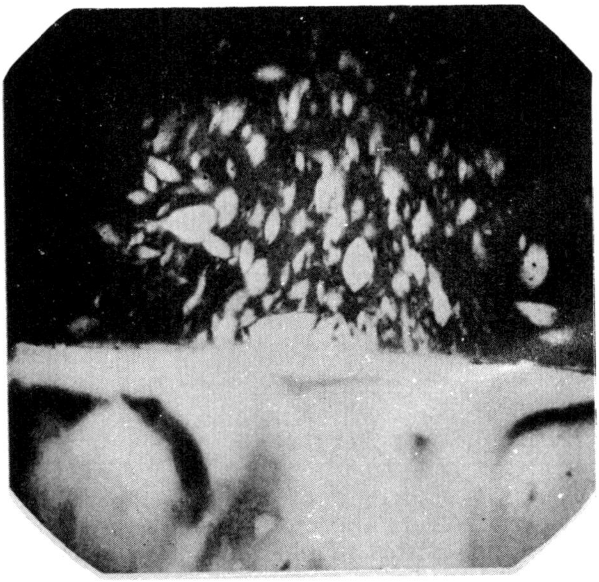

FIG. 40. — Photomicrograph, in polarised light, of the interface between top and bottom layers in a solution of tobacco mosaic virus separating in a cell 1 mm. deep. The photograph was taken after the fluid had stood for 6 hours, when birefringent spindle-shaped droplets were still forming and settling. The bottom layer when viewed is highly coloured. (BAWDEN, F. C. and PIRIE, N. W., 1937, Proc. R. Soc. B., *123*, 274).

The concentration at which neutral salt-free solutions of tobacco mosaic virus become liquid crystalline varies greatly with the method of purification. If procedures are used that cause extensive aggregation, such as heating the sap or incubation with trypsin, layers will separate from freshly made preparations at concentrations around 1.6%, but preparations made solely by ultracentrifugation or precipitation methods do not usually layer unless much more concentrated. On ageing, however, most preparations enter into a condition in which they form liquid crystalline layers at a concentration of 1% or less. This change is often accompanied by the separation of some insoluble material at the interface between the two layers, suggesting that the further aggregation of the virus particles, or the coalescence of separate tactoids, has been accompanied by the release of previously combined contaminants. The rate at which solutions layer, and the limiting concentration at which they will do so, are affected by temperature. Separation is faster at 37°C. than at 0°C., but solutions that are liquid crystalline at 0°C. may lose the property when warmed, because of increased Brownian movement. The presence of electrolytes also affects the formation of liquid crystals, and dilute solutions that are spontaneously birefringent when salt-free may lose this property if a little salt is added. BEST (1947) found that a preparation which produced a liquid crystalline layer at 0.9% in the absence of salt did not do so at 2% when the solutions contained 7 millimols of NaCl per litre.

From X-ray measurements on the liquid crystalline solutions, BERNAL and FANKUCHEN (1941) have confirmed the parallel alignment of the particles and found other regularities of structure previously unsuspected in liquids. They have measured the distance between particles in solutions under various conditions, and found this to be inversely proportional to the square root of the concentration. The particles are distributed uniformly through the liquid equidistant one from another and arranged in a hexagonal array so that they fill the available space as uniformly as possible (Fig. 41). An organised structure within the solutions is also suggested by the electron micrograms made by WYCKOFF (1947) from solutions dried when concentrated. In preparations from dilute solutions rod-like particles occur individually or in small clusters, but in more concentrated preparations, areas in which rods are aligned parallel to one another become increasingly common, and structures occur in which sheaf-like areas of parallel particles abut on one another at random, suggesting that these areas represent individual tactoids. In preparations made from solutions dried while frozen, the most striking feature is the occurrence of sheets of aligned particles, which cover extensive areas. There are often successive layers of such sheets, but whether these pre-existed in solution or are a product of drying while frozen is uncertain. The sheets are fragile and readily disintegrate when exposed to the beam of electrons, and WYCKOFF (1947) states that they not only look like connective tissue but break up in the same manner. A portion of such an organised sheet of aligned particles overlying a collodion film, which is itself strewn with individual particles, is shown in the frontispiece. The existence of an organised structure in the fluids is also indicated by their ability to show some rigidity when subjected to small forces, by their discontinuous flow in capillary tubes, and by the manner in which any pattern imposed, for example, by stirring, does not disappear spontaneously for some hours (PIRIE, 1945).

The birefringence of orientated preparations of tobacco mosaic virus is sufficiently high to suggest that the particles themselves might be optically anisotropic. Nevertheless, LAUFFER (1938) has obtained results suggesting that the particles are optically isotropic, and that the optical effect results from the orientation of isotropic rods in an isotropic medium of different refractive index. LAUFFER measured the refractive index of a solution of tobacco mosaic virus of a known concentration and from this calculated the refractive index of the virus itself to be approximately 1.6. He then measured the anisotropy of flow of tobacco mosaic virus in solvents of different refractive indices, such as glycerol-water mixtures and aniline-glycerol-water mixtures, and found that it decreased with increasing refractive index of the solvent and was lost in a solvent with a refractive index of 1.55. LAUFFER states that the activity of the virus was unaffected by the solvents used and that the reduction of anisotropy of flow was not an effect of viscosity differences between the different fluids. If this is so, it seems that the particles have little or no intrinsic double refraction, but the results await confirmation. BAWDEN and PIRIE (1940*a*) found that exposure to glycerol-aniline mixtures immediately reduced infectivity. Also, until the virus has been examined in solvents of still higher refractive indices and shown to regain anisotropy of flow, the possibility cannot be excluded that some phenomenon (for example, disaggregation of long particles) other than the changing refractive index of the solvent, was responsible for the loss of anisotropy of flow.

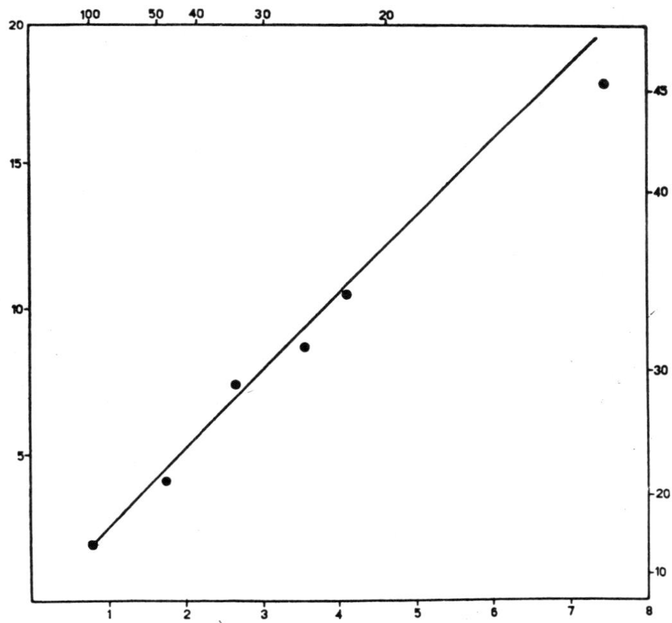


FIG. 41. — Graph showing the relationship between the distance separating virus particles and the concentration of tobacco mosaic virus solutions. (J. D. BERNAL and I. FANKUCHEN, unpublished).

When a beam of light is passed through virus solutions the light is scattered and they, like other colloidal solutions, show a Tyndall cone. LAUFFER (1938) has shown that the light scattering properties of liquid crystalline solutions of tobacco mosaic virus differ considerably from those of top-layer fluids. In the upper layer the scattered light is not depolarised appreciably, whereas in the bottom layer the scattered light is very largely depolarised, even when the incident light is polarised. The fact that the scattered light in the upper layer is not depolarised supports the view that the virus particles themselves have little intrinsic double refraction, for in suspensions of small isotropic particles the light scattered perpendicularly is completely polarised. The depolarisation of the light in the bottom-layer liquid does not result from the scattering by the virus particles themselves, but is produced by the birefringent tactoids composed of many parallel particles.

Only solutions of tobacco mosaic virus have yet been examined in any detail, but other plant viruses show similar phenomena. A number of different strains of tobacco mosaic virus (aucuba mosaic, enation mosaic, masked tobacco mosaic, mild tobacco mosaic virus, cucumber viruses 3 and 4) and three strains of potato virus X all show anisotropy of flow, and, when sufficiently purified and concentrated, give liquid crystalline solutions. At room temperature, solutions of potato virus X will give a liquid crystalline layer until diluted to below 1.5%, when they show anisotropy of flow. Solutions of virus X are more viscous than those of tobacco mosaic virus at the same concentration, so that the separation of originally homogeneous solutions into layers takes place more slowly. Solutions of severe etch, potato Y and henbane mosaic

TABLE 19. — *Anisotropy of flow of 0.108% solutions of tobacco mosaic virus in solvents of different compositions:—*

| *Composition of solvent in volume per cent* | | | *Refractive index of solvent* | *Amount of anisotropy of flow* |
|---|---|---|---|---|
| Water | Glycerol | Aniline | | |
| 100 | 0 | 0 | 1.334 | 54 |
| 80 | 20 | 0 | 1.362 | 42 |
| 50 | 50 | 0 | 1.402 | 27 |
| 10 | 90 | 0 | 1.454 | 14 |
| 2 | 68 | 30 | 1.503 | 4.5 |
| 2 | 28 | 70 | 1.55 | 0 |

The virus solutions were made to flow through a capillary between crossed Nicol prisms. The light transmitted by the second prism was passed through a photoelectric cell to a galvanometer, and the anisotropy of flow is expressed as millimetres of galvanometer deflection (LAUFFER 1938).

virus also show anisotropy of flow and give birefringent pellets when sedimented by high speed centrifugation, but these have not been obtained in sufficient quantities to determine whether they form liquid crystalline solutions. In the absence of any evidence to the contrary, it is reasonable to assume that all these viruses show these properties for the same reason as tobacco mosaic virus, but some may be plates instead of rods.

When TAKAHASHI and RAWLINS (1933) first demonstrated the occurrence of anisotropy of flow in sap from plants infected with tobacco mosaic virus, they used a microscopic technique. However, this is not necessary, for the phenomenon is readily observed simply by shaking sap in polarised light, and can be adapted to serve as a simple and rapid method of testing plants for the presence of this virus. The orientation of particles is more complete the greater their length to width ratio and the optical effects are more easily visible the fewer other substances there are present in the fluid. Thus, although the phenomenon can be seen in freshly expressed infective sap clarified by centrifugation, it is much more obvious if the virus particles are aggregated and the normal plant proteins removed. This can be done conveniently by heating the sap to 60-70° C. and then centrifuging for a few minutes at 3,500 r.p.m. The clear brown fluid shows the phenomenon of anisotropy of flow strongly if samples are examined in polarised light and tipped from end to end of half-filled test tubes 0.5 cm. in diameter. Examination in polarised light is not always necessary, for after heating and centrifuging, the presence of the virus is often easily detectable merely by examining the flowing fluids by ordinary reflected light, when they show an intense sheen. The test for anisotropy of flow in sap is not specific for tobacco mosaic virus, because potato virus X also occurs in sufficient concentration to show the phenomenon, particularly in tomato plants. However, the two can readily be distinguished by heating samples at 60 and 70°C.; if both samples show it equally, they contain tobacco mosaic virus, whereas if the phenomenon is seen only in the sample heated to 60°C., the cause will be potato virus X.

Dilute solutions of tobacco mosaic virus and potato virus X are orientated by alternating electric fields but not by magnetic fields up to 5000 gauss. Like the birefringence obtained by flow orientation, that produced in neutral solutions containing less than 2 g. of tobacco mosaic virus per litre is positive at all values of the field strength. In more concentrated solutions, or in those acidified to *p*H 5, on the other hand, the birefringence is negative in weak electric fields, passes through an inversion point as the field strength is increased, and becomes positive to reach a maximum as the field strength is further increased. The field strength at which the inversion from negative to positive birefringence occurs varies with virus strains, the age of the preparation and the concentration; it presumably depends on the degree of aggregation of the virus particles and the extent to which they are interacting with one another, but the interpretations are uncertain (LAUFFER, 1939).

**Crystals and Liquid Crystals:** — The name liquid crystal was first coined to describe liquids possessing the property of double refraction. The essential difference between a true crystal and a liquid crystal, however, is not that one is a solid and the other a liquid, but is the degree of regularity with which the constituent units are arranged. Normally, crystallinity presupposes an indefinite repetition of identical units in three dimensions, and, as an external indication of their internal regularity, true crystals are usually bounded by plane surfaces. In liquid crystals the regularity is incomplete. Between the perfect regularity of a true crystal and the completely random arrangement characteristic of amorphous materials, there are many possible arrangements showing intermediate degrees of regularity. Some of these are known as liquid crystals. The only one which need be considered here is one that had not previously been discovered but corresponds to the class RRD in HERMAN's (1931) classification of liquid crystal systems (BERNAL and FANKUCHEN, 1941). It is formed by the parallel orientation of particles at equal distances from one another, so that there is a two dimensional regularity at right angles to the direction of the particles but no regularity of arrangement in the direction of their length. Such a degree of orderly arrangement is confined to liquids and wax-like solids, the latter being described as paracrystalline or mesomorphic.

Solid forms of tobacco mosaic virus which are birefringent can be obtained in a number of ways. When neutral solutions of the virus are centrifuged at high speeds, birefringent gels are deposited. The solid content of the gels depends directly on the concentration of the fluid centrifuged, and varies between 10 and 35%. Similar gels are also formed when neutral solutions are concentrated by evaporation. The structure in the gels is essentially the same as that in the bottom layer solutions, *i. e.*, they are composed of regions of parallel orientation arranged at random. However, they can be further orientated by rolling between two glass slides or by being sucked into capillary tubes, so that the whole preparation behaves as a single liquid crystal and shows extinction positions when rotated in a beam of polarised light.

When solutions of purified tobacco mosaic virus are allowed to dry, a birefringent skin forms at the surface, showing that the particles are orientated relative to the air-water interface. If some solution is introduced between a microscope slide and a cover-glass, kept apart by two glass threads, and allowed to evaporate from one surface, well orientated solid preparations of the virus can be obtained. During the drying the virus solutions pass through a number of different conditions. As the solutions are concentrated their birefringence increases to a limiting value of about 0.001 (BERNAL and FANKUCHEN, 1937). The first distinguishable layer to be formed at the surface is a wet gel containing about 50% of water and with a higher birefringence of 0.006. On further drying, this layer shrinks by about 50% and passes into a dry gel with birefringence 0.003, intermediate between that of the liquid and the wet gel. The dry gel cracks if further dried and splits into single pieces of approximate 1 × 0.5 × 0.5 mm., which can be examined by X-rays as if they were ordinary crystals.

Orientated preparations of different concentrations give X-ray patterns showing some remarkable differences and similarities. The patterns can conveniently be considered in two parts: one of scattering at large angles, which corresponds to small spacings, and the other of scattering at small angles, which corresponds to large spacings. For the large angle scattering there is little difference between the patterns given by virus at all concentrations from the dry gel to top layer solutions orientated by flow through capillary tubes. As this pattern is independent of the distance the particles are apart, it can only result from regularities within the virus particles themselves and can be referred to as the intramolecular pattern. The pattern for the small angle scattering varies with concentration of the preparation, the more dilute the solutions the greater the spacings, and clearly results from the regular packing together of the virus particles. The virus preparations, therefore, possess two kinds of regularity. The individual particles are built up from subunits which are arranged with the perfect regularity of a crystalline lattice, while the particles may also associate together in a regular manner. The visible aggregates of tobacco mosaic virus, however, lack a complete three-dimensional regularity. All the intermolecular reflections observed in X-ray photographs lie in a plane at right angles to the length of the particles, that is, they refer to the relative sideway positions of the particles in the aggregates. A careful search has failed to reveal intermolecular reflections in a direction perpendicular to the orientation axis. Hence, there is no evidence of more than a two dimensional regularity in the arrangement with which the particles pack together. Similarly, the electron micrograms made by WYCKOFF (1947) show a parallel alignment of particles but reveal no orderly arrangement extending to the third dimension.

The most complete X-ray pattern is given by the dry gels. The reflections correspond to the first four planes of a two-dimensional hexagonal packing of side 15.2 μ (BAWDEN and others, 1936; BERNAL and FANKUCHEN, 1937). The lines are sharp, indicating that the areas of regular packing are large compared with the size of the particles, and agree perfectly with those calculated for a hexagonal lattice. The patterns for wet gel and orientated fluids are less complete, but again correspond to a hexagonal packing. In the wet gel the spacings correspond to an intermolecular distance of 21 mμ, and in the fluids the spacings increase with the dilution.

The addition of acid or ammonium sulphate to purified tobacco mosaic preparations produces an opaque precipitate that shows an intense sheen. It is composed of needle-shaped, birefringent bodies, tapering at

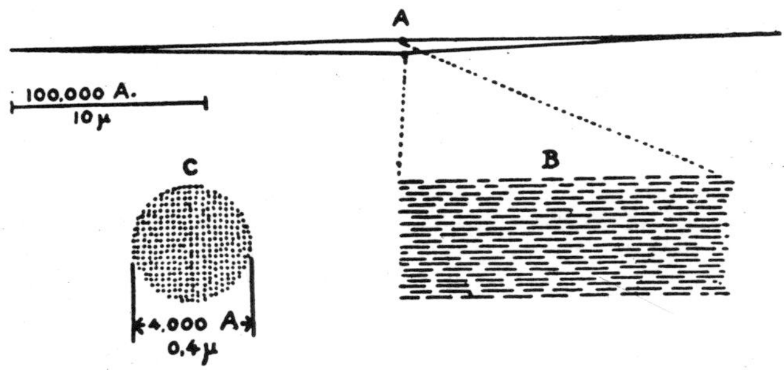


FIG. 42. — Diagrammatic representation of tobacco mosaic virus paracrystals. A: Typical paracrystal. B: Longitudinal section. C: Transverse section. Note regularity of arrangement in cross-section, but none in the direction of the length of the particles. (BERNAL, J. D. and FANKUCHEN, I., 1937, Nature *139*, 923).

both ends. They average about 40 μ long and 0.4 μ thick, look fibrous and microscope examination reveals no facets. STANLEY (1937*b*) and WYCKOFF and COREY (1936) concluded that they were crystals, but later work has shown that they are more accurately described as paracrystals or mesomorphic fibres. There is no evidence that the virus particles are arranged in the needles with a higher degree of regularity than that demonstrated in the preparations described as wet gel. That is, there is a regularity of arrangement in the cross-section, but no evidence of any along the length.

If the precipitate produced by the addition of acid is sedimented in an ordinary centrifuge tube, a dense white precipitate results containing about 75% of water. On the other hand, if it is centrifuged in the end of an L2 Chamberland filter candle, the acid precipitate is converted into a soft, translucent mass with a water content of about 50%, that is, similar to the water content of the wet gel. This translucent precipitate can be orientated completely by rolling between glass slides, when it becomes transparent and highly birefringent (BAWDEN and PIRIE, 1937*d*). In this condition its optical properties and X-ray pattern are indistinguishable from those of wet gel. However, when it is soaked in water and shaken it returns to a suspension of needles with the characteristic sheen. It seems, then, that the particles in the needle-shaped precipitates are also packed with a perfect two-dimensional regularity at right angles to their length, but again there is no evidence of regular arrangement in the direction of their length. Their most probable internal arrangement is that shown in Fig. 42 (BERNAL and FANKUCHEN, 1937).

Attempts have been made to increase the size of the paracrystals by slow precipitation methods (BAWDEN and PIRIE, 1937*a*). These failed, but suggested that the first step in the production of the paracrystals is the formation of a gel. If neutral solutions of the virus are kept still while the acidity or salt content is slowly increased, they gradually turn into limp jellies. When shaken or stirred these immediately break up to give a suspension of needles. The process of gel formation is most conveniently studied in mixtures containing 0.5-1.0% of the virus, 0.25-0.5% of glycine and 5% of neutralised ethyl formate. The *p*H is adjusted to about 5 and the fluid immediately placed into vessels in which it can be observed without being disturbed. As the ethyl formate undergoes hydrolysis, the *p*H is reduced and ultimately reaches the precipitation point 3.3. No change in the optical properties of such dilute solutions occurs, although the mixture turns to a fairly rigid jelly. If stirred it changes instantly to a suspension of needles with the typical sheen. It now behaves exactly like virus solutions to which acid has been added directly, and if allowed to stand undisturbed does not revert to the gel state. If more concentrated solutions are slowly acidified in this way they behave rather differently. The undisturbed gel breaks up to give a mosaic of birefringent spindles disposed at random, resembling bottom-layer fluids. When stirred, however, this also breaks up to give a suspension of needles.

These phenomena all suggest that the visible precipitates of tobacco mosaic virus are more accurately described as fibrous than as crystalline, and that their regular appearance under the microscope is to be attributed to the more or less regular breaking up of a gel rather than to the wholly regular process of crystal formation. Other properties of the needles also serve to identify them as pieces of gel. On pressing, they readily fuse together, so emphasising their partially liquid nature. Also, if the precipitate obtained at *p*H 4.2 in salt-free solutions is centrifuged off, it produces an opaque, white mass. When this is covered with an N/10 salt solution and allowed to remain undisturbed, the upper layers gradually change to a transparent, birefringent jelly. If slightly agitated, the jelly breaks up to give rise to long birefringent fibres, presumably similar to those described by BEST (1937*b*) in purified virus preparations around this *p*H value. When such suspensions are shaken vigorously, the long fibres break up and the virus apparently goes into solution. By transmitted light such solutions are transparent, but by reflected light they show an intense sheen. They are viscous and thixotropic; if allowed to stand undisturbed they turn into birefringent jellies but can be made to flow again if stirred or shaken.

BEST (1937*a*) has pictured long birefringent fibres that settle from infective sap when stored for some months at 0°C. (Fig. 43). When undisturbed, the fibres may be several centimetres long; they are flexible but usually break into fragments 2-5 mm. long when mounted under a cover glass. Gentle agitation breaks them into small needles similar to those obtained by precipitation with acid or ammonium sulphate, and violent shaking reduces them to a state in which they are no longer visible. When the shaken fluids stand undisturbed, the fibres reform, the time needed for their production decreasing with increasing age of the preparation. Similar birefringent fibres that are disrupted and dispersed by shaking are produced by purified virus preparations in the presence of clupein, nicotine, guanidine and arginine (BAWDEN and PIRIE, 1940*a*), and by exposure to concentrations of salt too dilute to cause rapid precipitation in the form of small needles (BEST, 1940, 1947).

Suspensions of precipitated tobacco mosaic virus show anisotropy of flow at higher dilutions than do neutral solutions, suggesting that the constituent particles are spontaneously orientated in the paracrystals. On the addition of acid or salt to neutral solutions, there is an increase in anisotropy of flow before conditions are such that a solid phase separates. This increase probably results from an end-to-end association of

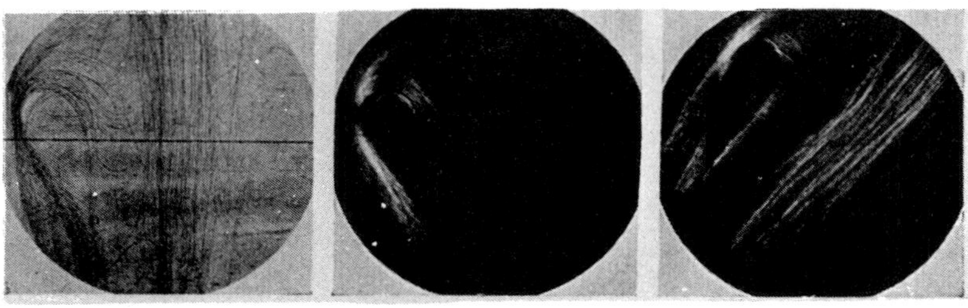

**FIG. 43. — *Left:* Photomicrograph of mesomorphic fibres produced in the clarified sap from plants suffering from mosaic after storing at 1° C. for several months. *Centre:* The same field as in A, photographed between crossed Nicol prisms. *Right:* The same field between crossed Nicols after rotating the stage through 45°. × 60. (BEST, R. J., 1937, Nature *139*, 628).**

individual particles to form aggregates with a still greater length to width ratio and which are therefore still more easily orientated. LAUFFER (1938*a*) has measured the specific viscosity of tobacco mosaic virus solutions at various hydrogen ion concentrations and obtained further evidence for the two stages in the production of paracrystals. Between *p*H 7 and 5.5 the viscosity is almost constant, but it rises sharply between *p*H 5.5 and *p*H 4, indicating an increase in the length to width ratio of the particles. Near the isoelectric point, where the virus is precipitated, the viscosity falls to a value approximately the same as that at *p*H 7. By contrast, the anisotropy of flow is almost unchanged between *p*H 5.5 and 3.3. The difference in the behaviour of anisotropy of flow and viscosity at the isoelectric point can possibly be explained by changes in the length to width ratios of the virus aggregates. According to LAUFFER, viscosity is a function of both the length and thickness of particles, whereas anisotropy of flow is determined chiefly by the length. Thus, if the first stage in the formation of visible aggregates is the formation of longer particles by a linear aggregation, both viscosity and anisotropy of flow would be increased. On the other hand, if the second stage is a side-to-side association of these long thin aggregates, viscosity would be greatly reduced while anisotropy of flow would be little affected.

From most points of view the distinctions between true crystals and solids in the liquid crystalline state are of little practical importance. But there are two reasons why it is of value to differentiate between the paracrystals of tobacco mosaic virus and true crystals. First, it prevents the apparent crystallinity of virus preparations being advanced as evidence of their purity. Were they true crystals their formation would by no means necessarily show that the preparations were homogeneous, for a number of crystalline protein preparations (*e. g.* trypsin) at first thought to be homogeneous have later been shown to be heterogeneous. If they are regarded as pieces of jelly containing 50% of free water between the particles, however, this danger is avoided, for it is obvious

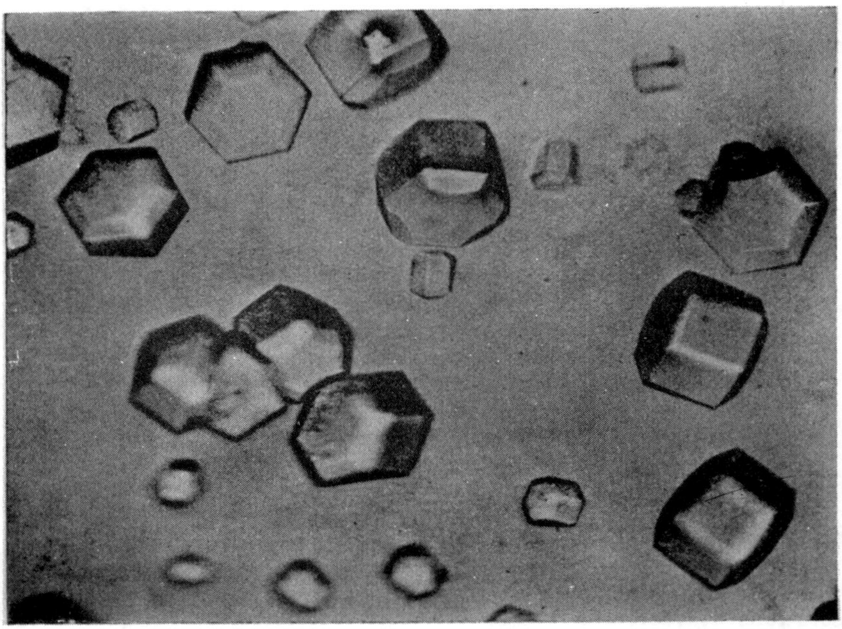

FIG. 44*a* and *b*. — Hexagonal birefringent prisms formed by a tobacco necrosis virus. — *a:* Photographed by transmitted light, and *b:* Photographed between crossed Nicol prisms.

that they can contain impurities and that the fact that repeated "recrystallisations" has no effect on activity cannot be taken as evidence of homogeneity. Secondly, the differentiation is important because it indicates that the virus particles as formed in the plant may be in a different physical state from those in the purified preparations.

No solid preparations of purified tobacco mosaic virus have yet been produced *in vitro* in which a three-dimensional regularity has been demonstrated. In the infected plants, however, true crystals occur in large numbers. For reasons discussed in Chapter 3, these crystals may not be pure virus, but that they are rich in virus is shown by their change when acid is added into needle-shaped fibres, microscopically indistinguishable from those produced by the addition of acid to solutions of the purified virus. Their relation to the virus is further indicated by their constant association with infection, by their hexagonal habit and positive birefringence. That they possess a full three-dimensional regularity, and hence may contain particles of identical length, is shown by the fact that they are bounded by plane surfaces symmetrically arranged. It is, of course, possible that in the plant the virus unites to form true crystals with a substance which has not been tested *in vitro*, or that in the plant a process of slow crystallisation goes on that has not been simulated *in vitro*. BERNAL (1938), however, considers that the crystals are typical of less asymmetrical particles than those found in purified preparations. Hence, it is possible that, in spite of their great asymmetry *in vitro*, the virus particles as produced in the plant are more uniform and less elongated. The ready transition of the material in the crystals to the paracrystalline needles suggests that these postulated primitive particles

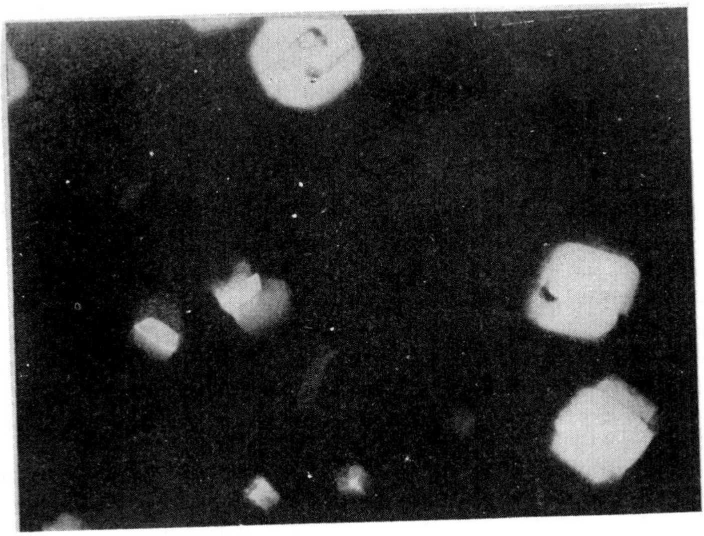

FIGURE 44*b* (*see* p. 214).

are already arranged in rows, and that the effect of acid is first to remove intercalary material, so allowing them to aggregate into separate long rods, and to form the visible needle-shaped aggregates. KAUSCHE (1939) claims that the virus can pass from the needle form to hexagonal crystals *in vitro*. He precipitated virus from sap with ammonium sulphate and then added clarified infective sap to the virus-containing precipitate. KAUSCHE examined this mixture microscopically and stated that the hexagonal crystals developed out of the precipitate of needles. At Rothamsted we have been unable to confirm this. When such mixtures of sap and precipitated virus were made, the precipitated virus slowly went into solution, unless the concentration of ammonium sulphate was too great when it remained unaffected. The only crystals formed were those of calcium sulphate, produced from the calcium salts in the sap and the ammonium sulphate in the precipitated material.

Other strains of tobacco mosaic virus that infect solanaceous plants behave, both in plants and *in vitro*, in much the same manner as tobacco mosaic virus itself. Cucumber viruses 3 and 4 also form the same types of paracrystalline precipitates *in vitro*, but no crystalline or paracrystalline inclusions have been seen in plants infected with them. Whether this indicates any differences in properties between these and other strains, or results from differences between solanaceous and cucurbitaceous plants, is unknown. The crystallinity of potato virus X resembles that of tobacco mosaic virus in some respects, but differs strikingly in others. Dilute solutions show anisotropy of flow to about the same extent, and liquid crystalline layers separate at about the same concentration; also when the virus is sedimented by ultracentrifugation, it forms birefringent gels. On the other hand, when it is precipitated from solution by acid, salts or other substances, the floccules are isotropic and amorphous. With quantities of salt or acid insufficient to cause flocculation, there is an increase in anisotropy of flow, suggesting that the particles, like those of tobacco mosaic virus, are first aggregating linearly, but on coming out of solution it seems that they associate totally at random. No reasons are known for the differences in behaviour, but electron micrograms of virus X suggest that the particles are more flexible than those of tobacco mosaic virus and that they intertwine laterally to a much greater extent (TAKAHASHI and RAWLINS, 1946; BAWDEN and CROOK, 1947). Thus it may be that the greatly elongated particles lack the necessary rigidity to maintain a parallel alignment. Potato virus Y and henbane mosaic virus behave in much the same manner as potato virus X, but have not yet been examined in such great detail.

The inability of tobacco mosaic virus and potato virus X to form true crystals *in vitro* may well be the result of changes in particle size that occur during precipitation and that produce heterogeneous products, for all preparations contain particles of widely different size and the production of true crystals demands some degree of uniformity. With viruses whose particles are uniform and remain discrete, crystals are readily producible, though different viruses form them in different conditions and adopt characteristic shapes. Slow precipitation from salt solutions in the cold produces crystals of several, but not all. In these conditions tomato bushy stunt virus crystallizes as isotropic rhombic dodecahedra (BAWDEN and PIRIE, 1938*b*), whereas when crystallized with heparin it forms isotropic prisms (COHEN, 1942). Several of the viruses that cause tobacco necrosis also crystallize from cold ammonium sulphate solutions. The serologically related strains called potato and tobacco VI by BAWDEN and PIRIE (1942) crystallize consistently as thin lozenge-shaped plates (Fig. 22a) and hexagonal prisms respectively; prisms with twenty facets are pictured in Fig. 44, which shows that they are birefringent when viewed along one axis but not along the other. A slight variant of this crystal form, hexagonal prisms with pointed ends and eighteen facets, is shown in Fig. 22b. The culture called tobacco I, which is not serologically related to the others, crystallizes in a variety of different forms. The two commonest are isotropic dodecahedra and birefringent bipyramids, but thin round laminae and elaborately twinned structures also occur (Figs. 33, 34, 35). The causes that determine the crystal form are unknown; if infective sap is divided and the two parts purified separately, one sometimes crystallizes in one form and the second in another. Similarly, if dodecahedral crystals are dissolved and then recrystallized, bipyramids or circular laminae may be produced, and attempts to influence the crystallization by "seeding" with different forms have failed.

Another tobacco necrosis virus, the Rothamsted culture, which is serologically unrelated to the others, has never crystallized from cold salt solutions. When sedimented from neutral solutions, instead of the isotropic gels produced by the others, the Rothamsted culture gives birefringent crystalline pellets (Fig. 45a). Crystallization occurs, too, when salt-free solutions more concentrated than above 0.5% are kept at 0°C.,

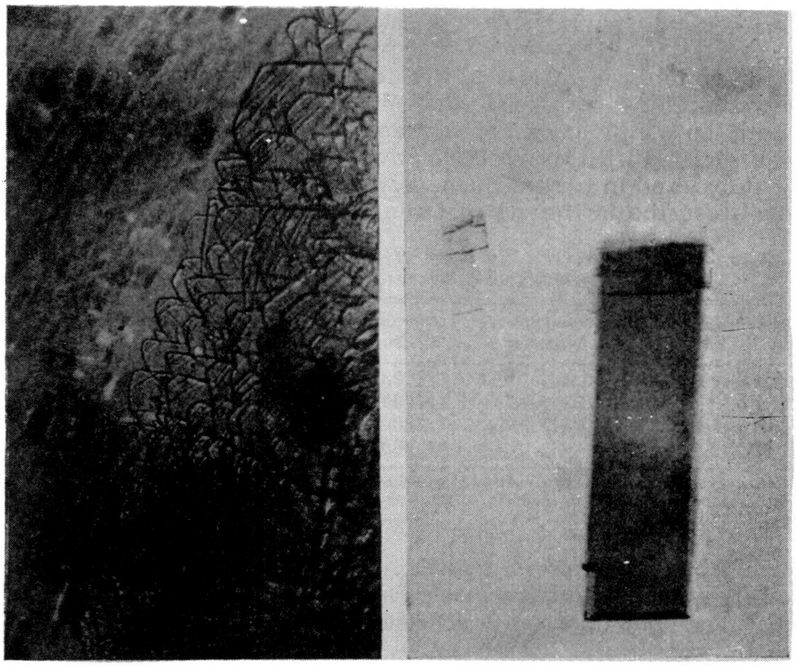

FIG. 45. — Crystals of the Rothamsted culture of tobacco necrosis virus. *Left:* The serrated edge of a crystalline pellet sedimented by ultracentrifugation. × 45. *Right:* Large crystals produced after crystal growth had proceeded in a salt-free solution for three months. ×9. (BAWDEN, F. C. and PIRIE, N. W., 1945, Brit. J. exp. Path., *26*, 277).

and produces thick triclinic prisms and thin hexagonal, or pseudohexagonal, plates. These begin to form after a few days, but growth continues for months and will give rise to single crystals several millimeters long (Fig. 45*b*). The relationship of this crystalline material to infective virus is uncertain, for by the time crystallization has proceeded the preparations have lost most of their infectivity. The crystalline pellets that are sedimented from fresh preparations are demonstrably inhomogeneous, in spite of their apparent uniformity and complete crystallinity. The pellets dissolve only slowly and by extracting them several times with successive lots of water, a series of fractions can be obtained containing material that dissolved at different rates. Material in the successive fractions may have the same sedimentation constant and serological activity, but it differs in infectivity, there being a steady reduction in infectivity in passing from the most to the least rapidly dissolving (TABLE 20).

The rate at which pellets dissolve decreases as the preparations age and lose infectivity, and though no positive interpretation of these results can be given, it seems likely that the infective particles pass into non-infective forms of the same size and serological activity, but which are less soluble and crystallize more readily. However, it is equally possible that infective particles form only a very minor part of even fresh preparations and that the crystallizable nucleoprotein is not directly related to the virus but is a secondary effect of infection (BAWDEN and PIRIE, 1945*a*).

Another tobacco necrosis virus, the potato culture, also crystallizes from water; this gives lozenge-shaped plates similar to those obtained by precipitation with ammonium sulphate. The crystallization of this virus is not associated with loss of infectivity, such as occurs with the Rothamsted culture, and in partially crystallized preparations there is no difference between the infectivity of the crystalline fraction and that of the

TABLE 20.—*Properties of successive extracts of a crystalline pellet of Rothamsted tobacco necrosis virus:—*

| *Extract* | *Sedimentation constant* | *Serological titre* | *Average lesions per leaf at* $10^{-4}$ | $10^{-5}$ |
|---|---|---|---|---|
| First | 49.9 | 1/300,000 | 430 | 230 |
| Second | 48.5 | 1/300,000 | 240 | 63 |
| Third | 49.7 | 1/300,000 | 89 | 17 |
| Fourth | — | 1/300,000 | 30 | 7 |

The first extract was made by stirring the pellet with water and centrifuging at once; for the second and third, the pellet was left to soak for 30 mins. each and for the fourth for 24 hours before centrifuging at 3,500 r.p.m.

fraction which remains in solution. Southern bean mosaic virus also crystallizes from salt-free solutions at *p*H 5.5 to produce various kinds of isotropic rhombic plates and prisms; these resemble the crystals obtained by treatment with ammonium or magnesium sulphates, but are larger, sometimes reaching 3-4 mm. in length (MILLER and PRICE, 1946*c*). Turnip yellow mosaic virus resembles the Rothamsted tobacco necrosis virus in forming a crystalline pellet when sedimented by ultracentrifugation; from salt solutions it crystallizes as isotropic octahedra, and from 20% alcohol at *p*H 3.7 as birefringent needles. As yet there is little information about the crystallinity of other viruses that have been purified. Tobacco ringspot virus forms true crystals in infected plants (BAWDEN and SHEFFIELD, 1939), but its behaviour *in vitro* is uncertain. STANLEY and WYCKOFF (1937) claimed to have isolated this virus as "a high molecular weight crystalline protein", but the meaning to be attributed to crystalline was not defined. Later it was stated to have rod-shaped particles and to show anisotropy of flow (STANLEY, 1938*b*; LAUFFER and STANLEY, 1938; STANLEY and LORING, 1939), but in the most recent paper on tobacco ringspot virus, STANLEY (1939) described it as having spherical particles which give isotropic solutions and amorphous precipitates.

In addition to its possible value as a fractionation procedure, the production of crystals, or of liquid crystals, is useful in making X-ray analysis a technique applicable to studying the size and structure of virus particles. Much information has already been gained about viruses from these techniques and more can be confidently expected. As a reverse benefit, the study of virus crystals also promises to provide a new approach to problems of crystal formation, for by the use of electron micro-

FIG. 46. — An electron microgram showing the regular array of particles on faces of several crystals formed by a tobacco necrosis virus. × 60,000. (MARKHAM, R., SMITH, K. M. and WYCKOFF, R. W. G., 1948, Nature *161*, 760).

scopy the manner in which particles arrange themselves to form crystals becomes directly observable for the first time. PRICE, WILLIAMS and WYCKOFF (1945; 1946) noted that electron micrograms of tomato bushy stunt, tobacco necrosis and southern bean mosaic viruses often showed an orderly arrangement of particles. Usually the arrangements were restricted to a single layer, so that only a two-dimensional regularity was observed, but in some, particularly with southern bean mosaic virus, orderly layers of particles were stacked one above another to produce the three-dimensional regularity of true crystallinity. More recently pictures have been obtained of single small crystals of a tobacco necrosis virus which show the regular distribution of particles over several faces of the crystal (Fig. 46) and suggest a symmetry that it at least pseudocubic (MARKHAM, SMITH and WYCKOFF, 1947, 1948). Similarly electron micrograms of turnip yellow mosaic virus indicate that the particles in the octahedral crystals are arranged in a diamond-type lattice, a structure that was suggested also from X-ray analysis (COSSLETT and MARKHAM, 1948).

*Chapter 12*

# THE ESTIMATION OF PARTICLE SIZES

Few problems have received more attention in the study of plant viruses than the estimation of particle size; certainly no other single problem has attracted such a wide variety of ingenious and impressive techniques or has compelled the use of so much costly apparatus and of so many mathematical formulae. Few problems, too, have produced so many conflicting conclusions. Tobacco mosaic has been studied more extensively than any other virus and the values recorded for its size are about as numerous and different as the workers engaged and the methods employed. Many accounts of viruses contain tables or diagrams in which viruses are listed in order of size, their sizes sometimes being given with an accuracy of two or three significant figures. It is a valuable corrective to the air of precision carried by such tables to compare the values recorded at different times over the last dozen years or so, when it is found that tobacco mosaic virus has moved from being the smallest to the largest of those that have been studied. Such variable conclusions were probably inevitable from the nature of the problem, but they might have caused less amusement and scepticism had they been recorded with rather more qualifications. Size could not be found directly, but only calculated from measurements on such physical properties as filterability, sedimentation, diffusion and viscosity, and the calculations necessitated many assumptions, the validity of which was far from certain. Unfortunately these assumptions have often been accepted unreservedly, and some workers have seemed in no way concerned to indicate the uncertainties inherent in their methods or to distinguish adequately between the measurements they made and the deductions they drew from those measurements. The methods of electron microscopy and X-ray crystal analysis allow a more direct approach to the problem and have already given us more precise results than were previously possible; they have also shown some of the reasons for the variable results obtained by other techniques. However, they in turn have their limitations, and, in spite of the immense amount of work that has been put into the problem, it has still to be admitted that we do not know with certainty the exact size of the particles of any plant viruses. Our knowledge can be briefly summarised. About the great majority, nothing is known except that they are too small to be resolved by microscopes using visible light. The sizes of a few are known within fairly narrow limits, and some are known to have particles of widely variable lengths. No attempt is made in this chapter to reach any final conclusions about the sizes of different viruses, but rather to discuss the methods used, in order to indicate their relative values and limitations.

**Filtration: —** The existence of viruses was first recognised because of their ability to pass through filters that stop bacteria. This showed that viruses were smaller or less readily absorbed than other pathogens, but gave no information about their actual sizes. The porosity of ordinary filter-candles is too uncertain and variable, and their capacity for absorbing virus is too great, for calculations of size to be possible from filtration results. Before this could be done, new types of filters had to be developed. The pioneer work of BECHHOLD (1907) showed that membranes with pores of different average sizes could be produced from gelatin in water and collodion in acetic acid, and since ELFORD (1931) introduced his "gradocol" membranes, filtration has been extensively used, especially with animal viruses, to estimate particle size. The membranes are produced by coagulating mixtures of ether, alcohol, acetone and collodion under carefully controlled conditions. Pore size is varied within narrow limits by altering the period of heating or the proportion of amyl alcohol, and within wider limits by adding water or acetic acid to the other constituents. The average pore diameter (A. P. D.) of the membranes is calculated by measuring the rate at which water flows through the membranes under known pressure, applying Poiseuille's viscosity law and assuming that the membrane has certain structural regularities.

Results on the filterability of viruses are usually given as filtration end-points, these being the A.P.D. of the most porous membranes that prevent the passage of the virus particles. The end-point is found by filtering preparations through a series of membranes with successively smaller A.P.D. until a non-infective filtrate is obtained. The size of the virus particles is then estimated by multiplying the end-point by a factor. ELFORD (1933) derived an empirical relationship between the A.P.D. and the size of particle retained and concluded that the relationship varied with the size of the pores. Thus the diameter of retained particles was considered to be from 0.33 to 0.5 of the A.P.D. with membranes of A.P.D. between 10 and 100 mμ, from 0.5 to 0.75 with membranes of A.P.D. between 100 and 500 mμ, and from 0.75 to 1 with membranes of A.P.D. between 500 and 1,000 mμ. MARKHAM, SMITH and LEA (1942) considered that incorrect sizes were attributed to some of the particles used by ELFORD to obtain his relationship between A.P.D. and size of retained particle and they found no systematic change between the ratio of particle size to A.P.D. as A.P.D. increases. These workers decided that similar relationships between A.P.D. and particle size hold over the whole range from 10 to 750 mμ and that, at all sizes, there are considerable variations between A.P.D. and size of retained particles. They suggest that experimental errors inherent in the method make it impossible to get a closer relationship than size of particle = A.P.D. × 0.55-0.95.

Clearly, even if it is assumed that the filter membranes have uniform pores, and there is evidence with the coarser membranes that this is not so (GRABAR, 1938), the method can give only approximations to sizes. Again it can do this only if the particles are spherical and all of the same size. If the preparation being filtered is polydisperse, the size estimated will be for the smallest particles and not an average. In spite of its limitations and uncertainties, the method has been applied with considerable success to many animal viruses, which give sharp and reproducible filtration end-points. The results suggest that different animal viruses have widely different particle sizes and cover a range from just below the limit of resolution by ordinary microscope to particles of about 10 mμ in diameter.

With plant viruses, filtration has been less successful and has given many confusing and conflicting results. A few viruses, for example, tomato bushy stunt, have behaved consistently and given reasonably sharp and reproducible end-points, but others have not. Some, although they filter fairly readily, give variable end-points, and much virus is stopped by membranes with A.P.D. much larger than the end-point. Others may fail to pass through membranes with A.P.D. of more than 500 mμ, a size that should permit the passage of particles large enough to be visible (SMITH and DONCASTER, 1936; SMITH and MACCLEMENT, 1941). Some of the difficulties experienced in applying filtration to the study of plant viruses are undoubtedly caused by the properties of plant saps. The presence of proteins and mucilaginous substances tends to block the pores and prevent the passage of virus particles. Also, the sap of most plants is acid, so that the particles are near to their isoelectric points, conditions which cause some to aggregate and in which all carry small charges and so are readily absorbed on to the filters. THORNBERRY (1935) showed that filterability was much improved by freezing leaves before extracting sap and by adding *p*H 8.5 phosphate buffer and nutrient broth. In these conditions tobacco ringspot and cucumber mosaic viruses, both of which when tested in sap usually fail to pass through coarse membranes, filtered readily and gave similar end-points, around 45 mμ, to tobacco mosaic and potato virus X.

The filtration end-points recorded for tobacco mosaic virus and potato virus X have varied from 13 to 450 mμ for the former and from 45 mμ to 450 mμ for the latter. The filterability of tobacco mosaic virus depends on the treatments to which the preparation has been subjected and on the method whereby filtration is done, almost certainly because the particles can aggregate linearly to produce rods of greatly differing lengths. With rod-like particles of variable lengths, the most that should be expected from filtration is an estimation of width, and this could be expected only if the particles were orientated so that they pass straight through the pores without jamming. The efficacy of orientating the particles has been shown by SMITH and MACCLEMENT (1941), who filtered tobacco mosaic virus in a cataphoretic cell. As normally filtered under pressure, the virus in sap gives end-points varying between 50 and 180 mμ, the variations presumably depending on the degree to which the virus particles have aggregated before being filtered. The passage of an electric current through a preparation orientates the virus particles, which because of their negative charge move towards the anode. By placing a filter membrane between the anode and cathode, so that the pores are parallel to the direction of flow of the particles, filtration end-points of 13 to 15 mμ can be obtained, giving a measurement of width that agrees remarkably well with those obtained by more exact methods.

SMITH and MACCLEMENT (1940) obtained an unusual result with sap from plants suffering from tobacco necrosis. The end-point was constant at about 40 mμ, but much virus failed to pass through membranes with considerably larger pores. There is no evidence that these particles aggregate and a different kind of polydisperse system is probably responsible for this result. It is now known that tobacco necrosis is caused by a number of viruses, some of which have particles of different sizes, and it seems likely that SMITH and MACCLEMENT were working with mixed cultures, membranes with A.P.D. of 40 mμ being needed to stop the smallest viruses present and the others being stopped by coarser membranes. By plotting the relative infectivity of filtrates against the A.P.D. of the membranes through which they have passed, filtration may provide information on factors other than size. When infectivity drops steeply near the end-point, it can be inferred that the preparation is reasonably homogeneous and the end-point can then be assumed, with some justification, to be related to particle size. When a gradually sloping curve is obtained, however, a polydisperse system should be suspected and no estimation of size from end-point should be made until some explanation for the heterogeneity is known. Such curves are particularly useful in showing the presence of small particles whose occurrence might be undetected by other methods, and filtration also gives a convenient method of obtaining such particles free from larger ones.

**Sedimentation and Diffusion:** — The fact that tobacco mosaic virus is sedimented by high speed centrifugation was first demonstrated by

TABLE 21. — *Sedimentation constants of various viruses and virus strains:—*

| VIRUS | $S_{20} \times 10^{-13}$ |
|---|---|
| Tobacco mosaic | 170, 200, 235 |
| Aucuba mosaic | 185, 220, 256 |
| Masked tobacco mosaic | 211 |
| Cucumber viruses 3 and 4 | 173, 200 |
| Potato virus X | 113, 131 |
| Tobacco ringspot | 115 |
| Tomato bushy stunt | 132 |
| Tobacco necrosis (Princeton culture) | 116 |
| (Potato culture) | 116 |
| (Rothamsted culture) | 50, 235 |
| Turnip yellow mosaic (nucleoprotein) | 110 |
| (protein only) | 51 |
| Southern bean mosaic | 114 |
| Alfalfa mosaic | 74 |

BECHHOLD and SCHLESINGER in 1933, and since then high speed centrifugation has been increasingly used for the purification of viruses and in attempts to estimate particle size. Size has usually been calculated from measurements on the rate of sedimentation. The details of the underlying theory, and the assumptions entailed in deriving the necessary formulae, are to be found in numerous publications (SVEDBERG and PEDERSEN, 1940; COHN and EDSALL, 1943; MARKHAM, SMITH and LEA, 1942; LAUFFER, 1946) and only a brief general outline need be given here.

Most determinations of sedimentation have been made in oil- or air-driven ultracentrifuges specially designed to allow sedimentation to proceed without interference by convection or vibration. Small samples of the solutions to be examined are enclosed in sector-shaped cells with plane parallel quartz windows, through which they can be observed and photographed as they are revolved. Under ordinary conditions viruses form stable solutions, the particles being uniformly distributed because the speed at which they diffuse is greater than that with which they fall under gravity. When the gravitational force is increased sufficiently by centrifuging, the mixing action of diffusion is overcome and the particles begin to sediment. A boundary then develops between the top of the fluid which has been freed from particles and the lower part in which they are still present. Uniform particles sediment at the same rate, and solutions of homogeneous particles give single sharp boundaries (Fig. 47). Particles of different sizes and weights sediment at different rates, and heterogeneous preparations are characterised by diffuse boundaries or by the production of more than one boundary (Fig. 48).

The movement of the boundary is observed in two ways, by making use of the fact that the solutions and the solvent differ in their ability either to absorb or to refract light. Neither method specifically identifies viruses, for the ability to absorb light of the wavelengths used, or to have a different refractive index from water, is shared by many substances. The refractive index method is now the more widely used, and it more readily detects the presence of particles sedimenting at different rates; with the light absorption method, the presence of particles heavier than the majority is particularly likely to pass unnoticed, because their pres-

FIG. 47. — A series of sedimentation pictures of tomato bushy stunt virus, 0.27% in 0.02 M acetate buffer. Centrifugal force 20,000 times gravity, 9 minutes between each exposure. Note perfectly sharp boundary in all exposures. (MCFARLANE, A. S. and KEKWICK, R. A., 1938, Biochem. J. *32*, 1607).

ence is easily masked by the great absorption of light by the major component. However, neither method can be relied on to detect the presence of components that do not constitute at least 10% of the whole preparation. This may explain why solutions of tobacco mosaic virus often give a single sedimentation constant though they demonstrably contain particles of many different sizes.

The rate at which the boundaries move under a known gravitational field is measured and from these measurements the sedimentation constant is determined. It is the rate of sedimentation in cm./sec. under an accelerating force of 1 dyne, conventionally converted to the frictional conditions represented by water at 20°C. The constant is expressed as $s^{w}_{20} = X \times 10^{-13}$ cm. sec.$^{-1}$ dynes$^{-1}$. The sedimentation constants determined for purified preparations of various plant viruses are listed in TABLE 21. The results show that some viruses give single constants, whereas others give several. The former are mostly viruses that have been obtained in crystalline forms; they have particles that are spherical or nearly so, and purified preparations usually give sharp sedimentation boundaries. The latter are those with anisometric particles that also give variable filtration end-points; boundaries obtained are usually rather diffuse.

From measurements of the sedimentation constant it is possible to calculate values for particle weight and size, provided the partial specific volume is known and enough assumptions are made. It must be assumed that Stokes' law holds for the sedimentation of virus particles, that these particles are smooth, unhydrated spheres which do not interact with one another, and that the solutions are structureless and have no anomalous physical properties. By making similar assumptions, particle weight and size can also be calculated from the diffusion constant ($D_{20}o$), which is the rate of transfer under a unit concentration gradient. Diffusion constants have been determined for tobacco mosaic, tomato bushy stunt and southern bean mosaic viruses. As is usual with measurements of most properties with tobacco mosaic virus, different workers have obtained different results, ranging from $5.3 \times 10^{-8}$ sq. cm. per sec. (LAUFFER, 1944), to $3 \times 10^{-8}$ (NEURATH and SAUM, 1938) and $4.5 \times 10^{-9}$ (FRAMPTON and SAUM, 1939). When particle weights are calculated from these values, the results obtained are about a hundred times as great as when calculated from sedimentation constants. Clearly the basic assumptions do not hold for this system. The reasons are not far to seek. The optical properties of solutions show the presence of anisometric particles, and there is abundant evidence that the particles are of various lengths and interact with one another over relatively large distances. Solutions do not show normal Brownian movement, they are thixotropic and have anomalous viscosity and diffusion (BAWDEN and PIRIE, 1937*a*; ROBINSON, 1939; FRAMPTON, 1939, 1940). FRAMPTON (1939, 1939*a*) has stressed the ambiguity of estimates of size of tobacco mosaic virus based on measurement of physical properties and considers that any agreement between different methods is purely fortuitous. He vividly illustrates the difficulties of working with such anomalous material by calculating a particle weight of infinity from sedimentation data and of zero from the diffusion constant. His prediction that solutions do not fulfil the requirements of Stokes' law, has been confirmed by LAUFFER's (1940) observation that the sedimentation constant given by the same preparations varies with the concentration. The $s_{20}$ increases by 2.9 for every decrease in concentration of 0.1 mg. per ml., and, by extrapolation to infinite dilution, LAUFFER suggests that the true sedimentation constants of the components that give constants of about 170 and 200 as usually measured are 193 and 216. In spite of the many uncertainties, LAUFFER (1944) concludes that, by correcting the laws of sedimentation to apply to ellipsoids of revolution instead of to spheres, and by combining data from diffusion, sedimentation and viscosity tests made on the same preparation, reliable estimates of weight and size can be obtained. He states that his calculation of a "molecular" weight between 31 and $40 \times 10^6$, and size between 13.6 by 276 m$\mu$ and 15.2 by 270 m$\mu$, agrees well with electron micrograms of the same preparation, though these show the presence of particles of many different sizes, both shorter and longer than the postulated "molecule". The presence of particles of many sizes in preparations of tobacco mosaic virus that give single, and reasonably sharp, boundaries is only one of the anomalies in the sedimentation of tobacco mosaic. This may be explained by PIRIE's (1945) suggestion that increases in the length of a uniform rod may not significantly affect sedimentation rate, or by FRAMPTON'S (1939*b*, 1942*a*) suggestion that preparations behave like thixotropic gels, so that the effect of centrifugation is more nearly comparable to expressing fluid from the interstices of a gel rather than to the independent movement of particles through the fluid. This type of behaviour would account for the effect of dilution in increasing both sedimentation rate and the water content of the sedimented pellet, but not for the occurrence in some preparations of double boundaries (Fig. 48), for these could probably only appear if particles were to some extent independent of one another.

There is much evidence that tomato bushy stunt virus has spherical particles and its solutions show none of the anomalous physical properties that complicate interpretations of results with tobacco mosaic virus. The particles also seem to be uniform in size, for LAUFFER (1942) found

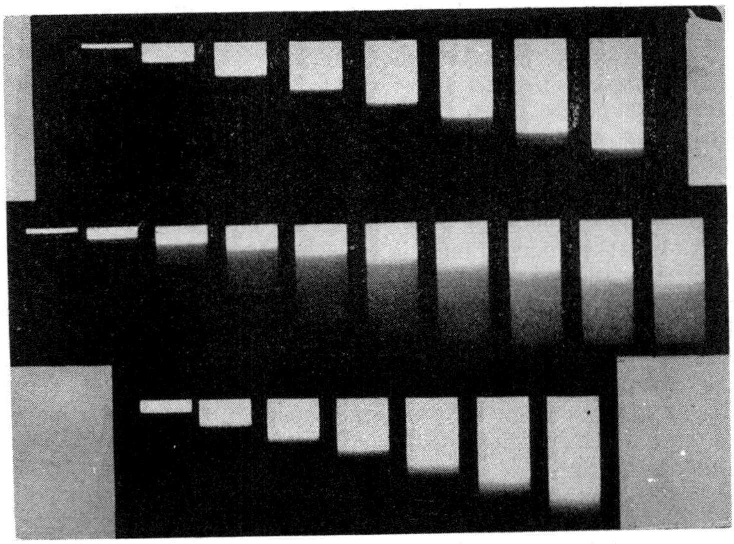

FIG. 48. — Sedimentation pictures of solutions of tobacco mosaic virus. *Top:* Virus purified by high speed centrifugation and dissolved in water. *Centre:* After the development of a second component caused by allowing the preparation to stand in the presence of salt. *Bottom:* After more extensive treatment with salt. (STANLEY, W. M., 1938, Harvey Lectures, 1937/38, 170).

that the spread in the sedimentation boundary could wholly be accounted for by diffusion. Nevertheless, when particle weight is calculated separately from sedimentation and diffusion constants, very different estimates are obtained; a sedimentation constant of $132 \times 10^{-13}$ suggests a weight equivalent to a molecular weight of $7.4 \times 10^6$, and a diffusion constant of $1.15 \times 10^{-7}$ cm.$^2$/sec. (NEURATH and COOPER, 1940) suggests one of $19 \times 10^6$. As the theory underlying the sedimentation and diffusion of unhydrated spheres is soundly based, the discrepancies clearly show that the assumption of unhydrated spheres is unjustified. The most likely explanation of the discrepancy is that the particles are hydrated. Knowing both sedimentation and diffusion constants, and the partial specific volume of the dry virus, estimates can be made of the weight and size of the particles and the degree of hydration. MARKHAM, SMITH and LEA (1942) calculate that 1 g. of dry virus is combined with 0.7 g. of water and on this basis they estimate diameters of 29 m$\mu$ and 37 m$\mu$ for the anhydrous and hydrated particles respectively, corresponding to equivalent molecular weights of $10.6 \times 10^6$ and $19 \times 10^6$. These conclusions, however, are at variance with the results obtained by McFARLANE and KEKWICK (1938) from studies on the sedimentation equilibrium of tomato bushy stunt virus, which suggested an equivalent molecular weight of $7.6 \times 10^6$ and that the particles were unhydrated. Whether these discrepancies arise from errors in the determination of sedimentation equilibrium or diffusion constant, or from some other cause, is unknown.

Similarly, with southern bean mosaic virus, because of hydration, the particle weight and size derivable from a sedimentation constant of 114 differ from the values obtained from a diffusion constant of 1.34 cm.$^{2}$/sec. The combined results of sedimentation and diffusion experiments suggest hydration to the extent of 0.7 g. or water per g. of dry virus, while viscosity experiments suggest rather more. By combining data from various physical properties, weights of $6.6 \times 10^6$ and $11.6 \times 10^6$ can be calculated for the dry and hydrated particles, corresponding to spheres of diameters of 24 and 31 m$\mu$ respectively, assuming uniform shrinkage on drying (MILLER and PRICE, 1946*a*).

Results on turnip yellow mosaic virus (MARKHAM and SMITH, 1949) afford a further illustration of the uncertainties entailed by calculating particle size and weight from sedimentation and diffusion constants when other properties are unknown, a practice that has been commonly adopted with other viruses. The two particles that occur in purified preparations both give the same diffusion constant of about $1.55 \times 10^{-7}$ cm.$^{2}$/sec. and both are known from other work to be approximately the same size and shape; yet, because of differences in density, the particles of nucleoprotein have a sedimentation constant of more than twice that of the particles consisting of protein only (MARKHAM, 1948).

**Microscopy and X-Ray Diffraction:** — The limiting resolving power of microscopes using white light is about 200 m$\mu$; particles smaller than this can be seen if the contrast between them and the background is increased by staining or by the use of dark-ground illumination, but their size cannot be measured. Resolution is limited by the numerical aperture of the objective lens and the wavelength of light and can be increased to about 100 m$\mu$ by substituting ultraviolet for white light. Using the ultraviolet microscope, particles of several animal viruses have been photographed and measured (Fig. 49), but the method has failed with the few plant viruses to which it has been applied, all these being too small to be resolved. They can, however, be resolved by the electron microscope.

In basic principles the electron microscope can be considered as analogous to ordinary microscopes, in which the light source and solid glass lenses are replaced by an electron source and magnetic field lenses respectively. The wavelength associated with a beam of electrons accelerated by a potential of 60 kilovolts is only 0.005 m$\mu$ compared with wavelengths of about 500 m$\mu$ for white light and 200 m$\mu$ for ultraviolet, and this fact gives the electron microscope its great resolving power. In theory it should be capable of resolving small molecules, but at its present state of development, because of deficiencies in the lenses and other technical difficulties, its actual performance falls far short of the potential, and particles smaller than 5-10 mμ in diameter are imperfectly resolved. Accurate calibration of the machine is far from easy, so that the exact magnification is not always certain, and when measuring particles near the limits of resolution the percentage error is high. These limitations, however, are minor compared with others imposed by the use of electrons. Electrons are readily scattered by air or water vapour, so that a high vacuum must be maintained throughout the electron path. To prevent over-heating, which might cause structural changes, the specimen must not be so thick that it absorbs excessive numbers of electrons. Thus specimens must be prepared as thin dry films, a treatment which obviously has many disadvantages for examining preparations of biological interest and which could introduce many artefacts. Further, specimens that are not inactivated by this preliminary treatment can certainly be expected to lose their biological activity when bombarded with electrons, although in measuring the sizes of virus particles this is likely to be of little importance compared with the errors that could occur during the mounting of specimens.

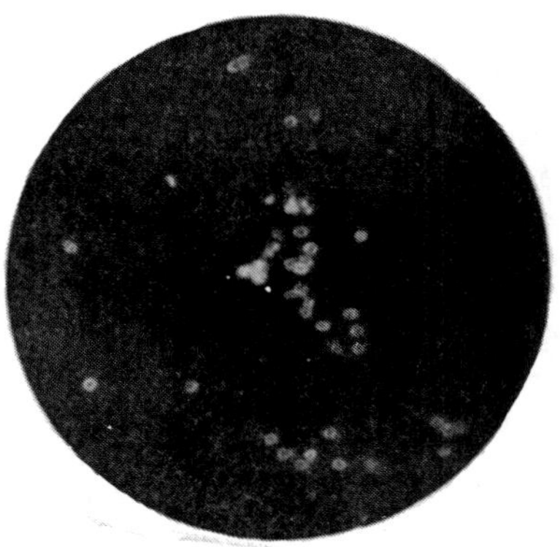

FIG. 49. — Virus of canary pox. Photographed with ultra-violet light, dark ground. × 3,200. (*Photograph by* J. E. BARNARD).

Specimens of viruses were first prepared for electron microscopy by drying drops of solutions on collodion membranes, but the pictures obtained were usually indistinct, as the viruses have little greater power of scattering or absorbing electrons than the supporting membranes. Hence the particles appeared merely as darker grey shadows on a grey background. Nevertheless, the pictures were adequate to confirm some of the conclusions on particle shape arrived at from studies of physical properties. For instance they showed that tomato bushy stunt virus has uniform, approximately spherical, particles and that tobacco mosaic virus occurs as rod-shaped particles of various lengths. The pictures were insufficiently sharp for any accurate measurements of width, but the length of the rods could be measured fairly accurately. As a result of such measurements different workers have drawn different conclusions. KAUSCHE, PFANKUCH and RUSKA (1939) gave the particle length as 150 and 300 mμ, MELCHERS *et al.* (1940) stated that the commonest lengths were 140 and 190 mμ, whereas STANLEY and ANDERSON (1941) found that 50 per cent. of the particles were 280 mμ long and concluded that this was the length of unaggregated particles, although their pictures showed many shorter than this. STANLEY and ANDERSON attributed the differences between their results and those of other workers to the fact that they were using different virus strains, but later comparisons of virus strains from different laboratories failed to support this conclusion, all the strains examined giving particles of 280 mμ as well as others both shorter and longer (OSTER, KNIGHT and STANLEY, 1947; KNIGHT and OSTER, 1947). After studying the distribution of particles of various lengths, FRAMPTON (1942) concluded that there was a tendency for particles to be multiples of 37 mμ, whereas RAWLINS (1942) could detect no sign of any single common factor.

The introduction of metallic shadow-casting (MÜLLER, 1942; WILLIAMS and WYCKOFF, 1944, 1946) has greatly enhanced the value of electron microscopy, virus particles being sharply defined in black and white instead of vague grey shadows (*cf.* Frontispiece and Fig. 27). The method consists of coating the specimen with a thin film of suitable metal (gold or palladium), by placing it in a vacuum at an oblique angle to a heated wire of the metal. The oblique deposition of the metal causes the higher elevations of the specimen to cast "shadows", regions free from metal, on the sides away from the wire. When the specimen is examined in the electron microscope, the scattering of electrons at different regions is proportional to the thickness of the deposited metal and great contrast is obtained between the particles covered with metal and the shadows they cast. In addition to increasing contrast and thereby increasing the accuracy with which measurements can be made, shadow-casting also permits calculations of the height or thickness of objects from measurements of the lengths of the shadows they cast. It brings small details into relief and affords contrast between regions of different slopes or at different elevations. The method can be used in two different ways. Specimens mounted on collodion or other substrate may be "shadowed" and photographed directly, or, if the specimen is too thick for direct examination, the shadowing can be done on replicas of the surface produced by strippings of Formvar or polystryrene-silica. These methods, largely developed by WYCKOFF, have produced pictures showing the manner in which virus particles are arranged in crystals and liquid crystals; examples of these are shown in Fig. 46 and the Frontispiece. Shadow-casting cannot, of course, obviate the main disadvantage of electron microscopy, namely that only dried specimens can be examined, but there is no reason to believe that the extra contrast and definition that it confers introduce any additional artefacts. The layer of metal deposited is too thin to affect measurements at present levels of magnification.

Preparations of only a few viruses have yet been examined in the electron microscope, and, as is usual, tobacco mosaic virus has been studied more than all the others put together. The certain conclusions that can be reached are that the particles from purified preparations are rod-like and have lengths varying from less than 100 to more than 1000 mμ, the average length being greater the more the preparation has been purified and the more intensively it shows the phenomenon of anisotropy of flow. Accurate measurements of width are difficult on single particles, even when shadow-cast, but from measurements across several particles in contact in a crowded field, KNIGHT and OSTER (1947) give a width of 15 mμ for tobacco mosaic virus and 13 mμ for cucumber viruses 3 and 4, which are serologically related to tobacco mosaic virus. Potato virus X also has rod-like particles that vary greatly in length; these seem to be more flexible than tobacco mosaic virus particles, and from measurements on two strains TAKAHASHI and RAWLINS (1946) give the mean width as 15.7 and 16.1 mμ respectively. A smaller value, about 10 mμ, has been reported by KLECZKOWSKI and NIXON (1950), who noted no differences in width between the various strains of virus X they examined. The virus that causes stem mottle of potato and rattle disease of tobacco has rod-shaped particles that resemble those of tobacco mosaic virus by being apparently rigid; they are about 20 mμ wide and vary in length between 100 and 700 mμ, the most frequent lengths being 150 and 300 mμ (VAN DER WANT and ROZENDAAL, 1948). The other purified viruses that have been examined appear to have spherical particles of a uniform size; approximate diameters are 26 mμ for tomato bushy stunt and southern bean mosaic (PRICE, WILLIAMS and WYCKOFF, 1946), 30 mμ for squash mosaic (TAKAHASHI and RAWLINS, 1947), 22 mμ for turnip yellow mosaic (COSSLET and MARKHAM, 1948), and one of the tobacco necrosis viruses (MARKHAM, SMITH and WYCKOFF, 1948) and 17 mμ for the Rothamsted culture of tobacco necrosis virus. All these values refer, of course, to dried particles and their relevance to particle size and shape of particles in solution is unknown. This can be deduced only by assuming uniform shrinkage and knowing the extent of hydration. The regular array of particles shown in electron micrograms of crystals shows that much of the original structure is preserved through drying, but distortions of the array suggest that disturbances are produced. From comparisons of spacings shown in electron micrograms and those obtained by X-ray diffraction, the extent of shrinkage and differences between hydrated and unhydrated particles may become clearer.

So far electron microscopy has been applied largely to purified virus preparations, but it may well prove to have wider applications and give* information on viruses not amenable to current purification techniques. Plant extracts contain much material that dries to produce particles covering a wide range of sizes and they are far from favourable subjects for examination by the electron microscope. Nevertheless, particles of tobacco mosaic virus and potato virus X are readily identifiable in specimens made by diluting clarified infective sap. Similarly other viruses whose particles differ in any recognisable manner from ordinary sap components may also be susceptible of examination. Already it has been shown that partially purified preparations of potato yellow dwarf virus contain short rods about 200 mμ long and 50 mμ wide, which have not been seen in preparations from uninfected plants (BLACK, MOSLEY and WYCKOFF, 1948). If these rods are the virus particles, potato yellow dwarf is much larger than any other plant virus so far studied, and in size, if in no other way, it seems to approximate more closely to some of the animal viruses. An unusually large size for this virus was also suggested by infectivity tests on centrifuged preparations, from which BLACK (1941) concluded that it sedimented much more rapidly than tobacco mosaic virus.

X-ray measurements provide the most accurate method of estimating particle size, and the width of 15 mμ, first proposed for tobacco mosaic

---

*Using the shadow-casting technique, rod-shaped particles, presumably the viruses themselves, have now been detected in clarified sap from plants infected with several viruses, including tobacco rattle, tulip break, pea streak, henbane mosaic, potato Y, tobacco etch, cabbage blackring, potato paracrinkle and sugar beet yellows. The first has rods that seem rigid, and the others seem more flexible and filamentous; all occur in rods or threads of variable lengths.

more than ten years ago by this method (BAWDEN *et al.* 1936; BERNAL and FANKUCHEN, 1938, 1941), is the only measurement on this troublesome virus that has gone undisputed. Unfortunately, the method is limited to those viruses that form crystals or liquid crystals, for it depends on the refraction of X-rays by regularly arranged layers of particles. When a beam of X-rays is passed through a substance the beam is deflected in various directions, the extent of the deflection being detected photographically. An orderly arrangement of constituent units in the specimen results in a regular diffraction pattern of spots, arcs or circles, at various distances from the centre of the photograph, depending on the angle of deflection (Fig. 50). From the pattern obtained, the arrangements of the constituent units and the distances separating them can be inferred. The pattern obtained with viruses falls into two parts, one of small spacings, which shows regularity of structural arrangement within the virus particles, and the other of larger spacings, which shows the manner in which the virus particles arrange themselves in crystals and liquid crystals. It is the larger spacings that are relevant to measurements of particle size; they are much larger than the spacings normally measured by X-rays and special cameras working at very low angles are needed to obtain clear patterns.

Low angle patterns of tobacco mosaic virus show that the particles are distributed in a hexagonal array in all kinds of orientated preparations. In the dried gel, in which it is to be expected that the rods will be packed tightly and touching, the particles have an effective width of 15.2 mμ and a cross-sectional area of 201 sq. mμ. X-ray measurements on virus precipitated as paracrystals by acid or salt give distances of 18.2 and 17.3 mμ between the centres of the rods. The rods are probably still in contact in these rigid orientated preparations so that these distances measure the effective diameter of the particles, showing an increase in cross-sectional area of 47% and 33% over that of the anhydrous particle, which implies association of the virus with a third to a quarter its weight of water. In solution the particles are likely to be associated with still more water and their effective diameters still further increased. No equivalent increase occurs in the measurements obtained in small spacings, *i. e.* in those produced by regularities within the body of the particles, from which BERNAL and FANKUCHEN (1941) conclude that the water does not penetrate into the body of the particle. Thus it seems that the particles in solution may be substantially dry internally but be surrounded by attached layers of water several molecules thick. No regular repeat pattern at large spacings has been found in the direction of the length of the particles, so that the method gives no indication of length. At small spacings, however, regularities are shown both across and along the particles, and the sharpness of the reflections show that the particles in aggregated preparations are sufficiently long for a large number of repeat units to be regularly arranged along the particle. This intramolecular pattern suggests a structure of the same order of complexity as in feather keratin; it is more complicated than that of myosin, the anisometric protein from muscle, but less so than pepsin (BERNAL, 1938). There is a repeat unit across the particle of approximately 2 mμ and the particle seems to be composed of piles of sub-molecules of dimensions 2.2 mμ × 2 mμ × 2 mμ. These units are rather smaller and less complicated than the molecules of many common proteins, and are themselves divided into nearly identical sub-units possessing half these dimensions.

Measurements on tomato aucuba mosaic and enation mosaic viruses (BERNAL and FANKUCHEN, 1941) and on the rib-grass strain (HOLMES, 1941) give the same width for the dry particles as for tobacco mosaic itself; two other strains, however, cucumber viruses 3 and 4, give spacings that suggest a width of 14.6 instead of 15.2 m$\mu$. All these strains give the same type of pattern at high angles. With potato virus X no pattern at low angles could be obtained, so that X-ray analysis has failed to give any evidence on particle width or length. The pattern at high angles shows a regularity within the particles, and the sub-units are similar to those in tobacco mosaic virus, but are arranged differently.

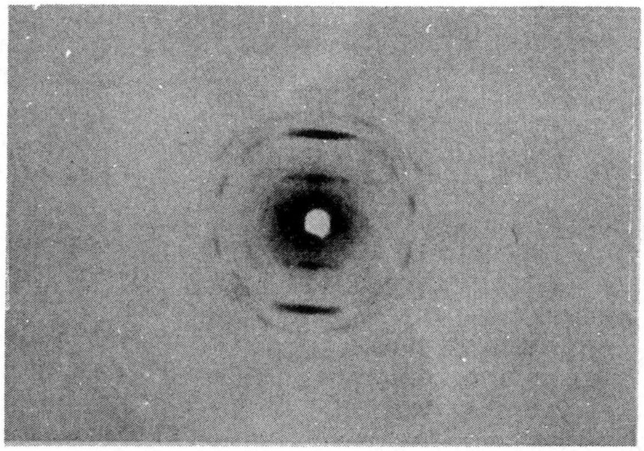

FIG. 50. — X-ray photograph made on a 15% solution of tomato enation mosaic virus, showing the pattern obtained by scattering of X-rays at wide angles. These spacings are almost independent of the solid content of the preparation, and the pattern results from regularities of arrangement within the virus particles. (*Photograph by* I. FANKUCHEN. Specimen 8 cm. from a copper anticathode, nickel filter, exposure 170 hours).

The spacings obtained at high angles with crystals of tomato bushy stunt virus indicate a body-centred cubic lattice with a side of 39.4 m$\mu$ for the wet crystals, which corresponds to a hydrated particle of effective diameter 34 m$\mu$. Drying causes the crystals to shrink by 20% and to give a cell with side of 31.8 m$\mu$, corresponding to a spherical anhydrous particle of diameter 27 m$\mu$. Equivalent molecular weights for these dimensions would be 24,000,000 for the hydrated and 12,800,000 for the dry particles. It is unknown in what manner the water is associated with the particles (BERNAL, FANKUCHEN and RILEY, 1938). The crystals that form in old purified preparations of the Rothamsted culture of tobacco necrosis virus are sufficiently large to be examined individually. X-ray patterns from a single crystal (CROWFOOT and SCHMIDT, 1945) show that the particles do not differ greatly from spheres and they give a minimum diameter for the wet particles of 16 m$\mu$ and a maximum of 20 m$\mu$. The relationship of these particles to infective virus is obscure, but if they are the same size then this is the smallest virus yet studied. Diffraction photographs made on the two kinds of particles occurring in preparations of turnip yellow mosaic virus show that both give wet crystals having a diamond-type lattice (BERNAL and CARLISLE, 1948). The interparticle distance, presumably the diameter of the hydrated particles, is 30.6 mμ for the nucleoprotein particles and possibly 1 mμ more for those free from nucleic acid. Assuming that the same lattice structure is maintained on drying, the diameters of the unhydrated particles are 22.8 for the nucleoprotein and 23.8 mμ for the particles free from nucleic acid. If, as seems possible, these two particles have the same protein constitution, it seems that the introduction of nucleic acid draws the particle into a smaller volume.

**Additional Methods:** — With tobacco mosaic virus the following estimates of particle length have been made from measurements on viscosity and on optical properties: 430 mμ by combining viscosity and sedimentation data (LAUFFER, 1938); 532 mμ from viscosity measurements with the width shown by X-ray diffraction (MARKHAM, SMITH and LEA, 1942); 580 mμ from the rotational diffusion constant in conjunction with a length width ratio estimated from viscosity measurements (MEHL, 1938); 1350 mμ from viscosity extrapolated to zero shear and assuming a width of 15 mμ (ROBINSON, 1939); 1400 mμ from the concentration at which solutions become spontaneously birefringent and assuming rotation in a flat cylinder (BAWDEN and PIRIE, 1937*a*). There is no reason to believe that any of these measurements were made on homogeneous preparations; all preparations contained aggregated particles, and the kindest interpretation that can be put on the different estimates of length is to assume that preparations of different states of aggregation were being studied. As FRAMPTON (1940, 1942*a*) has pointed out, most of the assumptions used in deriving these sizes do not allow for the thixotropic nature of tobacco mosaic virus solutions or their other anomalous characters, which may well invalidate calculations made from measurements of physical properties. Probably the most that should be deduced from such measurements is that, because of the variations in results on different preparations, the particles can vary in size.

Another property that has been studied to obtain the length of tobacco mosaic virus is light scattering (OSTER, DOTY and ZIMM, 1947). Measurements were made at two angles across the incident beam of light to determine the ratio of light scattered in the direction of the beam to that scattered in the opposite direction. Assuming there was no interference between individual particles, from the dissymetry of light scattered they calculated a length of 270 mμ. By using this result and applying it to the formula for calculating molecular weights from measurements of turbidity (light absorption) at different concentrations, they obtained a value of 42,000,000. OSTER, DOTY and ZIMM state that "these values are in complete agreement with sizes determined from electron microscope and viscosity studies made on the same sample". The uncertainties inseparable from viscosity studies with tobacco mosaic virus have already been sufficiently stressed, though the preparation used in this work was probably less aggregated than most that have been studied and so was probably less anomalous. However, the electron micrograms are hardly in "complete agreement", for more than 30% of the 500 particles measured were outside the range 260 to 300 mμ, the limits of measurement considered not to differ significantly from 270. From formulae relating the weight of spherical particles to the light absorption by solutions of different concentration, OSTER (1946) has calculated a value of 9,000,000 for the equivalent molecular weight of tomato bushy stunt virus, which agrees well with those obtained by other methods.

Calculations of particle sizes have also been attempted from measurements of the amount of inactivation caused by known doses of X-ray or other ionizing radiations. Various workers have found that inactivation follows a simple exponential curve and have concluded that inactivation is caused by the absorption of a single ionisation or a single cluster of ionisations (LEA, 1940). The method is, of course, subject to the uncertainties inseparable from infectivity tests, and it also assumes that the absorption of ionisation doses always cause inactivation, an assumption that is clearly unwarranted if irradiation causes mutations as sug-

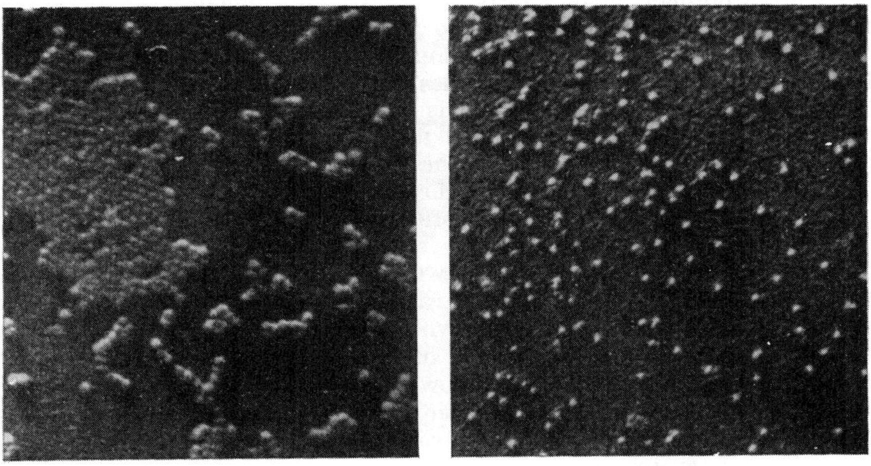

FIG. 51. — Electron micrograms of shadow-cast preparations of *(left)* tomato bushy stunt virus and *(right)* the Rothamsted tobacco necrosis virus. × 50,000. (*Photograph by* H. L. NIXON).

gested by KAUSCHE and STUBBE (1939), and GOWEN (1941). When applied to the spherical viruses, tomato bushy stunt, tobacco ringspot and one of the tobacco necrosis viruses, LEA and SMITH (1940, 1942) got reasonably consistent results with different kinds of irradiation, their results indicating "target volumes" of from one- to two-thirds the size of the particles estimated from other data. With tobacco mosaic virus, on the other hand, the "target volume" was less than one-tenth that of a particle 15 mμ wide and 280 mμ long. MARSHAK and TAKAHASHI (1942) found that sensitivity to X-rays is greater at *p*H 2.2 to 3.4 than at *p*H 3.4 to 7 and that different preparations also differ greatly in sensitivity. From their extreme values they calculate "target volumes" of 4,600 and 42,000 $m\mu^3$, while from measurements on another preparation GOWEN (1940) obtained a volume of 7,500 $m\mu^3$. These calculations are made assuming inactivation by one ionisation, but LEA and SMITH (1940, 1942) suggest that three are usually concerned, so that the actual "target volumes" are probably smaller than these figures. The most reasonable conclusion from these results is that the rods of 280 mμ long are aggregates in which the constituent units which can be inactivated independently, though the varying sensitivity of different preparations suggests that aggregation may increase total sensitivity, because energy released by ionisation absorbed by one particle may spread along the aggregate and inactivate contiguous ones (MARSHAK and TAKAHASHI, 1942). A similar explanation probably accounts for the observation that the "target volume" of potato virus X (LEA and SMITH, 1940), which is also known to have rod-shaped particles, is also much smaller than the volume of most of the rods photographed by the electron microscope.

**The Aggregation of Anisometric Viruses:** — Difficulties in arriving at any conclusion about the size of tobacco mosaic virus arise not only because of its anomalous physical properties, but also because there is a bewildering number of alternatives from which to choose. Some of the many different lengths attributed to particles have already been given in this chapter, but the range can be extended beyond these. It is generally accepted that the larger particles are aggregates, for the preparations measured were made by methods which, if they did not actively promote aggregation, at least included no precautions against it, but the status of the smaller particles and the minimum size of any basic infecive unit that may exist are both uncertain.

Linear aggregation was first postulated to account for the fact that purification reduced filterability and infectivity while simultaneously increasing anisotropy of flow, and BAWDEN and PIRIE (1937*a*) suggested that the greatly elongated rods characteristic of purified preparations were artefacts. Some rods were known to be present in freshly extracted sap, for it showed some anisotropy of flow, but their significance was uncertain. There was nothing to show that all the particles in sap were uniform in size, shape and activity, so that no definite conclusions could be drawn as to whether the rods were basic units or themselves aggregates. With a system so susceptible to aggregation, however, it was safest to assume heterogeneity, and BAWDEN and PIRIE suggested that particles as they are produced in the plant might be small and not greatly anisometric. This idea of rods of varying lengths built up by the union of different numbers of small units accords well with structure as shown by X-ray analysis (BERNAL, 1938; BERNAL and FANKUCHEN, 1941), and could also explain the occurrence of true crystals in infected cells though none has been produced *in vitro*. Evidence for the occurrence of small particles has come since from several different lines of work, but their origin and biological importance remain speculative.

LORING, LAUFFER and STANLEY (1938) agreed with BAWDEN and PIRIE that precipitating the virus caused aggregation, but they claimed that the particles in sap were uniform and that homogeneous preparaions could be made by ultracentrifugation. Mainly because "sharp" sedimentation boundaries were obtained, the claim for homogeneity has often been repeated (*e. g.* LAUFFER and STANLEY, 1939; LAUFFER, 1943), and STANLEY and his colleagues considered that aggregation is restricted to the end-to-end union of particles about 280 m$\mu$ long, which they have unhesitatingly called molecules. Every preparation of tobacco mosaic virus that has been examined critically, however, has been found to contain particles of diverse sizes, most containing particles both shorter and longer than 280 m$\mu$. Indirect evidence for small particles was early provided by the effect of dilution in procuring components which sedimented slowly (WYCKOFF, BISCOE and STANLEY, 1937) and filtered easily (SMITH and MACCLEMENT, 1940), as well as from interpretations of the effect of dilution on infectivity (BALD, 1937), but the first direct evidence came from electron microscopy. Almost all the published pictures showed small particles, though most workers ignored them and recorded only the more obvious rods. This may have been a sequel to the preconceived idea of uniform rod-like molecules, but FRAMPTON (1942) and RAWLINS (1942) drew attention to the presence in pictures published by STANLEY and ANDERSON (1941) of particles smaller than the postulated molecules.

The occurrence of small particles in electron micrograms is far from unequivocal evidence of their nature. The particles might well be impurities, totally unrelated to the virus, or they might be artifacts produced during the preparation of specimens. RAWLINS, ROBERTS and UTECH (1946), for instance, attempted to account for the small particles in their micrograms as fragments of rods broken by the drying during the preparation of the mount. Although this may account for some, further work makes it unlikely that many are produced in this manner. CROOK and SHEFFIELD (1946) found no increase in the proportion of small particles when the same specimen was dried and rewetted three times. The occurrence of small particles in plant extracts, and their independence of the effects of drying was demonstrated conclusively by studies on preparations of particles differentiated by sedimentation and then subjected to aggregating treatments, for these showed that sap contains particles of various shapes and sizes and that the small ones can be converted into long rods.

Preparations of particles of different sizes can be made by subjecting sap to differential centrifugation and separating the particles that sediment and compact readily from those that do not, but a larger proportion of the total virus nucleoprotein is obtained as small particles if some precautions are taken to prevent aggregation (BAWDEN and PIRIE, 1945*a*). Many different treatments cause aggregation; the most effective are incubation with trypsin or heating to 60°C. between *p*H 5 and 6, but ageing and exposure to many substances, including some constituents of sap, also do it. The juice that runs from leaves after freezing and thawing contains little virus but many of the constituents that cause aggregation; if this is removed before the leaves are minced to extract the virus, the proportion of small particles obtained is increased. A still larger proportion of small particles is obtained if the virus remaining in the fibre is extracted by fine grinding; whether this represents a qualitative difference between the two lots of virus as they occur in the plant, or is an effect produced by the method of extraction, is unknown. The particles in sap, and in these two kinds of extract, are far from homogeneous, and can be separated into different grades of size by ultracentrifugation. The method used by BAWDEN and PIRIE (1943*a*) separates four different fractions, two extremes, one of which sediments and compacts at 17,000g. and the other which sediments but does not compact at 80,000 g., and two intermediates; there is no sharp distinction between neighbouring fractions, but a steady graduation in properties from one to another. Some differences in behaviour between the particles present in the four fractions are summarised in TABLE 22, which shows that they differ in infectivity and serological activity as well as in sedimentation rate. Only about half the total material in the most slowly sedimenting fractions is virus nucleoprotein; these fractions show no anisotropy of flow, have serological behaviour resembling that of somatic type antigens and have low infectivity. The most rapidly sedimenting fractions contain little except virus nucleoprotein, show anisotropy of flow, have serological behaviour resembling that of flagellar antigens and have relatively high infectivity. Electron micrograms of the most slowly sedimenting fractions show particles most of which are little longer than they are wide, while with increasing sedimentation rate the numbers of obvious rods increase as also does their average length (Fig. 52).

All the fractions are unstable and readily pass into forms that sediment readily, show intense anisotropy of flow and behave serologically like flagellar type antigens, the changes being most striking in the preparations that previously sedimented most slowly. Electron micrograms show that these changes in properties are accompanied by an increase in the number of rods and in their average length, and when any of the fractions is subjected to effective forms of aggregation, rods up to and longer than 5 μ may be produced (CROOK and SHEFFIELD, 1946).

FIG. 52*a-c.* — Electron micrograms of shadow-cast tobacco mosaic virus particles of different average lengths separated by differential centrifugation. — *a:* most slowly sedimenting particles. — *b* and *c:* more rapidly sedimenting particles. × 50,000. (*Photograph by* H. L. NIXON).

In the preparations that contained small particles these changes in size and properties are accompanied by the destruction of material other than virus nucleoprotein. Although changing the small particles into rods produces physical and serological properties indistinguishable from those of infective preparations, it does not increase infectivity; on the contrary the aggregation of all fractions usually reduces infectivity by about one-half.

SIGURGEIRSSON and STANLEY (1947) have confirmed that fractional centrifugation will separate the particles originally present in sap into preparations of different average particle size, and that preparations of the larger particles are much more infective than preparations of the smaller ones. Indeed, they conclude that the infectivity is restricted to particles 280 m$\mu$ or longer, and they attribute the different infectivities of different preparations to the presence of varying numbers of particles of this size. Smaller particles they consider are merely fragments of such rods, produced by breakages because of thermal or mechanical

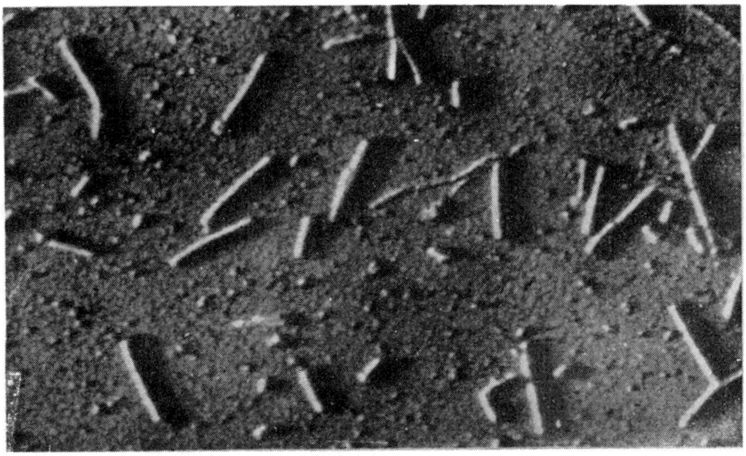

FIGURES 52 *b* and *c* (*see* p. 238).

stresses. From electron micrograms of particles in sap, SIGURGEIRSSON and STANLEY (1947) state that two-thirds of the rod-shaped particles are about 280 m$\mu$ long, and these particles they seem to conclude are uniform in both biological activity and size, that is, the equivalent of molecules. Similar proportions of particles about 280 m$\mu$ long have also been reported by OSTER and STANLEY (1946) to occur in juice pressed from cut hair cells. They consider that this treatment could not have led to any aggregation of particles and that such particles must therefore occur as such and that "they represent tobacco mosaic virus". The fact that fibrous inclusion bodies are sometimes formed in infected cells

(Fig. 12) (KASSANIS and SHEFFIELD, 1941), makes it probable that rod-like particles do occur in cells, but to assume that pressing juice from cut hair cells is without effect on the particles seems unwarranted. This treatment would not only expose the particles to sap constituents that can cause aggregation, but, from observations on the behaviour of crystalline inclusions, could also have other effects. SHEFFIELD (1939) has shown that the exertion of slight mechanical pressure on the outside of cells containing hexagonal type crystals, or the insertion of a needle into such cells, will cause the crystal to change into a suspension of para-crystalline needles, resembling those that are produced *in vitro* by precipitating purified virus. This cannot be interpreted with certainty, but suggests an alteration in the size and shape of the particles.

The concept of a basic particle which breaks into fragments of variable sizes, with fragments and particles capable of combining with one another linearly, could explain readily enough the occurrence of particles of very diverse lengths. However, our knowledge is still insufficient to accept it, and there are a number of phenomena that seem incompatible

TABLE 22. — *Properties of fractions of tobacco mosaic virus separated by differential centrifugation:—*

| *Sedimentation constants* | *Serological behaviour* Time for first precipitation | Antigen concentration at optimum | *Infectivity* No. of lesions per leaf at 0.02 | 0.002g/1 |
|---|---|---|---|---|
| 20, 28 | 3 hours | 0.015g/1 | 0 | 0 |
| 41, 63 | 1 hour | 0.03 | 3 | 0 |
| 122, 154, 190 | 15 minutes | 0.06 | 54 | 10 |
| 162, 190 | 5 minutes | 0.5 | 116 | 34 |

Antiserum used at a constant dilution of 1/400.

with it. For instance, fragmentation under thermal and mechanical stress might be expected to increase as the length of rods increases, yet high proportions of small particles occur only in freshly made extracts, and when virus preparations are subjected to additional treatments, the proportion of long rods increases. Also, as yet, there is no evidence that the infectivity does depend on the presence of rods with any one length; all that is known with certainty, is that with preparations of particles of different average lengths from sap, the infectivity increases as the average length of the particles increases up to about 280 m$\mu$. CROOK and SHEFFIELD (1946) found no rods as long as 280 m$\mu$ in some of the preparations separated by BAWDEN and PIRIE (1945*a*), though these had some infectivity, and TAKAHASHI and RAWLINS (1948) state that particles shorter than 225 m$\mu$ can be infective. TAKAHASHI and RAWLINS measured lengths of particles extracted from finely macerated pulp at different periods after infection and separated them into three groups, shorter than 225 m$\mu$, between 225 and 463 m$\mu$, and longer than 463 m$\mu$. They found that the total of middle and small group particles, but not of the long group, increased from the fourth to the sixteenth day; during this time the proportion of short particles increased from about 10 to 50% of the whole, while the proportion of long particles fell from more than 40% to less than 10%. They conclude that the short particles do not result from fracture of the long ones during extraction, but occur as such in the cells.

On present evidence no positive conclusions can be drawn about the significance of these different kinds of particles. As small particles can be made to aggregate into rods, without any increase of infectivity, clearly the small particles cannot be assumed to be identical units from which all others are composed. Also, as they give rods indistinguishable in appearance from those that seem to be more infective, there is obviously no reason to assume that all particles of approximately the same size and shape occurring in sap are uniformly infective. BAWDEN and PIRIE (1945*a*) have shown that, by various methods of purification, preparations can be made with similar physical, chemical and serological

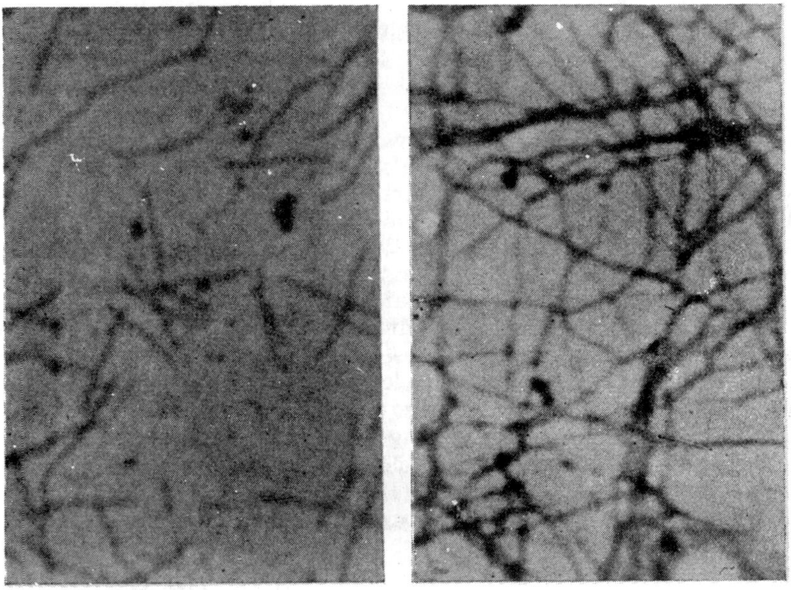

FIG. 53. — Electron micrograms of potato virus X particles in *(left)* clarified infective sap and *(right)* in purified preparation. × 30,000. (BAWDEN, F. C. and CROOK, E. M., 1947, Brit. J. exp. Path. *28*, 403).

properties but with widely different infectivities, whereas similar infectivities may be shown by preparations with very different other properties. They have suggested that the small particles may be virus that has become non-infective without losing serological activity, or that they are incompletely formed or mal-formed virus particles. There is some evidence (OSTER, 1948; TAKAHASHI and RAWLINS, 1948) that aggregation of small particles *in vitro* leads to rods that are longer, and of which fewer are about 280 mμ long, than those occurring in freshly prepared extracts. It may be, as has been suggested by BAWDEN and PIRIE (1945*a*), that in the cell the virus is synthesized as small particles with polar groups that tend to combine with other materials, or, in the absence of other suitable materials, with one another. Linear aggregation of small particles *in vitro* is accompanied by removal of other material, but nothing is known of what occurs within the cell. These small particles are obviously not fully active virus and it is possible that different ones carry different activities, several of which may need to be combined into one aggregate before infectivity is achieved. This may necessitate arrangement in an appropriate order, as well as inclusion of an adequate number, and infected cells may provide an orientating mechanism that is absent *in vitro*. On this assumption, the probability of a particle being infective would increase with increasing size, variations in the types of units incorporated would account for the production of mutant forms, and there is no need to postulate the existence of a single identical molecule that arises fully fledged and then fragments. All kinds of work with tobacco mosaic virus emphasises its variability and heterogeneity, and there is no method yet available for deciding what proportion of particles in any one preparation is infective. Until such methods are obtained, and we can test for other biological activities than ability to initiate infection, the significance of these particles of different sizes, and their relationships to one another, will probably remain unsettled.

Detailed experiments have so far been made only with tobacco mosaic virus, but it is likely that similar variability occurs in particle size of other anisometric viruses. Electron micrograms of sap from plants infected with potato virus X show particles of various sizes, and the rods tend to be longer than with tobacco mosaic virus. TAKAHASHI and RAWLINS (1946) concluded that the predominating length was 500-600 m$\mu$, but some are ten times as long as this. Aggregation seems to occur even more readily than with tobacco mosaic virus, and during purification the particles become so elongated and intertwined that it is often impossible to distinguish any separate entities (Fig. 53). No attempts have yet been made to fractionate the particles occurring in sap and to discover whether those of various sizes also vary in infectivity. Only a little virus X can be obtained by grinding the fibrous residues from infected leaves, but this is obtained in particles of a shorter average length than those in sap, few exceeding 250 m$\mu$ (BAWDEN and CROOK, 1947). As with tobacco mosaic virus, these small particles give precipitin reactions resembling those of somatic antigens. They can be aggregated into long rods by incubation with sap from healthy plants, when their serological behaviour changes to flagellar type. Critical infectivity tests are more difficult to make with potato virus X than with tobacco mosaic virus, but preparations of virus from fibrous residues seem, weight for weight, to be as infective as preparations of virus from sap.

*Chapter 13*

# TYPES OF INACTIVATION

Since JOHNSON (1927) pointed out the uncertainties of identifying viruses from symptomatology and that other characters often had greater diagnostic value, most workers giving first accounts of mechanically transmitted viruses have recorded data on such properties as dilution end-point, thermal-inactivation point, longevity *in vitro* and resistance to alcohol. Such properties are customarily determined using crude infective sap, and they have considerable value in showing differences between viruses that cause similar symptoms. For several reasons, however, they are of less value in suggesting relationships between viruses that cause different symptoms, or that occur in different hosts, and the relevance of such tests to the intrinsic properties of the viruses is uncertain. The virus concentration, the ratio of virus to other constituents, and the kinds of other constituents, may vary widely from host to host, and any such variations may well determine the result of tests. With cucumber mosaic virus, for example, depending on the strain used and the host in which it was propagated, different workers have reported dilution end-points varying from 1/1000 to 1/100,000 and thermal-inactivation points between 55 and 70°C. Even transmission by inoculation may be determined by the host, for cucumber mosaic virus is readily sap-transmissible between most hosts but not from sugar beet or *Phytolacca* sp. (HOGGAN, 1935; WELLMAN, 1935; BHARGAVA, 1948).

Values recorded for the dilution end-points of different viruses in sap range between less than 1/100 to over 1/1,000,000, thermal-inactivation points from 42°C. for tomato spotted wilt to 92°C. for tobacco mosaic virus, and longevity *in vitro* from a few hours to many years. Many viruses lose infectivity when heated for 10 minutes at around 60°C. or after standing for a few days at room temperature, conditions in which many normal plant proteins also denature and precipitate. Clearly tests of stability have little diagnostic value for these viruses, and it is also uncertain whether inactivation has occurred because of intrinsic instability on the part of the virus or because it has been affected by conditions obtaining in the sap; for instance it might be absorbed on to precipitating proteins or inactivated by changes occurring in the sap. In crude sap tomato spotted wilt is inactivated within a few hours at room temperature or by heating to 42°C. (SAMUEL, BALD and PITTMAN, 1930) but if the *p*H is adjusted to 7 resistance to both ageing and heating is increased. If a reducing substance is also added, or if oxygen is excluded, longevity is extended to 35 days and the thermal inactivation point raised to 52°C. (BEST, 1937; NORRIS, 1946). Thus the ready inactivation of tomato spotted wilt virus in sap does not seem to occur because the virus is intrinsically much more unstable *in vitro* than some other viruses, but because it is more affected by oxidation and slight acidity. Similarly, potato virus Y remains infective for longer *in vitro*, and is less readily inactivated by acid, when it is purified than when it is in sap. Because of the uncertainties in interpreting results obtained with sap, the discussion in this chapter deals largely with the inactivation of purified viruses.

When studying inactivation most workers have relied solely on infectivity tests and loss of infectivity has usually been taken as synonymous with destruction of the virus. This assumption, however, is often unjustified, for three different kinds of reaction that lead to loss of infectivity can be distinguished. The first is readily reversible, and is perhaps better described as inhibition or neutralisation of infectivity than as inactivation. The second is apparently irreversible, but is not accompanied by disruption of the particles, which retain their serological activity and many of their other characteristic properties. The third cause of loss of infectivity is also irreversible and is accompanied by disruption of the particles and loss of all characteristic properties. Some

TABLE 23.—*Recovery of infectivity by diluting a mixture of virus and inhibitor:—*

| ORIGINAL SOLUTION | INFECTIVITY Average no. of lesions per leaf with solution diluted | | | |
|---|---|---|---|---|
| Tobacco mosaic virus 0.05 g/1 in | 1/1 | 1/10 | 1/100 | 1/1000 |
| Water | 110 | 64 | 20 | 4 |
| Sap from *Phylacca esculenta* diluted 1/10 | 0 | 3 | 6 | 4 |

treatments cause loss of infectivity in only one of the three ways, but no rigid distinctions can be drawn, because others may act in more than one way, or act on one virus in one way and on a second in another.

**Inhibition of Infectivity:**—There are many substances that, with varying degrees of efficiency, reversibly inhibit the infectivity of tobacco mosaic virus. The effects of these substances on other viruses have been less studied, but the existing evidence suggests that there is little specificity in the reaction, and that a substance that strongly inhibits the infectivity of one virus will also inhibit others. Although many different types of substance act as inhibitors, there are several common features in their behaviour. Loss of infectivity usually occurs immediately the substances are mixed with the viruses and no additional effect is obtained by increasing the time of contact. The mixtures still react with virus antisera, and if the inhibitor is removed or destroyed the virus can be regained with its original activity. Also, when mixtures of concentrated virus and inhibitor, which may have little or no infectivity, are diluted, they usually regain some infectivity. This effect is illustrated in TABLE 23, which shows the increase in infectivity produced by diluting mixtures of purified tobacco mosaic virus and sap from *Phytolacca esculenta* (KASSANIS and KLECZKOWSKI, 1948).

Some inhibitors, *e. g.* sap from *Phytolacca* sp. (DUGGAR and ARMSTRONG, 1925), trypsin (LOJKIN and VINSON, 1931) and products from *Aerobacter aerogenes* and *Aspergillus niger* (JOHNSON and HOGGAN, 1937) were at first thought to destroy tobacco mosaic virus, but CALDWELL (1933) showed this to be unlikely by demonstrating that mixtures with trypsin regained infectivity when heated. STANLEY (1934*a*) confirmed this and further showed that loss of infectivity occurred immediately the virus and enzyme are mixed, that it did so at *p*H values at which trypsin is not proteolytically active, that there was no further loss of infectivity after incubation in conditions suitable for proteolysis, and that infectivity could be regained by digesting the trypsin with pepsin or by diluting the virus-enzyme mixtures.

Two hypotheses have been advanced to explain the inhibiting action of trypsin. From the results of infectivity tests with tobacco mosaic virus and trypsin mixed in various proportions, STANLEY (1934*a*) suggested that the enzyme had no direct action on the virus but produced its effect by increasing the resistance of the host. He attempted to see whether virus and enzyme combined by measuring the rate at which trypsin diffused alone and in the presence of the virus, and as he found no differences suggested there was no combination. Experiments with potato virus X, led BAWDEN and PIRIE (1936) to suggest that the loss of infectivity might result from combination between viruses and trypsin to form non-infective complexes that are dissociated by dilution. They found that the infectivity of virus X also is immediately reduced by the addition of trypsin, but, unlike tobacco mosaic virus, this one is hydrolysed by the enzyme. Thus when incubated in conditions favourable for tryptic activity, there is a second drop in infectivity which is also accompanied by corresponding destruction of the virus and loss of serological activity. As the first step in proteolysis is a combination between enzyme and substrate, this was good evidence that virus X and trypsin combine. Since then there has been much evidence showing that trypsin also combines with tobacco mosaic virus, though this virus is not a susceptible substrate. HILLS and VINSON (1938) first obtained evidence for combination by showing that the diffusion rate of both trypsin and tobacco mosaic virus differs when the two are mixed than when they are in solutions alone. Combination is also demonstrable by mixing the virus and enzyme, ultracentrifuging the mixtures and assaying the supernatant fluids and resuspended pellets for their enzyme contents. Using this method KLECZKOWSKI (1944) found that tobacco mosaic virus and potato virus X both combine with trypsin and that the combination is readily broken by dilution. Oddly enough, virus X, which is a substrate for the enzyme and is more inhibited by it, combines with less trypsin than does tobacco mosaic virus.

The ability to inhibit infectivity is not general to proteolytic enzymes. Pepsin and papain do not significantly inhibit tobacco mosaic virus or potato virus X, though they combine with the latter, for it is readily hydrolysed by them when incubated in conditions favourable for proteolysis. Papain also seems to combine with tobacco mosaic virus, for when the two are mixed an insoluble complex separates, but it does not hydrolyse the virus (BAWDEN and PIRIE, 1937*a*). STANLEY (1934*a*) and ROSS and VINSON (1937) stated that pepsin inactivated tobacco mosaic virus after incubation for many days at *p*H 3 and 37°C., but other workers have found it to have no proteolytic activity against tobacco mosaic virus and KLECZKOWSKI (1944) found that the two do not combine; pepsin combined with potato virus X and the denatured protein from tobacco mosaic virus, both of which are susceptible substrates, but not with active tobacco mosaic virus.

Other substances are now known which are much more active as inhibitors of infectivity than is trypsin. Of these, the ones that have been most studied are a polysaccharide from yeast (TAKAHASHI, 1942, 1946), ribonuclease (LORING, 1942; KLECZKOWSKI, 1946; BAWDEN and KLECZKOWSKI, 1948), and a glycoprotein from *Phytolacca esculenta* (KASSANIS and KLECZKOWSKI, 1948). Ribonuclease and the glycoprotein are about equally effective as inhibitors and, weight for weight, they are about a hundred times as effective as trypsin. They both combine with the viruses to form dissociable complexes. Combination with tobacco mosaic virus is evident from the production in salt-free solutions of insoluble paracrystalline fibres, which dissolve immediately on the addition of salt. Ribonuclease also precipitates potato virus X, but these precipitates do not dissolve in dilute salt solutions; the precipitates dissolve fairly rapidly when incubated with trypsin and slowly with borate buffer. The addition of ribonuclease to virus X causes a reduction in the precipitin titre, but the original titre is usually regained when the precipitates are re-dissolved by incubation; this re-solution sometimes also leads to a slight recovery of infectivity. Potato virus X does not appear to be hydrolysed by ribonuclease as it is by proteolytic enzymes; ribonuclease also has no enzymic action on tobacco mosaic and tomato bushy stunt viruses. Neither tobacco mosaic virus nor ribonuclease appear to suffer any change as a result of combining, for both can be recovered separately showing their original activities. Weight for weight potato virus X combines with less ribonuclease than does tobacco mosaic virus, but, as with trypsin, its infectivity is more reduced.

All the inhibitors so far studied combine with the viruses they inhibit. This is circumstantial evidence in favour of attributing inhibition to the production of non-infective complexes, in which groups essential for infectivity have become blocked. However, there are observations that do not fit too well with a simple quantitative neutralisation of virus particles and that suggest that the host plant is also concerned. For example, LORING (1942) found that the inhibitory effect of ribonuclease on tobacco mosaic virus was greater when *Phaseolus vulgaris* was used as a host plant than when mixtures were inoculated to *Nicotiana glutinosa,* though this could be interpreted by assuming that *N. glutinosa* is better able than *P. vulgaris* to split the complexes. KLECZKOWSKI (1946) has shown that ability to combine with, and precipitate, tobacco mosaic virus is not confined to substances that are powerful inhibitors. On the contrary, it seems to be a general property of all substances that carry an opposite electric charge from the viruses, yet some have little inhibiting power. Hence mere combination of a virus with another substance *in vitro* is inadequate to explain inhibition, unless inhibitors combine with and block specific groups. Inhibition is not proportional to the amount of combination *in vitro,* for this depends greatly on the *p*H of the mixtures and whether or not salt is present, but variations in *p*H or salt content have relatively little effect on the infectivity of mixtures (KLECZKOWSKI, 1946). However, the results of such tests are not readily interpretable, as it is unknown to what extent the host tissues alter the conditions of the inoculum. It may be that widely differing fluids are brought to similar conditions by contact with the host cells, and it is to be expected that the amount of inhibitor combined at that time would be the only relevant factor in determining whether or not infection occurs.

Were inhibition dependent on, and a consequence of, combination between the virus and inhibitor, it is to be expected, as with toxin-antitoxin unions, that for complete loss of infectivity a given weight of virus would need to combine with a certain minimum quantity of inhibitor. There is, however, no evidence for such a fixed neutralising ratio, and much suggesting that none exists. The development of infectivity when non-infective mixtures are diluted itself implies that there is no fixed ratio, for dilution leaves the relative proportions of virus and inhibitor unchanged, but this could be accounted for by the disruption of complexes to set free virus particles. Better evidence for the absence of any fixed neutralising ratio is provided by experiments on the amount of inhibitor needed to neutralise the infectivity of virus solutions of different concentrations. KASSANIS and KLECZKOWSKI (1948) found that the ratio of inhibitor to virus needed to neutralise all infectivity decreased as the concentration of virus increased. In other words, the higher the concentration of virus, the smaller is the amount of inhibitor needed to neutralise a given weight of virus. This differs strikingly from results with toxins and anti-toxins, in which equivalent quantities combine to give neutralisation, and suggests that combination between virus and inhibitor may not be an essential part in the inhibition. An effect on the host plant is therefore suggested, which is also implied by the fact that the number of lesions is reduced, not only by mixing virus with inhibitor, but if leaves are rubbed first with inhibitor, then washed and inoculated with virus alone.

Too little is known of the processes involved in the establishment of infection to allow any final interpretation of the mechanism of inhibition, or, with treatments that do not obviously change virus or host, to draw any clear distinctions between actions that affect solely one or the other. All that is known with certainty is that the viruses and inhibitors combine, and that the presence of inhibitors prevents infection. It is likely that both viruses and inhibitors can also combine with constituents of host cells, and combination with specific constituents may be an essential first step in virus multiplication. The inhibitors may act therefore by preventing this first step, either because their union with the viruses prevents this necessary combination, or because the inhibitors themselves combine with these cell constituents. As such a wide range of substances can act as inhibitors, clearly no great structural specificity is involved, and a lack of specificity is also apparent from the fact that inhibitors have been found to act generally against all the viruses so far tested. However, the problem has so far received little attention and inhibitors acting specifically against individual viruses may well exist. When such chemically different substances can act as inhibitors, it is to be expected that they will be widely distributed. Already they have been recorded in extracts of many different plants, in insects, milk, serum, and other animal products, and as growth products of fungi and bacteria (BLACK, 1939; JOHNSON, 1941; FULTON, 1943; KUNTZ and WALKER, 1947). Their occurrence complicates the interpretation of transmission and inactivation tests and it is important that, when making tests, attempts should be made more often than in the past to distinguish between virus-inactivation and the phenomenon of inhibition.

**Inactivation Without Loss of Serological Activity:** — Infectivity is usually the first character of virus to reflect changes in the particles and it is often lost by changes too slight to affect many other properties. Consequently, non-infective preparations can be obtained with the serological and physical properties of normal virus preparations. The ease with which the separation of infectivity from other properties can be made varies with different viruses; in this section the treatments described are those that do so readily with all the viruses to which they have been applied. Other treatments, described in the next section, readily produce a similar separation with some viruses but not with others.

The treatments most generally useful for destroying infectivity without affecting serological activity are exposure to formaldehyde,

TABLE 24. — *The effect of X-rays and ultra-violet radiation on 0.2% solutions of potato virus X:—*

| RADIATION | TIME | SEROLOGICAL TITRE | INFECTIVITY Mean number of lesions per leaf at 1/50 |
|---|---|---|---|
| X-rays | 100 mins. | 1/4000 | 22 |
| | 360 mins. | 1/4000 | 0 |
| Ultra-violet | 3.3 mins. | 1/6000 | 40 |
| | 10 mins. | 1/4000 | 1 |
| | 30 mins. | 1/6000 | 0 |
| Control | Not irradiated. | 1/6000 | 342 |

nitrous acid or hydrogen peroxide and irradiation with ultra-violet or X-rays. The non-infective preparations of tobacco mosaic virus and potato virus X show anisotropy of flow to the same extent as control preparations, and, if sufficiently concentrated, form liquid crystalline solutions. When precipitated by acid or salts, tobacco mosaic virus forms paracrystalline needles indistinguishable from those of active virus. Similarly, preparations of bushy stunt virus inactivated by these methods crystallize in the usual dodecahedra (Fig. 54). The loss of infectivity is proportional to the length of irradiation, or the strength of the reagents used. By suitable exposures, therefore, preparations can be made with varying degrees of activity, but which precipitate with antisera and salts in the same manner. Mixtures of infective and non-infective bushy stunt virus are not fractionated by successive crystallizations, for each crystalline fraction possesses the same infectivity. Also, as the sedimentation constant of the virus is unaffected by this type of inactivation, no fractionation is to be expected from high speed centrifugation. This being so, the physical uniformity of virus preparations giving high serological titres obviously cannot be taken as proof that they are homogeneous and consist solely of active virus.

Nitrous acid or hydrogen peroxide must be used with care, because in excess they denature and destroy serological activity. The denaturing concentration varies with different viruses. For example, tobacco mosaic virus is not denatured by 5% $H_2O_2$ acting for 5 hours, whereas potato virus X is denatured by concentrations of over 1%. Loss of infectivity without affecting the serological reactions of potato virus X occurs at concentrations between 0.2% and 1%. Excess of formaldehyde does not appear to cause denaturation. Nor do long periods of irradiation with X-rays or ultra-violet, provided that the solutions irradiated are kept cold.

TABLE 24 shows the results of irradiating potato virus X solutions. In the ultra-violet test a layer 1 mm. thick, under a quartz plate, was irradiated at a distance of 8 cm. from a 2 amp. lamp. In the X-ray test,

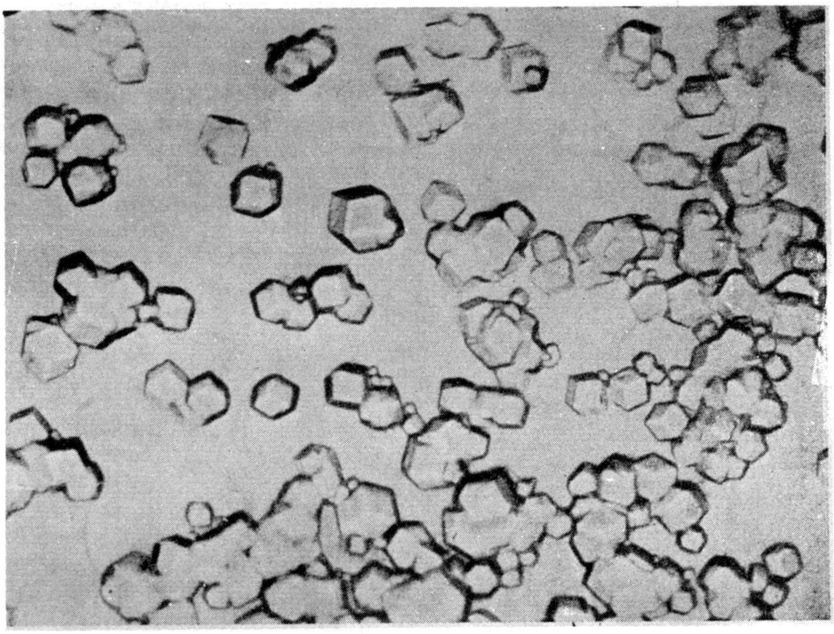

**FIG. 54. — Crystals of tomato bushy stunt virus inactivated by irradiation with ultra-violet light. The crystalline material is still serologically active. × 300. (BAWDEN, F. C. and PIRIE, N. W., 1938, Brit. J. exp. Path. *19*, 251).**

the solution was irradiated at 8 cm. from a copper anticathode of an X-ray tube run at 30 K.V. and 20 m.a. The wavelengths found by HOLLAENDER and DUGGAR (1936) to be most effective in destroying the infectivity of tobacco mosaic virus are those at which purified preparations have since been found to absorb most strongly. Wavelengths above 300 mμ have little effect, around 260 mμ, where the purines of the nucleic acid absorb strongly, there is an increase in efficiency, and lower wavelengths at which there is general absorption by the protein constituents of the virus are still more efficient. Inactivation by irradiation occurs over a wide *p*H range, but MARSHAK and TAKAHASHI (1942) find that the effect of X-rays on tobacco mosaic virus is greater at *p*H 2.2 than on the alkaline side of the isoelectric point.

The nature of the reactions responsible for inactivation are unknown, but irradiation is known to cause hydrolytic and oxidative changes and many such might occur without producing obvious effects on the gross structure of the particles. Serological activity and ability to crystallize are more closely linked with one another than with infectivity, and disruption of the particles, or at least some large structural change, is necessary for their destruction. Sufficient differences have been detected between infective preparations and those inactivated by some of these treatments to show that changes have occurred. STANLEY (1936*c*) has shown that the amino-nitrogen content of preparations treated with hydrogen peroxide or formaldehyde is lower than that of untreated controls, and that, as would be expected, preparations treated with nitrous acid contain almost no amino-nitrogen. Solutions treated with nitrous acid are more opalescent than untreated solutions, and the non-infective proteins are more easily denatured. The sedimentation constant of tobacco mosaic virus is unaffected by these treatments, but the boundary becomes more diffuse, especially after treatment with nitrous acid, showing that heterogeneity has been increased, although the large particles have not been disrupted (WYCKOFF, BISCOE and STANLEY, 1937). According to STANLEY (1936*c*) the isoelectric point is unaffected by irradiation with ultra-violet light, but shifts to the acid side after treatment with formaldehyde or nitrous acid.

ROSS and STANLEY (1938) reported that tobacco mosaic virus treated with formaldehyde at *p*H 7 could be partially reactivated by dialysing at *p*H 3. By exposure to formaldehyde for various lengths of time, they obtained preparations with approximately 10, 1 and 0.1% of the original infectivity, and after three days dialysing at *p*H 3 they found infectivity had increased respectively to 20, 10 and 1% of the original. During the treatment with formaldehyde at *p*H 7, there is a reduction in the amino-groups and groups that react with Folin's reagent, and ROSS and STANLEY state that the reactivation during dialysis is accompanied by a corresponding increase in these groups. KASSANIS and KLECZKOWSKI (1944) could not confirm these results and got no evidence that inactivation by formaldehyde was at all reversible. They found that tobacco mosaic virus was inactivated by 2% formaldehyde at all *p*H values between 3 and 7.5, the rate of inactivation being minimal at *p*H 3.5. At any stage during exposure to formaldehyde the process of inactivation could be stopped by dilution or dialysis, but neither of these treatments reactivated any virus. Inactivation does not depend on changes in groups giving the Van Slyke test for amino-nitrogen, because preparations inactivated at *p*H 3 give the same value as untreated virus. Nor does there appear to be a close correlation between inactivation and groups that give Folin's reaction, for treatment at *p*H 7 causes a greater reduction in this colour test than treatment at *p*H 3. Thus, by treatment for various times at different *p*H values, preparations are obtainable that have the same infectivity but give different colour values, or that give the same colour values but have different infectivities. KASSANIS and KLECZKOWSKI found no increase in the colour value when preparations inactivated at *p*H 7 were dialysed at *p*H 3. They suggest that the preparations used by ROSS and STANLEY may have contained some inhibitor of infectivity which was destroyed by dialysis at *p*H 3.

Although infectivity is the most sensitive indicator of changes in the virus particles, not all changes seem to destroy it. SCHRAMM and MÜLLER (1940) found that the amino groups of tobacco mosaic virus could be completely covered with acetyl or phenylureido groups without decreasing infectivity. Prolonged acetylation, however, did inactivate, which they suggested was because of phenolic groups becoming affected. MILLER and STANLEY (1941) could not get complete coverage without loss of infectivity, but they state that about 70% of the amino groups and 30% of the phenol plus indole groups can be covered without any inactivation. Similarly, ANSON and STANLEY (1941) oxidised the sulphydryl groups of tobacco mosaic virus with iodine without reducing infectivity. If the treatment was sufficient to change the tyrosine to di-iodotyrosine the virus was inactivated and denatured.

A further treatment that falls within this section is ultrasonic irradiation (treatment with high frequency sound waves), for though this does alter the physical and serological properties of tobacco mosaic virus, it destroys infectivity while still leaving preparations serologically active. The inactivation of tobacco mosaic virus by ultrasonic irradiation was first demonstrated by TAKAHASHI and CHRISTENSEN (1934), and confirmed by STANLEY (1934) who concluded that it was caused by cavitation of dissolved gas, rather than by direct action of the irradiation on the particles, for he got no inactivation when preparations were irradiated *in vacuo*.

When liquid crystalline solutions of the virus are treated, they immediately lose their birefringence, and as the treatment is continued their ability to show the phenomenon of anisotropy of flow also decreases. This change is accompanied by an increase in the number of small particles (KAUSCHE, PFANKUCH and RUSKA, 1941). OSTER (1947) has studied the changes by electron microscopy and finds that the rod-like particles break across the middle to give shorter rods. Using a preparation containing originally 60% of particles about 280 m$\mu$ long, the number of particles this length decreased with continued treatment; early in the treatment, particles of about 140 m$\mu$ increased in number, but later these also became fewer and were replaced by particles 70 or 35 m$\mu$ long, or even shorter. Infectivity and number of rods 280 m$\mu$ long both decreased exponentially with time, which OSTER concludes shows that only rods 280 m$\mu$ long have any infectivity. Although this may be so, the conclusion would seem hardly justified from the evidence advanced. There is no reason to think that the only effect of ultrasonic treatment is to break the rods, and all effects would be expected to increase exponentially with time. All that is known is that ultrasonic irradiation both breaks the rods and causes loss of infectivity, but there is nothing to show that the second effect is a direct consequence of the first. Irradiation with X-rays and ultraviolet destroys infectivity without affecting the rod-length or any other property yet studied; similarly ultrasonics inactivation may occur in some as yet undetermined manner, and the breaking of the rods may be simply an irrelevant extra effect. Indeed this is rather implied by the fact that aggregating the broken rods does not lead to any recovery of infectivity, for if the small particles are unaffected in any other way than length, they might be expected, when put together again, to possess at least some infectivity, but they do not. Only tobacco mosaic virus has yet been studied, but if spherical viruses such as tomato bushy stunt also lose infectivity when treated, but show no changes in particle shape or size, some other explanation will clearly be needed.

After treatment with ultrasonics, particles of tobacco mosaic virus structurally resemble the small particles that can be separated by differential ultra-centrifugation of sap. They also resemble them in their serological behaviour. Before treatment, the rod-like particles precipitate rapidly over a wide range of antigen/antibody ratios, but after treatment precipitation occurs slowly and there is an extended zone of antigen excess in which precipitation is inhibited. MALKIEL (1947) has shown that the disrupted particles combine with more antibody that the untreated rods, but could find no evidence that any new antigen groups were produced by the treatment.

**Inactivation and Denaturation:** — The denaturing treatments that have been most studied with plant viruses are heating, ageing, drying, and the addition of acids, alkalis, oxidising agents, and protein precipitants. Only a few viruses have been studied, but enough has been done to show that they vary widely in their ability to withstand these treatments. It is also clear that there is much interaction between denaturing treatments and that a term such as thermal inactivation point is largely meaningless unless the conditions of heating are clearly defined. For example, if tobacco mosaic virus is heated around *p*H 7, it is inactivated in 10 minutes at 75°C., whereas at *p*H 5.5 it is not completely inactivated after 10 minutes at 90°C. Whether this difference is regarded as a reduction in the resistance to heat by alkali, or an increase in the alkaline inactivation by heat, is a matter of convenience rather than any real distinction.

Work on inactivation by heat has largely been done to find the thermal inactivation point, *i. e.* the minimum temperature at which heating for 10 minutes destroys infectivity. With many viruses, such as tobacco mosaic and potato X, different workers have agreed closely on the thermal inactivation points in sap. On the other hand, with others, such as tomato bushy stunt and tobacco necrosis viruses, variable results are obtained. An explanation for this constant behaviour of some viruses and the variability of others has come from studies on the rate of inactivation at different temperatures and on the correlation between loss of infectivity and denaturation (PRICE, 1938, 1940; LAUFFER and PRICE, 1940; BAWDEN, 1941; BAWDEN and PIRIE, 1942*a*). With potato X and tobacco mosaic, loss of infectivity on heating is closely associated with denaturation. As with other examples of protein denaturation, the temperature coefficient, that is, the ratio between the rates of inactivation at temperatures separated by 10°C., is large. In the critical range, the increase in the rate of inactivation, with relatively small increase in temperature, is so great that variations in other factors, such as the virus content of sap or the susceptibility of the test plants, are unimportant. With tomato bushy stunt and tobacco necrosis viruses, on the other hand, the $Q_{10°}$ for thermal inactivation is small, for the loss of infectivity is not necessarily associated with denaturation. The range over which infectivity is lost fairly rapidly is wide, and increase in temperature is only a little more important than other factors in determining the precise inactivation point in any test.

The differences in the behaviour of the two kinds of viruses when heated for 10 minutes at various temperatures are shown in TABLE 25, the results of tests in which potato virus X and tomato bushy stunt virus were heated at $p$H 6. Loss of infectivity is slight with potato virus X at 59°C., but complete at 9 degrees higher, whereas with tomato bushy stunt virus there is a range of over 30°C. in which heating for ten minutes causes loss of infectivity. Loss of infectivity with potato virus X is correlated with a fall in serological titre, whereas with bushy stunt virus there is no reduction in serological titre unless solutions are heated to above 80°C. The loss of serological activity with both viruses occurs over a range of only a few degrees, indicating a large $Q_{10°}$ for this phenomenon and that it is closely linked with denaturation.

There are similar differences between the behaviour of tobacco mosaic and tobacco necrosis viruses when heated for 10 minutes at various temperatures and at $p$H 6. The serological activity of both is affected only at about 90°C., and the infectivity of tobacco mosaic virus is reduced only a little more than its serological activity. By contrast, tobacco necrosis viruses, which may not be completely non-infective after 10 minutes at 80°C., have their infectivity considerably reduced after 10 minutes at 50°C. Preparations of tomato bushy stunt and tobacco necrosis viruses with no demonstrable infectivity can be produced by heating at temperatures well below the thermal inactivation points. A day at 50°C., a few hours at 60°C. or 1 hour at 70°C. give non-infective preparations, which crystallize well and are indistinguishable from infective preparations in their serological and physical properties. Such complete separation of infectivity from serological activity is unobtainable with potato virus X and tobacco mosaic virus, although there is sometimes slight separation of the two properties. Potato virus X when heated for 16 hours at 50°C. at $p$H 6 may lose three-quarters of the infectivity but only one-quarter of the serological activity. Heated preparations of tobacco mosaic virus also usually lose rather more infectivity than would be expected from measurements on the amount of denatured protein produced. Thus with all the viruses heat apparently produces a series of reactions and infectivity is lost because of changes early in the series. In some these first changes are rapidly followed by others leading to disruption of the particles, whereas with bushy stunt and tobacco necrosis viruses loss of infectivity apparently occurs because of internal changes too slight to affect the stability of the particles. Not all the original antigens of tobacco mosaic and bushy stunt viruses are destroyed by denaturation. If rabbits are injected with the coagulum of heat-denatured viruses, they produce weak antisera that flocculate normal virus solutions. When heating is a little less than that needed to produce

TABLE 25.—*Effect of heating potato virus X and tomato bushy stunt viruses for ten minutes at various temperatures and pH 6*:—

| POTATO VIRUS X | | | BUSHY STUNT VIRUS | | |
|---|---|---|---|---|---|
| Temperature | Infectivity | Serological titre | Temperature | Infectivity | Serological titre |
| Unheated | 108 | 1/256 | Unheated | 153 | 1/500 |
| 59°C. | 94 | 1/256 | 50°C. | 79 | 1/500 |
| 62°C. | 29 | 1/128 | 60°C. | 20 | 1/500 |
| 65°C. | 2.5 | 1/8 | 70°C. | 9 | 1/500 |
| 68°C. | 0 | No ppt. | 80°C. | 2 | 1/500 |
| | | | 85°C. | 0 | No ppt. |

a coagulum, bushy stunt virus preparations become opalescent and material can be sedimented by centrifugation at 8,000 r.p.m. This material can be resuspended in saline and flocculated specifically by virus antiserum. The amount of heating needed to cause the formation of this easily sedimentable material, or to cause the separation of a coagulum, depends on the *p*H. At *p*H 4, solutions become opalescent after 10 minutes at 60°C. and coagulate at slightly higher temperatures, whereas at *p*H 6, 10 minutes at 80°C. is needed (BAWDEN and KLECZKOWSKI, 1942).

High pressures inactivate tobacco mosaic virus and in some respects resemble the effects of heating. BASSETT *et al.* (1938) found that pressures up to 6,000 atmospheres have no effect on infectivity, serological behaviour or crystallinity, but above 8,000 atmospheres all these properties are destroyed. Between 6,000 and 8,000 a coagulum of nucleic acid-free protein separates from the treated solutions, but at higher pressures the products of denaturation remain in solution. As with heating, coagulation follows the course of a first order reaction, the rate of denaturation being proportional to the virus concentration (LAUFFER and DOW, 1941), but several different effects on the virus are probably involved, for loss of infectivity proceeds faster than the separation of a coagulum.

As with heat, different viruses behave differently when they age *in vitro*. Viruses such as potato X and Y, which lose infectivity and serological activity at about the same rates when heated, also lose these properties at about the same rate on standing at room temperature. Infectivity is lost a little more rapidly than serological activity so that non-infective preparations can sometimes be obtained which react with antisera, but they give only low titres. By contrast, bushy stunt or tobacco necrosis viruses, which also become non-infective in a few weeks or months, retain their full serological activity. Non-infective crystalline nucleoproteins can be isolated from old sap that are indistinguishable in their properties from normal virus preparations. This type of inactivation seems to occur more rapidly in sap that in purified preparations. Differences of the type found between potato X and tobacco mosaic on the one hand, and tomato bushy stunt and tobacco necrosis on the other, may explain why heating and ageing produce successful vaccines with some animal viruses but not with others.

The ability of tobacco mosaic virus to withstand drying has been long known and appreciated as a property that separates it from most other viruses. The ability to be infective after drying, however, does not mean that the treatment has had no effect. When purified preparations of tobacco mosaic virus are dried and redissolved, their birefringence and infectivity are both reduced and their serological behaviour altered. Repeated drying and re-solution affects the properties still further. BAWDEN and PIRIE (1937*a*) found that a preparation that was liquid crystalline at 1.8% lost half its infectivity after one drying and when redissolved gave only a trace of a liquid crystalline layer at 5.4%. After a second drying it failed to give any liquid crystalline layer, though it still showed anisotropy of flow, and after seven successive dryings the infectivity was reduced to about 1% of the original. The non-infective particles produced by drying do not break down to liberate nucleic acid and, if incubated with trypsin, preparations can be regained with the normal optical properties and serological activity.

When dried in leaves or sap, most other viruses have been found to be inactivated. By cutting leaves into small pieces and drying rapidly at 1°C., however, McKINNEY (1947) has found that such unstable viruses as cucumber mosaic, tobacco ringspot, potato Y and tobacco etch may remain infective for more than a year, if they are stored in moisture-free conditions. The variations in the effects of drying produced by other factors is also shown by the behaviour of potato virus X. When leaves or sap are dried slowly, infectivity is lost; when purified preparations are dried, some insoluble material (denatured protein, free from nucleic acid) is produced. The amount depends on the *p*H of the solutions. At *p*H 6 about half the preparation is denatured, whereas at *p*H 4 and 8 the loss may be more than 90%. Fully active virus can be recovered from solutions of the dried material by ultracentrifugation. Some of the tobacco necrosis viruses withstand drying from sap, but not from purified solutions. Purified solutions when dried suffer a greater reduction in infectivity than in serological activity, but the effect of drying depends greatly on whether preparations are dried from liquid or when frozen. The dried material from unfrozen preparations dissolves readily, but when dried frozen it is insoluble and inactive.

Freezing and thawing, without drying, can also denature viruses, though the effects depend greatly on the fluids dried. Freezing leaves or sap from infected leaves has little or no effect on the activity of most viruses. It facilitates the clarification of sap from tobacco and many other plants, and is widely used as a first step in purification. It leads to loss of infectivity with some viruses, such as tobacco etch and alfalfa mosaic, probably because of adsorption onto the coagulum of host protein rather than to destruction of the viruses by freezing. Viruses that will withstand freezing in sap, however, may be denatured when purified preparations are frozen. Tomato bushy stunt is a good example. BAWDEN and PIRIE (1938*b*) showed that solutions of purified virus were denatured by freezing in conditions that had no effect on tobacco mosaic and potato virus X, but STANLEY (1940) found that freezing leaves caused no inactivation. Further work showed that the freezing of purified preparations produces widely different results depending on the condition of the fluid when frozen (BAWDEN and PIRIE (1942*a*). The amount of bushy stunt virus inactivated is increased by increases in the concentration of the virus in the solutions frozen, by the addition of acid and by increases in the length of time the fluids are held frozen. The virus is protected against the inactivating effects of freezing by the presence of glucose, proteins, salts and other substances. The protection afforded by different salts depends on their salt:ice:water eutectic temperature. Those with a high eutectic temperature do not protect if freezing is carried out at a lower temperature. The concentration of virus in sap is sufficiently small and the *p*H is far enough from the isoelectric point to account for the failure of freezing to inactivate. Added to this, the salts, sugars and colloids in the sap supplement the other protective effects. If infective sap is dialysed and brought to *p*H 3, then freezing causes inactivation. In general, loss of infectivity in frozen samples of bushy stunt virus is accompanied by a corresponding loss of serological activity and by the precipitation of denatured protein. However, inactivation by freezing, as by heating, is a complex process and loss of infectivity is produced by the earliest changes, so that frozen samples are sometimes less infective than would be expected from their serological titres. Occasionally, completely non-infective preparations with full serological activity can be prepared by freezing 0.2% virus solutions at *p*H 4.25 for 2 hours at −10°C.

Tobacco ringspot virus behaves in much the same way as bushy stunt virus when frozen (STANLEY, 1939). When purified preparations are frozen and thawed, they are inactivated and 75% of the original material separates as a precipitate. This inactivation is also prevented by the presence of salts and other materials from plant sap. A rather different effect has been described with purified preparations of tobacco necrosis viruses (PIRIE and others, 1938). After freezing, solutions lose their characteristic slight opalescence; there is loss of infectivity and a reduction in the precipitability with antiserum, especially in the region of antigen excess. Whether freezing in other conditions would cause changes more comparable with those with bushy stunt and tobacco ringspot viruses has not been determined. It is probable that other viruses now thought to be unaffected by freezing and thawing would also be denatured if frozen under different conditions. Preparations of tobacco

TABLE 26.— *The effects of alkali on tobacco mosaic virus:—*

| pH | TIME | SEROLOGICAL TITRE | AVERAGE NUMBER OF LESIONS PER LEAF AT | |
|---|---|---|---|---|
| | | | $10^{-4}$ | $10^{-5}$ |
| 11.0 | 24 hr. | 1/40,000 | 0 | 0 |
| 10.5 | 24 hr. | 1/3,000,000 | 12 | 2 |
| 9.3 | 24 hr. | 1/4,000,000 | 106 | 21 |
| 8.5 | 24 hr. | 1/4,000,000 | 74 | 9 |
| 8.5 | 5 min. | 1/4,000,000 | 65 | 11 |
| Control | — | 1/4,000,000 | 73 | 10 |

mosaic virus, for example, are unaffected by repeated freezings and thawings at neutrality, but freezing at *p*H 3 causes loss of infectivity and serological activity.

Like other proteins, viruses are stable over only a limited range of hydrogen ion concentration, but different viruses inactivate at different *p*H values, and vary more in their resistance towards acid than towards alkali. For example, potato virus Y is inactivated at *p*H values below 4.5, potato X and tobacco ringspot below *p*H 4, whereas tobacco mosaic and tomato bushy stunt viruses are stable down to below *p*H 2. As with other treatments, it is probable that infectivity is the first property to be lost on treatment with acid, but no careful tests have been made, and loss of infectivity is closely associated with disruption of the particles and loss of serological activity.

Tobacco mosaic virus appears to be no more resistant to alkali than are many other viruses, but different workers have given different values for the pH at which inactivation becomes appreciable. BEST (1936) found that loss of infectivity occurs at about *p*H 7.8, and that the amount lost in 12 hours at room temperature was progressively larger with increase in *p*H, until at *p*H 10.2 more than 99% was inactivated. Between *p*H 8.0 and 8.9, giving 20% and 90% inactivation respectively, the ratio of the concentration of hydrogen ions to remaining active virus was a constant (Fig. 55). BEST concluded that inactivation was produced by neutralisation of acidic groups, and with the discovery that the virus was a nucleoprotein he later (1937*b*) suggested that these might be the nucleic acid fractions. BAWDEN and PIRIE (1940*a*) found their preparation to be more resistant to alkali, and exposure at *p*H 10 or so was necessary to cause appreciable inactivation. It is possible that these discrepancies arise because of the use of different strains or because of different treatments to which the preparations had been previously exposed.

The results of exposing a purified preparation of tobacco mosaic virus, made by precipitation methods, to various degrees of alkali are shown in TABLE 26. It is clear that the action is complex, for the variations in infectivity and serological titre are not correlated. At *p*H 10.5 there is loss of infectivity without a corresponding fall in serological activity, and at *p*H 11 there is still some serological activity, though the preparations are not infective.

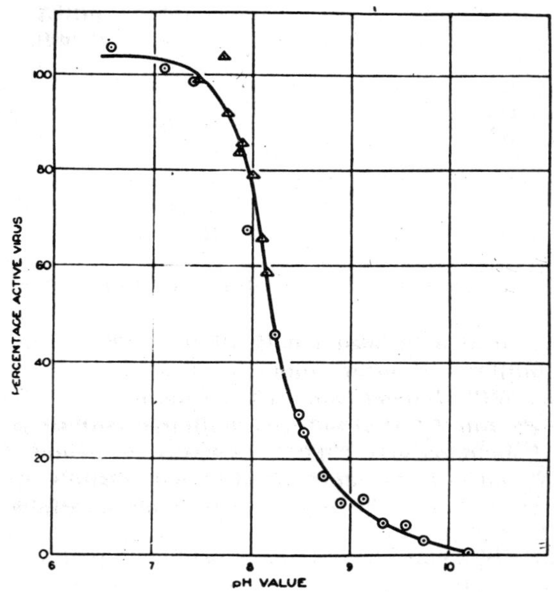


FIG. 55. — Showing the percentages of tobacco mosaic virus remaining active after 12 hours' storage at various *p*H values. Triangles represent values obtained with purified virus. Circles represent values obtained with clarified infective sap. (BEST, J. R., 1936, Austral. J. exp. Biol. Med. Sci. *14*, 323).

Exposure at *p*H 9.3 shows an effect frequently obtained, an apparent increase in infectivity with no change in the serological titre. This activation can be regarded as a partial reversal of the fall in infectivity produced by purification, which is believed to be caused by the linear aggregation of virus particles, and the simplest explanation is that gentle treatment with alkali disaggregates the long rods. Such an effect might be more than sufficient to mask any small amount of inactivation occurring simultaneously. Thus alkali can have at least three successive effects on purified preparations of tobacco mosaic virus. First, there is a disaggregation into smaller active particles, secondly a change within the particles, making them non-infective without destroying their structure or serological properties, and thirdly a disruption of the particles leading to loss of all characteristic properties. The course of the third stage is also influenced by the temperature, for at low temperatures the denatured products remain soluble whereas at high temperatures they precipitate. The centrifugal studies of ERIKSSON-QUENSEL and SVEDBERG (1936) and WYCKOFF (1937) suggest that the third stage is also complex, for the extent to which the virus is split into products of smaller molecular weight is increased by increased exposure to alkali.

The *p*H range over which bushy stunt and tobacco necrosis viruses lose infectivity without undergoing denaturation is wider than that of tobacco mosaic virus, and, as with heating, it is correspondingly easier to produce non-infective preparations that are fully active serologically. Exposure of these viruses to *p*H 9 or so causes fairly rapid loss of infectivity, but serological activity and ability to crystallize are not affected until the *p*H is raised to 10 or higher. Even after denaturation sets in and preparations of bushy stunt virus lose their homogeneity, they can still react with virus antiserum. For example, after 2 hours' exposure at 18°C. to *p*H 11, a part of the preparation precipitates on neutralisation, but the remaining soluble material still reacts specifically with virus antiserum. The behavior of virus X has not been studied in detail, but there does not seem to be a wide *p*H range in which loss of infectivity occurs without denaturation, for LORING (1938*b*) states that at *p*H 10 and above, where infectivity was lost, the homogeneity of the preparations was also destroyed.

Of the various substances that have been found to inactivate viruses, urea has been the most studied. BAWDEN and PIRIE (1937*a*) found that three strains of tobacco mosaic virus were denatured and lost their liquid crystallinity after some hours exposure to saturated urea, and MEHL (1938) also described loss of anisotropy of flow after treatment with urea. FRAMPTON and SAUM (1939) and FRAMPTON (1939) reported that dissolving tobacco mosaic virus caused a hundred-fold increase in the diffusion constant with no change in infectivity. They interpreted this as indicating that urea disaggregated the virus particles into molecules with a molecular weight of about 100,000 which they suggested were the true virus molecules, the larger particles being simply aggregates of these. However, no increase in infectivity was produced such as might be expected from such a disaggregation, and later work has not confirmed it, but has shown that urea disrupts the virus. STANLEY and LAUFFER (1939) got no evidence suggesting that the proteins with smaller molecular weights produced by the action of urea possessed any infectivity. They showed that the residual activity in partially inactivated preparations was associated with residual, unchanged virus and not with the small break-down products. The size of the degradation products varies with the extent of the treatments. MARTIN (1939), from measurements in the ultracentrifuge, gave a molecular weight of 400,000 for inactivated virus, but by measurements of osmotic pressure STANLEY and LAUFFER (1939) showed that on continued action of urea it fell to 40,000. They also found that the disintegration depended on the concentration of the urea, the type and concentration of the electrolyte, the *p*H and the temperature.

BAWDEN and PIRIE (1940) found that the four viruses, tobacco mosaic, potato X, tomato bushy stunt and tobacco necrosis are all irreversibly denatured by urea. The denaturation is closely linked with loss of infectivity and serological activity. For each virus there is a critical concentration of urea below which there is no irreversible effect on infectivity; this concentration is smallest for potato virus X and greatest for tomato bushy stunt virus. In addition to these irreversible effects, however, it is probable that urea can cause changes that can be readily reversed and have no effect on activity. For example, FRAMPTON (1939) found that the viscosity of tobacco mosaic virus solutions was greatly reduced by the addition of 1 M urea. As inactivation occurs only slowly in much more concentrated urea solutions, it is probable that this immediate effect on viscosity is caused by changes in hydration of the virus particles and not by denaturation. The rate of inactivation for all four viruses is greatly increased by the presence of alkali. The rate is minimum at about 20°C. and is much increased by cooling to —10°C. The viruses differ in the manner in which they break down as well as in their resistance to urea. The inactivation of potato virus X and tobacco mosaic virus is accompanied by the separation of the nucleic acid from the protein and the products of denaturation are soluble in urea solution, whereas the inactivated bushy stunt and tobacco necrosis viruses are insoluble in urea solutions and the precipitates contain nucleic acid.

Other simple organic substances are more efficient inactivators than urea or than the usual solvents such as alcohol and acetone, whose precipitating and inactivating actions are more widely known. Of 15 substances tested ranging from urethane, guanidine and related substances to pyridine, benzoate, salicylate and phenol, all except arginine and nicotine cause inactivation when used in neutral solutions at concentrations of 4 M or less (BAWDEN and PIRIE, 1940*a*). Arginine and nicotine give reversible fibrous precipitates with tobacco mosaic virus; other substances also often act as precipitants without causing inactivation when used diluted, but when used concentrated they denature the viruses and dissolve the products of denaturation (BEST, 1940). For each substance there is a threshold concentration below which no irreversible changes are produced. As with urea, this threshold is lowest with potato virus X, intermediate with tobacco mosaic virus and highest with tomato bushy stunt virus. With none of these other substances is there an increase in the rate of denaturation by cooling below 20°C., as there is with urea. In general, the course of denaturation seems similar to that with urea, leading to a separation of nucleic acid from the protein with tobacco mosaic virus and potato virus X but not with bushy stunt virus. Sodium dodecyl sulphate, a surface active agent, however, splits off the nucleic acid from all three viruses.

Viruses can also be affected in various ways by exposure to inorganic salts. Many workers have observed that the presence of salt in inocula affects the numbers of lesions produced, but neutral salts appear to have no permanent effect on most viruses, except to cause those with anisometric particles to aggregate. Tobacco ringspot (STANLEY, 1939) is exceptional in being inactivated by ammonium sulphate, which separates nucleic acid. The Rothamsted culture of tobacco necrosis virus (BAWDEN and PIRIE, 1945*a*) is also rendered non-infective by exposure to low salt concentrations, but the treatment does not denature this virus and it leaves the serological activity unaffected. Work with potato virus X (BAWDEN and CROOK, 1947) shows that the type of preparation treated with salt may determine whether or not inactivation occurs. Infective sap is unaffected by incubation for some hours at 37°C. with *p*H 7 0.1 M/phosphate buffer, but purified preparations are denatured; in the presence of chloroform, denaturation is more rapid and shows by the separation of a protein coagulum free from nucleic acid. Inactivation depends on *p*H and temperature, and on the presence of other materials. At *p*H 5.5 0.1 M/phosphate has no effect on infectivity or serological activity. Also, at *p*H 7, the inactivating effect of phosphate can be prevented by the presence of sap from healthy leaves and some other protein-containing solutions. At room temperature or below, phosphate has little effect on virus X, at 37°C. more than 90% is denatured in 16 hours and at 50°C. denaturation is almost complete in 2 hours. At 37°C. the reaction occurs at a fairly uniform velocity, but it almost ceases after 16 hours, usually leaving a small residuum of serologically active material with little infectivity.

Claims have been made that tobacco mosaic virus (WENT, 1937) and a tobacco necrosis virus (MANIL, 1938) can be inactivated by mercuric chloride and then at least partially reactivated by dilution or by precipitating the mercuric ions by adding suitable salts. KASSANIS and KLECZKOWSKI (1944) found that mercuric chloride had various effects on tobacco mosaic virus, but obtained no evidence for any reversible inactivation. At *p*H values above 6, mercuric chloride caused denaturation; as with exposure to alkali, short exposures often produced an initial rise in the numbers of lesions produced, but longer exposures led to loss of both serological activity and infectivity. This reaction could be stopped at any stage by dilution, acidification, or the addition of appropriate salts, some such as NaCl and KCl, which neither precipitate mercuric ions nor change *p*H, but the inactivated virus could not be reactivated by these treatments. Below *p*H 6 mercuric chloride has no denaturing action, but at concentrations above 0.01% it behaves as an inhibitor of infectivity.

### *References to Chapters 9-13:-*


ADAIR, G. S. (1937): Ann. Rev. Biochem. *6*, 163.
ANSON, M. L. and STANLEY, W. M. (1941): J. Gen. Physiol. *24*, 679.
BALD, J. G. (1937): Austral. J. Exp. Biol. and Med. Sci. *15*, 211.
BASSET, J., GRATIA, A., MACHEBOEUF, M. and MANIL, P. (1938): Proc. Soc. Exp. Biol. and Med. *38*, 248.
BAWDEN, F. C. (1939): *Plant Viruses and Virus Diseases* (1st Edition). Chronica Botanica, Leiden, Holland.
—— (1941): Brit. J. exp. Path. *22*, 59.
——, and CROOK, E. M. (1947): Brit. J. exp. Path. *28*, 403.
——, and KLECZKOWSKI, A. (1942): Brit. J. exp. Path. *23*, 178.
——, and —— (1945): J. Pomology *21*, 2.
——, and —— (1948): J. Gen. Microbiol. *2*, 173.
——, and PIRIE, N. W. (1936): Brit. J. exp. Path. *17*, 64.
——, and —— (1937*a*): Proc. Roy. Soc. B. *123*, 274.
——, and —— (1937*b*): Brit. J. exp. Path. *18*, 275.
——, and —— (1937*c*): *ibid. 18*, 290.
——, and —— (1937*d*): Réunion Intern. de Physique, Chimie et Biologie, Paris, p. 382.
——, and —— (1938*a*): Brit. J. exp. Path. *19*, 66.

——, and —— (1938*b*) : *ibid.* *19*, 251.
——, and —— (1938*c*) : *ibid.* *19*, 264.
——, and —— (1939) : *ibid.* *20*, 322.
——, and —— (1940) : Biochem. J. *34*, 1258.
——, and —— (1940*a*) : *ibid* *34*, 1278.
——, and —— (1942) : Brit. J. exp. Path. *23*, 314.
——, and —— (1943) : Biochem. J. *37*, 66.
——, and —— (1944) : Brit. J. exp. Path. *25*, 68.
——, and —— (1945*a*) : *ibid.* *26*, 277.
——, and —— (1945*b*) : *ibid.* *26*, 294.
——, and —— (1946) : *ibid.* *27*, 81.
——, PIRIE, N. W., BERNAL, J. D., and FANKUCHEN, I. (1936) : Nature *138*, 1051.
——, and ROBERTS, F. M. (1947) : Ann. appl. Biol. *34*, 286.
——, and SHEFFIELD, F. M. L. (1939) : *ibid.* *26*, 102.
BECHHOLD, H. (1907) : Z. phys. Chem. *60*, 257.
—— and SCHLESINGER, M. (1933) : Phytopath. Z. *6*, 627.
BERNAL, J. D. (1938) : Proc. Roy. Soc. B. *125*, 299.
—— and CARLISLE, C. H. (1948) : Nature, *162*, 139.
—— and FANKUCHEN, I. (1937) : *ibid.* *139*, 923.
—— and —— (1941) : J. Gen. Physiol. *25*, 111.
——, ——, and RILEY, D. P. (1938) : Nature *142*, 1075.
BEST, R. J. (1936) : Austral. J. Exp. Biol. & Med. Sci. *14*, 323.
—— (1937) : Austral. Chem. Inst. J. and Proc. *4*, 375.
—— (1937*a*) : Nature *139*, 628.
—— (1937*b*) : *ibid.* *140*, 547.
—— (1940) : Austral. J. Exp. Biol. & Med. Sci. *18*, 307.
—— (1940*a*) : Nature *145*, 627.
—— (1947) : Austral. J. Exp. Biol. & Med. Sci. *25*, 283.
—— and LUGG, J. W. H. (1944) : *ibid.* *22*, 247.
BHARGAVA, K. S. (1948) : Unpublished.
BLACK, L. M. (1939) : Phytopath, *29*, 321.
—— (1941) : *ibid.* *31*, 3.
——, MOSLEY, V. N., and WYCKOFF, R. W. G. (1948) : Biochim. Biophys. Acta *2*, 121.
BUTENANDT, A., FRIEDRICH-FRESKA, H., HARTWIG, S., and SCHIEBE, G. (1942) : Z. physiol. Chem. *274*, 276.
CALDWELL, J. (1933) : Ann. Appl. Biol. *20*, 100.
CHIBNALL, A. C. (1923) : J. Biol. Chem. *55*, 333.
—— (1924) : Biochem. J. *18*, 395.
COHEN, S. S. (1942) : J. Biol. Chem. *144*, 352.
—— and STANLEY, W. M. (1942*a*) : J. Biol. Chem. *142*, 863.
—— and —— (1942*b*) : *ibid.* *144*, 589.
COHN, E. J. and EDSALL, J. T. (1943) : *Proteins, Amino Acids and Peptides.* Reinhold, New York.
COSSLETT, V. E. and MARKHAM, R. (1948) : Nature *161*, 250.
CROOK, E. M. (1946) : Biochem. J. *40*, 197.
—— and SHEFFIELD, F. M. L. (1946) : Brit. J. exp. Path. *27*, 328.
CROWFOOT, D. and SCHMIDT, G. M. J. (1945) : Nature *155*, 504.
DUGGAR, B. M. and ARMSTRONG, J. K. (1925) : Ann. Mo. Bot. Gard. *12*, 359.
ELFORD, W. J. (1931) : J. Path. Bact. *34*, 505.
—— (1933) : Proc. Roy. Soc. B. *123*, 384.
ERIKSSON-QUENSEL, I. B. and SVEDBERG, T. (1936) : J. Amer. Chem. Soc. *58*, 1863.
FRAMPTON, V. L. (1939) : J. Biol. Chem. *129*, 233.
—— (1939*a*) : Science *90*, 305.
—— (1939*b*) : Phytopath. *29*, 495.
—— (1940) : *ibid.* *30*, 666.
—— (1942) : Science *95*, 232.
—— (1942*a*) : Arch. Biochem. *1*, 83.
—— and SAUM, A. M. (1939) : Science *89*, 84.
FULTON, R. W. (1943) : Phytopath. *33*, 674.
GOWEN, T. W. (1940) : Proc. Nat. Acad. Sci. Wash. *26*, 8.
—— (1941) : Cold Spring Harbour Symposium *9*, 187.
GRABAR, P. (1938) : *ibid.* *6*, 252.
HERMANN, C. (1931) : Z. Krist. *79*, 1.
HILLS, C. H. and VINSON, C. G. (1938) : Res. Bull. Mo. Agric. Exp. Stat. No. 286.
HOGGAN, I. A. (1935) : Ann. appl. Biol. *22*, 27.
HOLLAENDER, A. and DUGGAR, B. M. (1936) : Proc. Nat. Acad. Sci. Wash. *22*, 19.
HOLMES, F. O. (1934) : Phytopath. *24*, 845.
—— (1941) : *ibid.* *31*, 1089.
JENSEN, J. H. (1937) : *ibid.* *27*, 69.
JOHNSON, J. (1927) : Wis. Agric. Expr. Stat. Res. Bull. 63.
—— (1941) : Phytopath. *31*, 679.

—— and Hoggan, I. A. (1937) : Phytopath. *27*, 1014.
Kassanis, B. and Kleczkowski, A. (1944) : Biochem. J. *38*, 20.
—— and —— (1948) : J. gen. Microbiol. *2*, 143.
—— and Sheffield, F. M. L. (1941) : Ann. appl. Biol. *28*, 360.
Kausche, G. A. (1939) : Naturwissenschaften *27*, 77.
——, Pfankuch, E. and Ruska, E. (1939) : *ibid. 27*, (18), 292.
——, ——, —— (1941) : *ibid. 29*, 573.
—— and Stubbe, H. (1939) : *ibid. 27*, 77.
Kleczkowski, A. (1944) : Biochem. J. *38*, 160.
—— (1946) : *ibid. 40*, 677.
—— and Nixon, H. L. (1950) : J. gen. Microbiol. *4*, in press.
—— and Watson, M. A. (1944) : Ann. appl. Biol. *31*, 116.
Knight, C. A. (1942*a*) : J. Biol. Chem. *145*, 11.
—— (1942*b*) : J. Amer. Chem. Soc. *64*, 2734.
—— (1943) : J. Biol. Chem. *147*, 663.
—— (1947) : *ibid. 171*, 297.
—— (1949) : J. Amer. Chem. Soc. *71*, 3108.
—— and Lauffer, M. A. (1942) : J. Biol. Chem. *144*, 411.
—— and Oster, G. (1947) : Arch. Biochem. *15*, 289.
—— and Stanley, W. M. (1941) : J. Biol. Chem. *141*, 29.
—— and —— (1941*a*) : *ibid. 141*, 39.
Kuntz, J. E. and Walker, J. C. (1947) : Phytopath. *37*, 561.
Landsteiner, K. (1936) : *The specificity of serological reactions.* C. C. Thomas, Baltimore, Md., U. S. A.
Lauffer, M. A. (1938) : J. Physical. Chem. *42*, 935.
—— (1938*a*) : J. Biol. Chem. *126*, 443.
—— (1939) : J. Amer. Chem. Soc. *61*, 2412.
—— (1940) : J. Phys. Chem. *44*, 1137.
—— (1942) : J. Biol. Chem. *143*, 99.
—— (1943) : *ibid. 151*, 627.
—— (1944) : J. Amer. Chem. Soc. *66*, 1188.
—— (1946) : *Viruses.* Edwards Bros., Ann Arbor, Michigan.
—— (1948) : J. Biol. Chem. *174*, 481.
—— and Dow, R. B. (1941) : *ibid. 140*, 509.
—— and Price, W. C. (1940) : *ibid. 133*, 1.
—— and —— (1947) : Arch. Biochem. *15*, 115.
—— and Stanley, W. M. (1938) : J. Biol. Chem. *123*, 507.
—— and —— (1939) : Chemical Reviews *24*, 303.
Lea, D. E. (1940) : Nature *146*, 137.
—— and Smith, K. M. (1940) : Parasitology *32*, 405.
—— and —— (1942) : Parasitology *34*, 227.
Lojkin, M. E. and Vinson, C. G. (1931) : Contrib. Boyce Thompson Inst. *3*, 147.
Loring, H. S. (1938*a*) : J. Biol. Chem. *123*, lxxvi.
—— (1938*b*) : *ibid. 126*, 455.
—— (1939*a*) : *ibid. 130*, 251.
—— (1939*b*) : *ibid. 128*, lxi.
—— (1942) : J. gen. Phys. *25*, 497.
—— and Stanley, W. M. (1937) : J. Biol. Chem. *117*, 733.
—— and Wyckoff, R. W. G. (1937) : *ibid. 121*, 225.
——, Lauffer, M. A. and Stanley, W. M. (1938) : Nature *142*, 841.
Loring, H. S., Osborn, H. J. and Wyckoff, R. W. G. (1938) : Proc. Soc. Exp. Biol. and Med. *38*, 329.
McFarlane, A. S. and Kekwick, R. A. (1938) : Biochem. J. *32*, 1607.
McKinney, H. H. (1947) : Phytopath. *37*, 139.
Malkiel, S. (1947) : J. Immunol. *57*, 55.
Manil, P. (1938) : C. R. Soc. Biol. Paris *127*, 1464.
Markham, R. (1948) : Private communication.
—— and Smith, K. M. (1946) : Nature *157*, 300.
—— and —— (1949) : Parasitology *39*, 330.
——, Matthews, R. E. F. and Smith, K. M. (1948) : Nature *162*, 88.
——, Smith, K. M., and Lea, D. E. (1942) : Parasitology *34*, 315.
——, ——, and Wyckoff, R. W. G. (1947) : Nature *159*, 574.
——, ——, and —— (1948) : *ibid. 161*, 760.
Marshak, A. and Takahashi, W. N. (1942) : Proc. Nat. Acad. Sci. *28*, 211.
Martin, L. F. (1939) : Rep. of 3rd Inter. Cong. Microbiol. 281.
——, Balls, A. K. and McKinney, H. H. (1938) : Science *87*, 329.
——, ——, and ——(1939) : J. Biol. Chem. *121*, 205.
Mehl, D. (1938) : Cold Spr. Harb. Symposia *6*, 226.
Melchers, G., Schramm, G., Trurnit, H. and Friedrich-Freska, H. (1940) : Biol. Zentr. *60*, 524.
Miller, G. L. and Price, W. C. (1946*a*) : Arch. Biochem. *10*, 467.

——, and —— (1946*b*) : *ibid.* *11*, 337.
——, and —— (1946*c*) : *ibid.* *11*, 329.
——, and STANLEY, W. M. (1941) : J. Biol. Chem. *141*, 965.
MÜLLER, H. O. (1942) : Kolloid-Z. *99*, 6.
MULVANIA, M. (1926) : Phytopath. *16*, 853.
NEURATH, H. and COOPER, G. R. (1940) : J. Biol. Chem. *135*, 455.
—— and SAUM, A. M. (1938) : *ibid.* *126*, 435.
NORRIS, D. O. (1946) : Coun. Sci. Ind. Res. (Australia), Bull. 202.
OSTER, G. (1946) : Science *103*, 306.
—— (1947) : J. Gen. Physiol. *31*, 89.
——, and STANLEY, W. M. (1946) : Brit. J. exp. Path. *27*, 261.
——, DOTY, P. M. and ZIMM, B. H. (1947) : J. Amer. Chem. Soc. *69*, 1193.
——, KNIGHT, C. A. and STANLEY, W. M. (1947) : Arch. Biochem. *15*, 279.
PFANKUCH, E. (1940) : Biochem. Z. *306*, 125.
PIRIE, N. W. (1940) : Biol. Rev. *15*, 377.
—— (1945) : Advances in Enzymology *5*, 1.
—— (1948) : Unpublished.
——, SMITH, K. M., SPOONER, E. T. C. and MACCLEMENT, W. D. (1938) : Parasitology *30*, 543.
PRICE, W. C. (1938) : Amer. J. Bot. *25*, 603.
——, (1940) : Arch. ges. Virusforsch. *1*, 373.
—— (1945) : Science *101*, 515.
—— (1946) : Amer. J. Bot. *33*, 45.
—— and WYCKOFF, R. W. G. (1938) : Nature *141*, 685.
—— and —— (1939) : Phytopath. *29*, 83.
—— and —— (1946) : Nature *157*, 764.
——, WILLIAMS, R. C., and WYCKOFF, R. W. G. (1945) : Science *102*, 277.
——, ——, —— (1946) : Arch. Biochem. *9*, 175.
RAWLINS, T. E. (1942) : Science *96*, 425.
——, ROBERTS, C., and UTECH, N. M. (1946) : Amer. J. Bot. *33*, 356.
RAYCHAUDHURI, S. P. (1947) : Current Science *16*, 26.
ROBINSON, J. R. (1939) : Proc. Roy. Soc. A. *170*, 519.
ROSS, A. F. (1940) : J. Biol. Chem. *136*, 119.
—— (1941) : Phytopath. *31*, 394.
—— (1941*a*) : *ibid.* *31*, 410.
—— (1941*b*) : J. Biol. Chem. *138*, 741.
—— (1942) : *ibid.* *143*, 685.
—— and STANLEY, W. M. (1938) : J. Gen. Physiol. *22*, 165.
—— and —— (1939) : J. Amer. Chem. Soc. *61*, 535.
—— and VINSON, C. G. (1937) : Res. Bull. Mo. Agri. Exp. Stat. 258.
SAMUEL, G., BALD, J. G., and PITTMAN, H. A. (1930) : Coun. Sci. Ind. Res. (Australia), Bull. 44.
SCHACHMAN, H. K. and LAUFFER, M. A. (1948) : *Quoted by* STANLEY (1948), American Scientist *36*, 59.
SCHRAMM, G. (1941) : Ber. chem. Ges. *74*, 532.
—— and MÜLLER, H. (1940) : Z. physiol. Chem. *266*, 43.
—— and —— (1940*a*) : Naturwissenschaften *28*, 223.
SCHWERDT, C. E. and LORING, H. S. (1947) : J. Biol. Chem. *167*, 593.
SHEFFIELD, F. M. L. (1939) : Proc. Roy. Soc. B. *126*, 529.
SIGURGEIRSSEN, T. and STANLEY, W. M. (1947) : Phytopath. *37*, 26.
SMITH, K. M. and DONCASTER, J. P. (1936) : Rap. III. Intern. Cong. Comp. Path. Athens.
—— and MACCLEMENT, D. (1940) : Parasitology *32*, 320.
—— and —— (1941) : *ibid.* *33*, 320.
SREENIVASAYA, M. and PIRIE, N. W. (1938) : Biochem. J. *32*, 1707.
STANLEY, W. M. (1934) : Science *80*, 339.
—— (1934*a*) : Phytopath. *24*, 1055.
—— (1934*b*) : *ibid.* *24*, 1269.
—— (1935*a*) : Science *81*, 644.
—— (1935*b*) : Phytopath. *25*, 475.
—— (1936*a*) : *ibid.* *26*, 305.
—— (1936*b*) : J. Biol. Chem. *115*, 673.
—— (1936*c*) : Science *83*, 626.
—— (1937*a*) : Ergebnisse der Physiologie *39*, 294.
—— (1937*b*) : Amer. J. Bot. *24*, 59.
—— (1937*c*) : J. Biol. Chem. *121*, 205.
—— (1937*d*) : *ibid.* *117*, 325.
—— (1937*e*) : Phytopath. *27*, 1152.
—— (1938*a*) : J. Physical Chem. *42*, 55.
—— (1938*b*) : Handbuch der Virusforschung, Springer, Wien.
—— (1939) : J. Biol. Chem. *129*, 405.

—— (1939*a*) : *ibid.* *129*, 429.
—— (1940) : *ibid.* *135*, 437.
—— (1943) : In *Virus Diseases* Cornell University Press. Ithaca.
—— and ANDERSON, T. F. (1941) : J. Biol. Chem. *139*, 325.
—— and LAUFFER, M. A. (1939) : Science *89*, 345.
—— and LORING, H. S. (1938) : Cold Spr. Harb. Symposium VI, p. 341.
—— and —— (1939) : Proc. IV. Intern. Cong. Comp. Path. (Rome) p. 45.
—— and WYCKOFF, R. W. G. (1937) : Science *85*, 181.
STOKES, J. L., GUNNESS, M., DWYER, I. M. and CASWELL, M. C. (1945) : J. Biol. Chem. *160*, 35.
SVEDBERG, T. and PEDERSEN, K. O. (1940) : *The Ultracentrifuge*. Clarendon Press, Oxford.
TAKAHASHI, W. N. (1941) : Phytopath. *31*, 1117.
—— (1942) : Science *95*, 586.
—— (1946) : *ibid.* *104*, 377.
—— (1948) : Amer. J. Bot. *35*, 243.
—— and CHRISTENSEN, R. J. (1934) : Science *79*, 415.
—— and RAWLINS, T. E. (1933) : Science *77*, 26.
—— and —— (1933*a*) : Science *77*, 26.
—— and —— (1933*b*) : *ibid.* *77*, 284.
—— and —— (1946) : Amer. J. Bot. *33*, 740.
—— and —— (1947) : *ibid.* *34*, 271.
—— and —— (1948) : Phytopath. *38*, 279.
THORNBERRY, H. H. (1935) : Phytopath. *25*, 931.
—— and MCKINNEY, H. H. (1939) : *ibid.* *29*, 250.
VINSON, C. G. and PETRE, A. W. (1929) : Bot. Gaz. *87*, 14.
—— and —— (1931) : Contrib. Boyce Thomp. Inst. *3*, 131.
VAN DER WANT, J. P. H. and ROZENDAAL, A. (1948) : Tijdschr. Pl. Ziekten *54*, 134.
WELLMAN, F. L. (1935) : Phytopath. *25*, 377.
WENT, J. C. (1937) : Phytopath. Z. *10*, 480.
WILLIAMS, R. C. and WYCKOFF, R. W. G. (1944) : J. appl. Phys. *15*, 712.
—— and —— (1946) : *ibid.* *17*, 23.
WOODS, W. M. and DUBUY, H. G. (1941) : Phytopath. *31*, 978.
WYCKOFF, R. W. G. (1937) : J. Biol. Chem. *121*, 219.
—— (1947) : Biochem. Biophysica Acta *2*, 139.
—— and COREY, R. B. (1936) : Science *84*, 513 .
——, BISCOE, J. and STANLEY, W. M. (1937) : J. Biol. Chem. *117*, 57.

*Chapter 14*

# TAXONOMY OF VIRUSES

**Nomenclature and Classification:** — The urgent need for some agreed methods by which viruses shall be identified and classified is most evident from the confusion that exists in virus nomenclature. Not only do names provide no information about either relationships between viruses or any of their intrinsic properties, but different workers commonly use different names for the same virus and similar names for unrelated ones. Some confusion was probably unavoidable because of the early methods whereby viruses were named and described, but it has grown and persisted unnecessarily because most workers seem in no way concerned to relate what they consider to be new viruses with those already recorded. It will continue unless workers show much more interest in taxonomy than they have hitherto, and the various schemes of classification so far proposed will have justified themselves if they do no more than stimulate this interest.

During the early work, plant viruses were generally thought to be stable, to produce constant symptoms and to have limited host ranges. Hence it was reasonable to give them names derived from the host plants in which they were encountered and from the main type of symptom caused. It was also reasonable to assume as "new" any virus that was found causing a clinical condition different from those previously described or that occurred infecting a plant not already recorded as a host for viruses. As knowledge accumulated, the original assumptions became untenable; some viruses were found to have wide host ranges or to exist in strains that cause widely different symptoms; it was also found that different viruses may cause similar symptoms and that the reactions of infected plants often vary strikingly with changes in the environment. Nevertheless, the old practice continued unaltered, and most workers appear to have been unaffected by the warnings of the few who were conscious of the confusion inseparable from its continuance. First and foremost among those to realise the need for a new system of nomenclature was JOHNSON, who in 1927 pointed out the inadequacy of attempting to identify and name viruses on the dual basis of common host and symptoms caused. At that time and since (JOHNSON, 1935; JOHNSON and HOGGAN, 1931, 1935), he has stressed the need for a systematic approach towards nomenclature and has urged the need for separating virus names from names of diseases. Although he realised the difficulties of attempting a system of classification with the little information then existing on intrinsic properties of the viruses themselves, he pointed out that tests of stability *in vitro* are often more reliable for virus identification than host plant reactions. As an interim measure, he suggested that viruses should be named simply by adding the word virus and a number to the common name of the host in which they were first discovered. JOHNSON did not propose a classification, but with HOGGAN (1935) he developed a descriptive key based on five characters by which about fifty viruses might be identified and placed into groups at least as homogeneous as those set up in some schemes of classification since proposed.

The continued use of the name of the host plant for the most prominent part of virus names has led almost inevitably to a grouping of viruses because they are associated with a particular plant. It has culminated in a scheme described by SMITH (1937) as a classification, in which viruses are divided into 51 groups. Each group takes the generic name of a host plant, for example, there is a *Cucumis* group, a *Nicotiana* group, a *Solanum* group, and so on, the individual members of each group being distinguished by different numbers. The name received depends merely on the host in which a virus is first found and on how many have previously been found in that host; except for a few viruses which may have limited host ranges, the grouping is largely a matter of chance, and has no bearing on any natural affinities between viruses. The viruses that cause aster yellows, tomato spotted wilt, and cucumber mosaic, to quote only a few with wide host ranges, might almost equally as well be placed in any other of the groups as in the *Callistephus, Lycopersicum* and *Cucumis* groups, where they now respectively find themselves.

A grouping based on such arbitrary grounds and which entirely neglects natural relationships is not a classification; it provides a system of nomenclature and of indexing but has no other significance. The taxonomic implications of such a method of grouping are well illustrated by applying a similar one to a few fungi. *Ophiobolus graminis* would be placed, together with *Erisyphe graminis* and *Puccinia graminis* in the *Triticum* group, and *Phytophthora infestans, Rhizoctonia solani* and *Synchytrium endobioticum* would all come together in the *Solanum* group. By contrast, fungi as closely related as *Puccinia Pruni-spinosae* and *Puccinia graminis* would be as widely separated as is the *Rosacea* from the *Gramineae* in HUTCHINSON's classification of plants.

Grouping and naming by hosts is a straight-forward procedure and undoubtedly provides a convenient method of indexing or cataloguing viruses, but more than this is needed to justify the term classification. Classification implies a systematic arrangement based on natural relationships and should result in uniform groups containing only individuals that have many features in common. It is impossible to work with a large number of viruses without appreciating the fact that they show the various grades of relationships implied by the terms family, genus, species and variety. Recognising this, HOLMES (1939) broke away from the previous methods of nomenclature and introduced a Latin binomial-trinomial system based on a proposed classification in which 89 viruses of flowering plants were separated into ten families. The host species plays no part in determining groups and at first sight the scheme appears taxonomically sound. Closer inspection, however, shows that natural relationships are often not shown and that some groups are very heterogeneous, widely-differing viruses often receiving the same generic name. The basis for separation is largely host plant reactions, supplemented by methods of transmission, so that the result is a classification of diseases rather than of viruses. The most frequent symptoms are probably leaf mottling and necrosis, and it is therefore not surprising that 53 of the 89 viruses treated by Holmes fell in the one genus *Marmor*. Viruses that are known to differ in so many ways from one another as do tobacco mosaic, potato X and Y, tomato bushy stunt and tobacco necrosis, to name only a few, all receive the same generic name because in some host or another they cause necrotic or chlorotic spotting, so that their names suggest that they are all as closely related as are tobacco mosaic virus and cucumber virus 3, though these two have many common attributes and share common antigens.

Although it is debatable whether the time is yet ripe for taking a decision as to whether or not viruses should be treated as organisms, the Latin nomenclature introduced by HOLMES, possibly because it is one familiar to all biologists, is proving attractive to many workers (VALLEAU, 1940; THORNBERRY, 1941; MCKINNEY, 1944*a*, 1944*b*; LIMASSET, 1946), and it has received the approval of the Nomenclature Committee of the American Phytopathological Society (BENNETT *et al.*, 1943). This type of nomenclature, however, carries with it definite implications of relationships and these are not borne out by the groupings in HOLMES' scheme. In attempts to remedy some of the anomalies, supporters of the new nomenclature have suggested amendments to the original scheme. Major shortcomings are the absence of genera within families and the setting up of families on insufficient and artificial grounds. VALLEAU (1940) has pointed out the erroneous relationships suggested by including so many different viruses in the genus *Marmor* and that there is no valid reason to set up the *Annulaceae* as a separate family. He proposed the abolition of trinomials for strains and classified some of the commonly occurring tobacco viruses into 7 genera. His scheme has the merit of maintaining uniformity within each genus and of suggesting no false relationships, but is limited to only a few viruses and some genera contain only one species.

An excellent historical account of nomenclature and classification has been given by MCKINNEY (1944*a*) who also makes many valuable suggestions to be followed in naming and classifying viruses. He points out that HOLMES' scheme is unprecedented in consisting exclusively of monogeneric families and in containing descriptions of varieties without descriptions of the genera or species. He proposes a scheme in which the ten families proposed by HOLMES are consolidated into two, *Marmoraceae* and *Rugaceae*. The division between the two is made by symptoms. *Marmoraceae* includes all viruses that cause mosaic symptoms and most of those that cause necrosis of parenchymatous tissues. *Rugaceae* comprises those that produce malformations without mosaic symptoms, those that produce the yellows type of chlorosis and most of those that cause phloem necrosis. Within the *Marmoraceae* he includes six genera, *Marmor, Lethum, Poecile, Fractilinia, Ochrosticta* and *Flavimacula,* while the *Rugaceae* contains twelve genera, *Chlorogenus, Chlorophthora, Blastogenus, Polycladus, Corium, Savoia, Galla, Ruga, Carpophthora, Rimocortius, Acrogenus* and *Minuor.* Descriptions are given for each genus and for one type species for each genus, but no attempt is made to fit all viruses into these genera. With such a scheme, extended as necessary for new genera, viruses could be named without the misleading implications of relationships that occur in HOLMES' scheme. However, in another paper MCKINNEY (1944*b*) describes some viruses which he proposes to place in the genera *Marmor, Fractilinea* and *Galla,* and from this it is clear that the genus *Marmor* at least will still be remarkably heterogeneous. The four viruses tobacco mosaic, tobacco mild dark-green mosaic, tobacco ringspot and wheat rosette are all described as species of the genus *Marmor,* implying the same degree of relationship between all four, though what is known of their properties suggests that the first two may be related strains and that the last two are unrelated either to one another or to tobacco mosaic virus.

As long as primary divisions for classification rest on symptoms, it is unlikely that such anomalous groupings will disappear. Some viruses may produce only one type of reaction in all hosts and this reaction may reflect some intrinsic character of the viruses themselves. Then, in default of any knowledge of the virus particles themselves, symptoms may have to be used as a criterion for grouping and they may give groups containing viruses that share several common features. The viruses that cause the yellows types of chlorosis, for example, could be provisionally grouped in this way, for they also resemble one another in their methods of transmission and there is nothing known about the structure or properties of the particles themselves to conflict with such grouping. Grouping on symptoms alone, however, should be no more than a temporary expedient and never employed if information exists on intrinsic properties of the pathogens themselves. The errors that can result from doing so are vividly illustrated by the manner in which typhoid and typhus remained inextricably confused until etiology replaced symptomatology. Symptoms reflect a host-parasite interaction and are often determined predominantly by the host. It would seem as reasonable to expect to classify flowering plants because of their reactions to a given number of viruses as it is to classify viruses by the symptoms they produce in a given number of hosts. Such a test would almost certainly result in placing different potato varieties in different genera or families and might well put tobacco and cucumber plants in the same genus. There is no reason to assume that similar anomalies will be avoided when the test is used to classify viruses.

Indeed, many such anomalies can be found in the most recent version of HOLMES' scheme, which appears as a supplement to the *Sixth Edition* of *Bergey's Manual of Determinative Bacteriology* published in 1948. The scheme is extended to cover animal viruses as well as plant and bacterial viruses. A new order is set up, *Virales,* with three suborders, *Phagineae* (bacteriophages), *Phytophagineae* (plant viruses) and *Zoophagineae* (animal viruses). Within the *Phytophagineae* separation is again mainly on symptoms and insect vectors, with five suggested families, *Chlorogenaceae, Marmoraceae, Annulaceae, Rugaceae, Savoiaceae,* and *Lethaceae,* of which the first two each contain six genera and the last four are all monogeneric. Of the 122 viruses dealth with, no less than 67 are given the same generic name of *Marmor,* so that it is hardly surprising that much heterogeneity is still evident within this genus.

The arbitrary decisions that may be needed in allocating viruses to genera, or even to families, when symptoms are used as criteria, are vividly illustrated by HOLMES' treatment of potato virus X, which causes different symptoms in different hosts and exists in many clinically-distinct strains. In 1939 it was placed in the genus *Marmor,* family *Marmoraceae,* and from what is known of the intrinsic properties of virus X it is clear that a better case can be made for grouping it with tobacco mosaic virus than for most of the others included in that genus. However, on the criteria used for this classification, some strains should have been placed in the *Annulaceae,* for they cause temporary ringspot diseases in tobacco, and recovery from ringspot is the hall-mark of the *Annulaceae.* Presumably because of this, in 1948 HOLMES has transferred virus X to the genus *Annulus,* though in fact many strains do not cause ringspot diseases from which plants recover. The net result of this transfer is that *Annulus,* which previously seemed a reasonably homogeneous genus, now resembles *Marmor* in containing both spherical and rod-shaped viruses, a situation that would be paralleled in bacterial classification by including cocci with either bacilli or spirochaetes in one genus.

In HOLMES' revised scheme, as in the earlier one, the most useful result is the grouping of related strains under the same specific name, though sometimes this is done only by ignoring the very criteria that are supposed to determine classification. Were these criteria followed with any rigidity, there would be no escape, with such viruses as potato virus X, from the farcical situation of having to place some strains in one genus or family and other strains in another. This provides perhaps the strongest evidence that can be brought against attempts to use host reactions as primary bases for setting up major groups. Unfortunately, the value of grouping strains under one specific name is largely destroyed by using the same generic name for viruses which there is no reason to consider are in any way related. The arbitrary grounds on which higher categories such as families are set up are relatively unimportant, for they do not show in the binomial nomenclature, which is concerned only with groupings at the generic and specific levels. If the nomenclature is to be extended to viruses, it is important that it should retain its valuable feature of showing relationships at these levels. To do this, much more knowledge of the intrinsic properties of viruses will be needed and more attention to these properties must be paid before they are assigned to genera.

In the introduction to Bergey's Manual it is stated that "No organism can be classified before we have determined its morphological, cultural, physiological and pathogenic characters". Whether deliberately placed in that order or not, it is worth noting that the pathogenic characters come last. It is odd that similar considerations are apparently not considered to apply equally to viruses, for there is no *a priori* reason why they should not. In adopting HOLMES' scheme, however, it is assumed that viruses are classifiable when their intrinsic properties are wholly undetermined and that they can be assigned to genera and families because of suspected clinical similarities that would be given little weight in classifying bacteria. Morphology, which is placed first for bacterial characters, is wholly neglected, and it is not only that groups and names used in the scheme suggest relationships between viruses about which little is known, but often they definitely mislead by associating under one generic name viruses with different shapes and widely different other characters. Such misalliances give no reason to consider that classification by host reactions will produce any more homogeneous groups with viruses than it would with other kinds of pathogens. Because viruses are smaller than bacteria, does not mean that they have no morphology, but only that different methods are needed to determine it. Morphology tends to merge into physico-chemical properties and a lasting classification is more likely to result from using the properties described in chapters 9-13 than from any others.

**The Grouping of Virus Strains:** — The main reason why symptoms are unsuitable for attempting to relate viruses is that different strains of one virus often cause strikingly different symptoms. This fact, too, is probably the main cause of confusion in virus nomenclature, for it has often led to related strains receiving very different names. The most valuable immediate advance in classifying viruses would undoubtedly be to establish accepted criteria whereby symptomatologically distinct viruses could be identified as variants of type species. Fortunately, it seems that this should not be difficult. From studies on viruses thought to have arisen as variants from one another, two phenomena have emerged that seem to be generally applicable in indicating close relationship. These are *(1)* the sharing of common antigens and *(2)* mutual antagonism when present together in the same host. Identification as related strains can therefore be made by testing either the ability of viruses to react with each other's antisera or to protect plants against one another. From applications of these two methods to viruses whose properties are already reasonably well known, it is clear that they group together only those that share many common characters and differ in only one or a few. Indeed, so far, most have been found to differ solely in host range or symptoms caused, though this is probably a transient phase and other differences will no doubt emerge with further work. Already differences in such properties as isoelectric point and aminoacid constitution (KNIGHT, 1947) are known between strains of tobacco mosaic virus; potato yellow dwarf virus (BLACK, 1941) and potato virus Y (BAWDEN and KASSANIS, 1947) occur in strains that have different insect vectors, and some tobacco necrosis viruses occur in strains that crystallize in different forms (BAWDEN and PIRIE, 1942).

Serological tests have much to recommend them; they give rapid results and unknown viruses are often immediately identifiable as strains of a type virus merely by testing whether or not crude sap is flocculated by an antiserum to that virus (CHESTER, 1937). However, they have their limitations and dangers. Attempts to apply serological methods to several viruses have been unsuccessful. The reasons for this may differ with different viruses, but there is little doubt that failure often results from a too low virus content of infective sap. With purified preparations of different viruses, the minimum concentration required for a positive precipitin reaction varies between 10 and 1 mg./l., the actual value depending at least partly on the shape of the virus particles, and sap from plants infected with many viruses contains less than the minimal amount. Nevertheless, some of these may still be amenable to serological techniques for the purpose of determining relationships, though the methods needed would be too laborious to use for the routine diagnosis of infected plants. For example, complement fixation might give results when the precipitin test fails, for it is often demonstrable with one-tenth the amount of antigen needed to give a visible precipitate. Neutralisation of infectivity might also be used, though such tests will need careful controlling to distinguish between the specific effects of antibodies and other unspecific inhibition. The precipitin test itself may be made more sensitive by adding suitable substances that would precipitate with the virus-antibody complex and so increase the size of floccules, or the viruses may be concentrated sufficiently to give reactions. Sap from plants infected with an avirulent strain of henbane mosaic virus does not flocculate with antisera prepared against a virulent strain, but it does after being concentrated ten-fold by precipitation with salts or by ultracentrifugation. Some viruses that occur too dilute to give the precipitin reaction nevertheless produce antisera when injected into rabbits. Severe etch virus, for example, is precipitated by

TABLE 27. — *Viruses grouped by serological reactions:*—

The viruses within each group precipitate with each other's antisera, whereas those in different groups do not.

GROUP

1. *Tobacco mosaic.*
   Common and masked tobacco mosaic viruses, tomato aucuba mosaic, enation mosaic and streak viruses; Japanese petunia virus; Holmes' rib-grass virus; cucumber viruses 3 and 4.
2. *Potato X.*
   Salaman's H, G, L, S and N strains; potato viruses B and D; potato mottle and ringspot viruses; Hyoscyamus virus 4.
3. *Potato Y.*
   Potato leaf-drop streak, rugose mosaic and veinbanding viruses; potato virus C; tobacco veinal necrosis virus; Hyoscyamus virus 2.
4. *Tobacco etch.*
   Severe and mild etch viruses; Blakeslee's Z-mosaic virus of Datura.
5. *Henbane mosaic.*
   Severe and mild strains of Hyoscyamus virus 3.
6. *Cucumber mosaic.*
   Price's isolates of cucumber virus 1; Valleau's delphinium virus.
7. *Soybean mosaic.*
8. *Pea mosaic.*
   Osborn's pea viruses 2 and 3.
9. *Tulip mosaic.*
10. *Sugar beet yellows.*
11. *Tobacco ringspot.*
    Wingard's and yellow and green tobacco ringspot viruses; tobacco ringspot virus 2.
12. *Tobacco necrosis A.*
    Potato, Princeton and Tobacco VI cultures.
13. *Tobacco necrosis B.*
    Tobacco I and II cultures.
14. *Tobacco necrosis C.*
    Rothamsted culture and bean stipple-streak.
15. *Tomato bushy stunt.*
16. *Southern bean mosaic.*
17. *Turnip yellow mosaic.*

serum prepared against sap from plants infected with mild etch virus, showing that the two are serologically related, though no reaction is obtained with the homologous antigen (BAWDEN and KASSANIS, 1941).

There is little doubt that serology can be more widely used for showing relationships than it has so far, but seventeen antigenically distinct types have already been identified, each of which might be regarded as a separate species. These are listed in TABLE 27, which is partly based on CHESTER'S (1937) results, and shows that some types comprise many strains whereas others seem as yet to be single units. The greatest value of the method is the grouping of clinically distinct strains around a type specimen, but it can also be useful, as in the study of tobacco necrosis and potato top-necrosis, in showing that the same disease is caused by distinct viruses (BAWDEN and PIRIE, 1942).

As with any other techniques, serological tests can give misleading results unless carefully applied. A negative result with a heterologous antiserum, for example, is valueless unless a positive result is obtained in similar conditions between the antigen under test and its homologous serum. On the other hand, when two viruses react with their homologous sera, but not with the heterologous, the result is better evidence that the two are serologically unrelated than a positive result may be of relationship. The reliability of positive results depends on the certainty that the antigens used both for immunising and testing are pure cultures, and entirely erroneous relationships may be suggested by the use of mixed cultures. The cultures under test should therefore be propagated from single local lesions, passed through some suitable set of differential hosts or transmitted in some specific manner to eliminate any likely contaminants. Mixtures of viruses, such as often occur in tobacco necrosis plants, should be suspected if double zones of precipitation occur, for preparations of single viruses usually show only one optimal precipitation point.

Tests of mutual antagonism give slower results than serological tests, but they are applicable to many more viruses. They do not depend on the presence of a high virus content in sap and can be used to study relationships between viruses whose intrinsic properties are totally unknown because they are not transmitted by sap inoculation. For instance, by intergrafting between peach trees suffering from yellows, little peach and rosette, KUNKEL (1936) showed that the first two are caused by related strains whereas the rosette virus does not interfere with the others and so presumably is not closely related to them. Using similar techniques with the clinically distinguishable viruses involved in causing swollen shoot of cacao, CROWDY and POSNETTE (1947) showed that some are mutually antagonistic and others not.

Plant protection tests are made most simply and give most definite results with two strains only one of which produces local lesions; if leaves are first infected with the strain that causes no local lesions and inoculations with the second strain are made at intervals, the protection against the second is clearly shown by the progressive fall in numbers of lesions as the first strain multiplies. Provided that the strains under test produce sufficiently distinctive symptoms, however, the method can be used reliably when both produce only systemic symptoms and when they are introduced by grafting or insects instead of by sap inoculation, though protection is often less complete and lasting when grafting is used. Even so, the delay in development of symptoms characteristic of infection with the second strain is usually adequate to detect that its progress has been interfered with. MCKINNEY (1941) has pointed out that the extent to which different strains dominate one another varies somewhat. He considers this a limitation in the technique for assessing strain relationships, but provided tests are adequately controlled with simultaneous infections of healthy plants with the second strain, it is rarely difficult to decide whether or not the two are antagonistic. This is especially so because two unrelated viruses often produce different and more severe symptoms than either produces alone, whereas with related strains, even though the first may not entirely prevent the establishment of the second, the second will produce its effects only slowly and the symptoms will pass gradually from those characteristic of the first strain, through intermediate stages, to those characteristic of the second.

One obvious limitation of tests of mutual antagonism is that they cannot be applied to viruses that have no common hosts, nor to those that cause closely similar symptoms or only local lesions. The relationship of cucumber viruses 3 and 4 with tobacco mosaic virus, for example, or between some of the viruses that cause tobacco necrosis, are readily shown serologically, but would not be demonstrable by plant protection tests. When both serological and antagonism tests have been made with the same viruses, they have almost always given the same results. That is to say, viruses that share antigens are also mutually antagonistic, and working with potato virus X, MATTHEWS (1948) found that the extent to which strains protected plants against one another increased with increasing degrees of serological relationship. Agreement between the two methods is not quite perfect. CHESTER (1937) stated that potato virus Y and cucumber mosaic virus, which are not mutually antagonistic, are serologically related, though this has not been confirmed. At Rothamsted we have always failed to get any reaction between antisera to virus Y and sap from plants infected separately with many different strains of cucumber mosaic virus, but this is inconclusive because we have also failed to get specific reactions between such extracts and the sera of rabbits that have been injected with them. Other workers (BIRKELAND, 1934; CHESTER, 1937; NIENOW, 1948) appear to have produced antisera against cucumber mosaic virus without difficulty, and this anomaly remains unexplained. We have, however, met one example of disagreement between serological and plant-protection tests; tobacco veinal necrosis (SMITH and DENNIS, 1940) regularly precipitates with virus Y antisera, though the two are not mutually antagonistic in plants.

Further work will, no doubt, reveal other disagreements between the two tests, but a positive result from either can probably be taken as reliable evidence of close relationship. Both tests presumably work because they depend on structural similarities in the particles that determine ability to combine *in vitro* with antibodies and *in vivo* with specific host constituents, with which combination may be an essential first step towards multiplication. There is no obvious reason why the same structural groups should be concerned with both *in vitro* and *in vivo* tests, but the more closely related two strains are, the greater number of groups they are likely to have in common and the greater will be the chances of both tests applying. Strongly positive results in both tests probably indicate many structural similarities, whereas a negative result in one but not the other, incomplete protection, or a weak serological reaction, may all indicate fewer common structures. Wholly negative results in both tests will not necessarily demonstrate complete lack of relationships, because viruses may share common structures not involved in either test. Nevertheless, such viruses are clearly distinguishable from those that give positive results, so that until better and more direct methods of assessing relationships are developed, it seems desirable to group as strains viruses that either share antigens or are mutually antagonistic, and to regard those that do neither as separate species.

**The Formation of Larger Groups:** — The immunological tests are limited in scope to the grouping of closely related strains around a type species. By refining serological methods, different strains can sometimes be shown to possess different degrees of structural relationships; for example, by doing cross-absorption tests, it can be shown that tomato aucuba mosaic and tobacco mosaic viruses share more antigens with one another than either does with cucumber virus 3 or HOLMES' rib-grass virus. These differences, however, are all at what, for convenience, may be called the variety level, and the techniques are inapplicable for forming associations of viruses into the equivalents of genera, families and orders. Such a limited grouping would fall short of a full classification, but its application to all plant viruses would be a major advance, for many that are now thought to be unrelated would probably be found to be strains, and the number of specific names needed might not then be cumbersomely large.

Indeed, determining the immunologically related viruses might so reduce the apparent number of distinct viruses that the need for a broader system of classification might seem less pressing that it now does. However, attempts at grouping on a wider basis are desirable, if only because they should stimulate work to gain further evidence of suspected relationships and also point to promising lines of research. For example, such groupings might well help to narrow the search for insect vectors and might indicate which viruses are most likely to be amenable to study by different techniques.

In attempting to set up groups wider than collections of strains, however, we are faced by the problems of what criteria shall be used and the relative values of different criteria. These, of course, are not problems peculiar to viruses, but the invisibility of viruses does present some unusual problems and provides some justification for attempts to classify by host reactions. Nevertheless, whenever possible it would seem best to avoid using similarities of symptoms as evidence of relationships, and to use criteria that more certainly represent intrinsic properties of the virus particles themselves. Shape, constitution, stability, manner of inactivation and disruption, any such characters would seem to offer more opportunity for a lasting taxonomy than symptoms, with which it is always possible that the host is more directly concerned than the virus.

Methods of transmission, too, might play a part, though these have their limitations. We know little about the factors that prevent the mechanical transmission of viruses, but they probably differ with different viruses and it would be rash to assume that the many viruses which we can now transmit only by grafting share any fundamental properties. It has been suggested (ELZE, 1931; STOREY, 1931) that viruses might be grouped on the basis of their insect vectors, but as some are transmitted by many species and the single species *Myzus persicae* transmits many viruses, some of which are clearly unrelated, a grouping on species of vector would have little taxonomic value. A grouping on type of insect vector would seem more helpful; except for squash mosaic virus, which is apparently transmitted by both beetles and aphids (FREITAG, 1941), it seems that the vectors of individual viruses are confined to nearly related species. For example, aphid-transmitted viruses do not appear to be transmitted also by leaf-hoppers or thrips, though it would be too soon to decide that this is always so

A further feature that might be used is whether or not the vectors undergo a latent period and the viruses persist for more than a limited time in the vectors, two phenomena that seem to be correlated and to be determined by the viruses rather than by the vectors (WATSON and ROBERTS, 1940). With the limited information available on most viruses, tentative groups may be desirable for such reasons as similarities in transmission or in causing some specific type of symptom. As with bacteria and fungi, however, the main use of pathogenic differences should lie in making distinctions within species, that is to say, in showing that two strains that may otherwise be indistinguishable are in fact not identical. Much of the confusion now existing in virus nomenclature has arisen because too much value has been placed on symptoms and host ranges in describing new viruses, and it would be unfortunate to carry this source of confusion into any systems of classification. With bacteria and fungi the type species were recognised first and the pathogenic variations which were discovered later could be grouped around them fairly simply. With viruses, on the other hand, symptomatology rather than the study of the viruses themselves has been the dominant subject; in the absence of any concept of type species, pathogenic variants in great numbers have been described as unrelated viruses.

The first essential step is to bring together these pathogenic variants under specific names, which should be possible by immunological tests. The next is to arrange these species in genera and families, and it is here that the difficulties will begin. At present most viruses are clearly unclassifiable; we know too little about their intrinsic properties. Nevertheless, it can be seen that some viruses have more in common than others, and a few natural groups could be set up, free from the heterogeneity that accompanies systems based on symptoms. Whether these should rank as families or genera is less obvious, and is perhaps unimportant, except for implications in nomenclature. Agreement on such distinctions will come only with increasing information about the attributes of viruses that are most significant taxonomically. Meanwhile, without conflicting with any existing knowledge, the many strains of tobacco mosaic virus and of potato virus X might be grouped together; a second group might contain the strains of potato virus Y, henbane mosaic, tobacco etch, and soybean mosaic viruses, all of which have rod-shaped particles and share many properties; with them too might be included sugar beet mosaic, pea mosaic, cauliflower mosaic, and perhaps some others that seem to inactivate under similar conditions *in vitro*, are aphid-transmitted and fail to persist in their vectors, though some of these may need moving to other groups as information is gained about their structure; a third could contain tomato bushy stunt, tobacco necrosis, southern bean mosaic, turnip yellow mosaic and squash mosaic viruses, all of which crystallize, have spherical particles, similar gross constitutions and seem to inactivate in many similar ways; the viruses that cause the "yellows" type of disease might be placed in a fourth, for this seems a distinctive type of symptom caused only by viruses that are transmitted similarly and behave similarly in their insect-vectors; the strains of tomato spotted wilt virus, which have received so many different names in the past, might well be segregated into another group on their own, for this virus seems distinctive in many of its characters; another group might centre around tobacco ringspot virus, which has been found to differ considerably in constitution from other viruses that have been purified.

With the limited amount of information now available, no groupings can be expected to remain unaltered, but to reduce changes and consequent synonymity to a minimum, viruses should not be assigned to genera and families unless there is good reason to consider that they fit there comfortably and that they have no properties which conflict with those of their associates. This means that the grouping should be based on fundamental characters of the virus particles themselves. To accept this means also to accept that many viruses will remain unclassifiable for a long time, but this is not a valid reason for its rejection. The taxonomy of fungi was attempted long before all the members could be assigned to genera or families, the uncertain ones being placed in a section apart and clearly labelled *Fungi imperfecti.* Similarly with viruses, to avoid a spurious air of knowledge with its possible repercussions on research, we should classify only those about which we have a reasonable amount of information; the rest should be set apart, provided with names suitably chosen to stress our ignorance of their properties, and they should remain in limbo until there is enough knowledge to justify their removal.

*References:-*


BAWDEN, F. C. and KASSANIS, B. (1941): Ann. appl. Biol. *28*, 107.
—— and —— (1947): *ibid. 34*, 503.
—— and PIRIE, N. W. (1942): Brit. J. exp. Path. *23*, 314.
BENNETT, C. W. *et al.* (1943): Phytopath. *33*, 424.
BIRKELAND, J. M. (1934): Bot. Gaz. *95*, 419.
*Bergeys Manual of Determinative Bacteriology,* Sixth Edition, (1948). Williams & Wilkins Co., Baltimore, Md.
BLACK, L. M. (1941): Amer. Pot. J. *18*, 231.
CHESTER, K. S. (1937): Phytopath. *27*, 903.
CROWDY, S. H. and POSNETTE, A. F. (1947): Ann. appl. Biol. *34*, 403.
ELZE, D. L. (1931): Phytopath. *21*, 675.
FREITAG, J. H. (1941): *ibid. 31*, 8.
HOLMES, F. O. (1939): *Handbook of Phytopathogenic Viruses.* Burgess, Minneapolis, Minn.
JOHNSON, J. (1927): Wis. Agric. Exp. Stat. Res. Bull. 76.
—— (1935): Rep. to Intern. Bot. Conf. Amsterdam.
—— and HOGGAN, I. A. (1931): Science, *73*, 29.
—— and —— (1935): Phytopath. *25*, 328.
KNIGHT, C. A. (1947): J. Biol. Chem. *171*, 297.
KUNKEL, L. O. (1936): Phytopath. *26*, 201.
LIMASSET, P. (1946): Ann. Epiphytes *12*, 317.
MCKINNEY, H. H. (1941): Amer. J. Bot. *28*, 770.
—— (1944*a*): J. Wash. Acad. Sci. *34*, 139.
—— (1944*b*): *ibid. 34*, 322.
MATTHEWS, R. E. F. (1948): Ph. D. Thesis, Cambridge University.
NIENOW, I. (1948): Phytopath. *38*, 62.
SMITH, K. M. (1937): *Textbook of Plant Virus Diseases.* Churchill, London.
—— and DENNIS, R. G. W. (1940): Ann. appl. Biol. *27*, 65.
STOREY, H. H. (1931): 2nd Cong. Inter. Path. Conf. *2*, 471.
THORNBERRY, H. H. (1941): Phytopath. *31*, 23.
VALLEAU, W. D. (1940): Phytopath. *30*, 820.
WATSON, M. A. and ROBERTS, F. M. (1940): Ann. appl. Biol. *27*, 227.

*Chapter 15*

# VIRUS DISEASES AND HOST-PLANT PHYSIOLOGY

**Host-Plant Metabolism and Susceptibility:** — Viruses are unique in being the only plant pathogens that are both wound parasites and obligate parasites, and their ability to enter and multiply in all the vegetative parts of many plants is a further unusual character. With such a combination of properties, there are clearly many different ways in which changes in the physiology of host plants might influence susceptibility to viruses. Age of plant, supply of nutrients and water, illumination, temperature, presence of another pathogen, and any other factor that influences host metabolism, might be expected *a priori* also to influence the ability of viruses to enter, move, multiply and cause symptoms. Most workers will appreciate that these expectations are realised, for there must be few who are not familiar with the fact that susceptibility and reactions to viruses may depend greatly on the physiological state of the test plants. Nevertheless, there has been little experimental work on such phenomena, and most mentions of them in the literature are qualitative observations on the manner in which environment or nutrition affect symptoms. Some of these are described in Chapter 2 and need no further consideration now, when factors affecting the entry, multiplication and movement of viruses will be discussed. One of the most important factors, whether or not another virus is already present, also need not concern us here, for the resistance conferred on plants by recovery from the acute stages of some diseases or by infection with avirulent strains, as well as the synergistic interactions of some unrelated viruses, are described in Chapters 2 and 6.

As viruses are wound parasites, any change in a plant that renders it more liable to injury could be expected also to increase its likelihood of becoming infected. The increased resistance to infection by many sap-inoculable viruses that occurs as host plants age, may be largely attributable to the hardening of epidermal layers. Similarly, the increase in susceptibility occasioned by raising plants under conditions of low light intensity, may simply reflect the fact that shading increases succulence and fragility, with a consequent increase in the likelihood of suffering injuries when rubbed. It seems unlikely that this is the whole explanation, for shading host plants enables some viruses to reach a higher virus content and cause more severe symptoms, not only in leaves that are directly inoculated, but also in those that become infected by systemic spread. It is difficult to attribute this effect to a softening of the epidermal layers, and it seems more likely that products of photosynthesis, in some unknown manner, interfere either with the passage of viruses from cell to cell or with their ability to become established in the cells. This explanation could also account for the increased susceptibility produced by placing plants in the dark for a short time immediately before they are inoculated. Such treatments have no obvious effect on the hardness of the leaves and do not appear to make the epidermal layers more susceptible to wounds, but, by removing most of the carbohydrates, they may facilitate the entry of virus particles either by reducing cell turgor or by reducing the number of cell components that interfere with the establishment of virus particles at sites where multiplication occurs (BAWDEN and ROBERTS, 1947, 1948).

When susceptibility is assessed by local-lesion counts, it is always difficult to ascertain whether differences arise because leaves vary in their resistance to injuries during inoculation, or because of more subtle physiological differences within the injured cells. It seems likely that infection occurs in two distinct stages, first, entry into cells and the attachment of virus particles to sites essential for multiplication, and secondly, multiplication leading ultimately to the development of lesions. If this is so, there is no reason why conditions that favour one phase should also favour the other, and there is evidence that physiological conditions of the host that predispose it to becoming infected are not necessarily those also most favourable for virus multiplication. For instance, if French bean plants are placed in the dark for 24 hours before they are inoculated with tobacco necrosis viruses, their susceptibility is increased, but if placed in the dark after inoculation, they usually produce fewer lesions than if kept in the light continuously. The susceptibility of beans is also reduced if the plants are exposed to atmospheres containing from 30-60% carbon dioxide, provided the exposure is made either immediately before or immediately after inoculation. The same reduction occurs with a two-hour exposure after inoculation as with one before inoculation, so that the degree of wounding during inoculation is apparently not involved and the effect seems to result from changes within the host cells (KALMUS and KASSANIS, 1944). The effect of exposure to carbon dioxide decreases with increasing time after inoculation and if applied after four hours the treatment does not reduce the numbers of lesions. Similarly, plants treated four hours or more before the time of inoculation produce as many lesions as untreated controls. It seems, then, that the treatment temporarily influences the physiology of the host cells, so that particles are less readily established at their sites of multiplication, but that it is without effect if they are so established, which apparently occurs within four hours. Exposure to carbon dioxide, like exposure to bright light, apparently reduces susceptibility by creating conditions that interfere with the first stage of infection. Some treatments possibly affect the second stage and not the first; immersing leaves in dilute solutions of sodium cyanide may be one. WOODS (1940, 1943) has found that this delays the production of lesions by both tobacco mosaic and tobacco ringspot viruses and that, after a given period, cyanide-treated leaves contain less virus than controls. As this treatment does not reduce the number of lesions produced by tobacco mosaic virus, but inhibits their development whether applied immediately after inoculation or some hours later, it seems not to interfere with the entry and establishment of the virus but to prevent its multiplication. WOODS attributes the effect to a reversible inhibition of respiratory systems on which the virus depends for multiplication, but too little is known about either virus multiplication or the action of cyanide for any certain conclusions. It has also been reported that various other narcotic treatments (RISCHKOV and GROMIKO, 1941), as well as treatment with thiamine, aniline and certain other substances (RISCHKOV, SMIRNOVA and GORODSKAJA, 1946), and malachite green (TAKAHASHI, 1948), inhibit the multiplication of tobacco mosaic virus, but the mechanism is also uncertain.

Nutrition so profoundly affects the growth and constitution of plants that it would be surprising had it no effects on susceptibility to viruses, but little work has yet been done to elucidate the many factors involved. The only extensive study has been with tobacco mosaic virus, and the results are rather inconsistent. SPENCER (1935*a*, 1935*b*) found that more local lesions were produced in tobacco, *Nicotiana glutinosa*, and French bean, by increasing supplies of nitrogen, even when the amounts given were in excess of those needed for optimal growth and were reducing the rate of growth of the plants. In tobacco, phosphorus increased susceptibility as long as plant growth was benefiting from extra phosphorus but not beyond, whereas small amounts of potassium increased susceptibility and larger amounts decreased it without having any apparent effects on plant growth. BAWDEN and KASSANIS (1948) found that potash had little effect on resistance to tobacco mosaic virus or on the growth of tobacco plants in soil; both nitrogen and phosphorus affected resistance, and their effects were correlated with the growth response of the plants. Phosphorus was more important than nitrogen in increasing susceptibility and, in the absence of adequate phosphorus, increasing the nitrogen supply reduced the numbers of local lesions. Susceptibility was increased by additional nitrogen only when the host plants also responded by increase in growth. The magnitude of the effects, however, depends greatly on the manner in which the results are expressed. Using the customary method, *i. e.* number of lesions per leaf, the differences are large because of the effects of the nutrients on leaf size, but if lesion counts are expressed as number per unit area the differences are much less striking. For instance, using the leaf as a basis, phosphorus increased the numbers of lesions by 500%, whereas it increased the number of lesions per 100 sq. cm. of leaf by only 50%.

Local-lesion counts measure only the effects of nutrition on the entry and establishment of viruses and may give no information about the manner in which it influences virus multiplication. To gain information on this point, the virus-content reached in the infected tissues needs to be measured. It would seem preferable, too, that this should be measured on tissues infected by systemic spread rather than on inoculated leaves, for differences with the latter might arise from different numbers of entry points rather than from differential multiplication of the virus in infected cells. Again, the basis for assessing susceptibility needs consideration, because the sizes of differently treated plants may vary widely and different conclusions might be reached if amount of virus per ml. of sap or per unit weight of leaf is considered than if the total virus contents of the plants are compared. Only with tobacco mosaic virus have any experiments been recorded bearing on this problem, and those deal only with nitrogen supplies. RISCHKOV and SMIRNOVA (1939) reported that the concentration of the virus in nitrogen-deficient tomato plants equals that in those adequately supplied. SPENCER (1939), on the other hand, reported that virus concentration in tobacco was directly correlated with nitrogen supply, and stated that, 35 days after infection, the virus was 80 times as concentrated in sap from plants receiving abundant nitrogen as in sap from nitrogen-deficient plants. He concluded that additional nitrogen encouraged virus multiplication even when it reduced plant growth and also stated that, in plants receiving inadequate supplies of nitrogen, the virus content of sap decreased with increasing duration of infection. These conclusions were reached from infectivity measurements, on the assumption that local-lesion counts could be translated into relative virus concentrations. Many things other than virus concentration can influence infectivity, so that this assumption is by no means necessarily valid, but in a later paper SPENCER (1941*a*) considered his conclusions to be justified because he found that differences in infectivity were correlated with the virus concentration as estimated from measurements on the amount of protein sedimented by ultracentrifugation.

In other papers, however, SPENCER (1941, 1941*b*, 1942) appears to reach different conclusions and claims that the infectivity per unit weight of tobacco mosaic virus is influenced by both host nutrition and the duration of infection. He reported (1941) that the total virus content of nitrogen-deficient leaves remained constant for 16 days while the infectivity fell to less than one-half, whereas in nitrogen-fed leaves the virus increased five times in weight and suffered no loss in infectivity. The results quoted suggest little increase in virus concentration per volume of sap by increasing nitrogen, but that increase depended on leaf growth, a conclusion at variance with SPENCER's earlier statement. In inoculated leaves, SPENCER (1941*b*, 1942) concluded that the virus content and the infectivity per unit weight of virus both increased up to 20 days after infection; the virus reached a higher concentration in nitrogen-fed than in nitrogen-deficient leaves, and he concluded that stopping supplies of nitrogen inhibited the development of infectivity per unit weight sooner than it inhibited virus multiplication. As with all experiments on inoculated leaves, the interpretation of quantitative differences is uncertain, because it is unknown to what extent differences at the time of assay reflect different numbers of entry points or differential multiplication at each point.

SPENCER's claim that host nutrition affects tobacco mosaic virus qualitatively as well as quantitatively has so far been unconfirmed, and BAWDEN and KASSANIS (1948) were unable to demonstrate any consistent differences between the infectivity of virus obtained from plants supplied with widely different quantities of nitrogen, phosphorus and potassium. As obtained in expressed sap, the virus is always heterogeneous and can be separated into fractions containing particles of different average sizes and infectivities (BAWDEN and PIRIE, 1945), but increasing the supply of nitrogen has not resulted in any increase in average infectivity. Differences in infectivity per unit weight of virus from differently fed plants are always small and seem in no way connected with nutrition. SPENCER's conclusions that nitrogen specifically increases infectivity appear to rest on several assumptions; first, it is assumed that his method of isolating virus recovers all the virus present in the different lots of sap; secondly, that the method isolates nothing but virus; thirdly, that the method does not cause aggregation of virus particles, or at least, aggregates all samples equally. As SPENCER provides no information on either the percentage virus recovered or the homogeneity of his final products, there is no reason to assume that any of these assumptions is justified. The method used was to centrifuge frozen sap twice at 66,000 g. for 1 hour, and centrifuge the resuspended pellets at low speed. Apparently the supernatant fluid from the second low-speed centrifugation was then assumed to contain only virus and the yield was calculated from micro-Kjeldahl determinations of nitrogen. It is noticeable that the products stated to have low infectivities were also those which gave the smallest yields; thus, in the absence of any evidence of purity, the differences could be explained by the occurrence, in all preparations, of a constant impurity, the effects of which would become proportionally smaller as the quantity of virus increased.

Our results on the effects of host nutrition on virus multiplication are also at variance with those described by SPENCER, because virus multiplication has been closely correlated with rate of plant growth, nutrients that encourage one also encouraging the other. Plants respond differently to nutrients at different times of the year, and correspondingly

TABLE 28. — *The effect of host nutrition on the production of tobacco mosaic virus in systemically infected leaves:—*

| NUTRIENT ADDED | WET WEIGHT OF PLANTS (g.) | VIRUS CONTENT OF SAP | |
|---|---|---|---|
| | | *Concentration mg. virus/ml. sap* | *Total virus in sap (mg.)* |
| Nothing | 9 | 2.3 | 13 |
| Nitrogen | 4 | 2.1 | 4 |
| Phosphorus | 22 | 4.3 | 54 |
| Potassium | 8 | 2.4 | 11 |
| N + P | 52 | 5.0 | 125 |
| N + K | 12 | 1.8 | 13 |
| P + K | 28 | 3.0 | 43 |
| N + P + K | 72 | 4.4 | 170 |

Tobacco plants grown in mixture of sand, soil and peat; leaves picked one month after infection (BAWDEN and KASSANIS, 1948).

nutrients may also have different effects on virus production. For example, in summer, when tobacco plants respond rapidly to nitrogen, nitrogen also affects virus production more than in winter when a corresponding amount of nitrogen may inhibit plant growth. In our experiments nitrogen has increased the amount of virus only when plants were also receiving adequate amounts of phosphorus and when the plants have responded by increased growth. Even then the effect is small, the main increase being produced by phosphorus, which has also produced the greatest increase in growth. The results of one experiment to find the effect on virus content in systemically infected leaves of adding nitrogen, phosphorus and potassium in various combinations to tobacco plants grown in a mixture of 50% sand, 25% infertile soil and 25% peat are recorded in TABLE 28. The virus present in sap from the minced leaves was purified by precipitation methods and weighed, and the table shows both the concentration and total weight of virus in the sap from the differently treated plants. When total yields of virus are compared, the effects of nutrients are much greater than when amounts of virus per ml. of sap are considered. This is because the nutrients affect plant size, but comparing the figures in the two columns shows that, in general, conditions that increased size also increased virus concentration.

although sometimes potassium apparently reduced the concentration without any corresponding effect on plant growth. Essentially similar results were also obtained with inoculated leaves, except that phosphorus, and phosphorus and nitrogen together, gave slightly larger increases in virus concentration than in systemically infected leaves, presumably because these elements also increased susceptibility to infection and so provided more multiplication foci.

**Effects of Viruses on Host-Plant Metabolism:** — After reviewing the many references in the literature to metabolic effects of viruses on their host plants, WYND (1943) ventured on only two generalisations; one, that respiration is increased early in infection, but not necessarily later, and, two, that infection alters the permeability of cytoplasm so that soluble substances diffuse less readily. In the present state of knowledge, however, it is doubtful whether even these conclusions are justified, for only a few diseases have been studied and with these different workers have often recorded divergent results. Also, many of the experiments have been insufficiently replicated and inadequately controlled; much more detailed and systematic studies will be essential before some of the recorded results can be accepted as fully established.

The various symptoms, changes in colour, shape and size, produced by infection with viruses are, of course, all outward manifestations of disturbed metabolism, but the direct causes of the various phenomena remain obscure. KRAYBILL *et al.* (1932) claim that tomato plants infected with tobacco mosaic virus contain a toxin that alone causes leaf deformities characteristic of tomato mosaic. It is heat stable, is not precipitated by protein precipitants, and does not occur in tomatoes with fern-leaf symptoms caused by cucumber mosaic virus. When injected into healthy tomato plants it causes leaf deformity but no mottling, and it does not interact with potato virus X to cause streak as does tobacco mosaic virus. THORNTON and KRAYBILL (1934) state that plants severely deformed by injecting toxins resemble virus-infected plants in that both contain less carbohydrate (reducing sugars, sucrose, starch and hemicellulose) but more nitrogen than normal plants.

Largely because symptoms usually occur only on actively growing leaves, some workers have concluded that chlorosis results solely from inhibition of chloroplast development and that viruses have no action on mature chloroplasts. This seems improbable, because mature leaves often become yellowed if they are rubbed with concentrated inocula. The chlorosis probably results from a combination of two different actions, the importance of each varying with different viruses or virus strains. One is a stimulation of the normal cell enzymes that attack chlorophyll and the other may be a diversion of plastid protein into virus. PETERSON and MCKINNEY (1938) have described how three strains of tobacco mosaic virus and an unrelated virus, probably potato virus Y, affect plastid pigments and chlorophyllase, and their results are summarised in TABLE 29. In healthy plants chlorophyllase was directly proportional to chlorophyll content, whereas in mottled leaves chlorophyllase was greatest in the most chlorotic leaves which contained least chlorophyll. Similarly, yellow areas from mottled leaves contained more chlorophyllase and less chlorophyll than green areas from the same leaves. Carotene and xanthophyll were reduced proportionally to the chlorophyll, so PETERSON and MCKINNEY suggest that chlorosis results not from an increase in yellow pigments, but because the normal masking action of chlorophyll is partially removed. These results, however, differ from ELMER'S (1925), who reported that tobacco mosaic doubled the carotene content of leaves while both chlorophyll and xanthophyll were reduced.

Increased enzymic activity has been recorded as a result of infection with many viruses. WOODS (1899, 1900) first commented on disturbances of oxidase and peroxidase systems in plants with tobacco mosaic, and BUNZEL (1913) reported that sugar beet leaves suffering from curly top had more than twice the normal oxidase activity. BEST (1937) found that sap from tomato leaves infected with spotted wilt virus contains an oxidase that is undetectable in normal leaves, although it is present in healthy tomato roots. The oxidase was not found in all susceptible species when infected. ROUZINOFF (1930) has also recorded increased oxidase activity in potatoes as a result of infection with leaf roll virus. WOODS and DUBUY (1942) describe complex effects of tobacco mosaic

TABLE 29. — *Relative chlorophyllase and plastid-pigment content of healthy and mottled tobacco leaves:—*

| ENZYME OR PIGMENT | HEALTHY PLANTS | TOBACCO MOSAIC VIRUSES | | | |
|---|---|---|---|---|---|
| | | *Common mosaic* | *Yellow mosaic* | *Mild dark green mosaic* | MILD MOSAIC (? *Potato Virus Y*) |
| Chlorophyllase | 100 | 94.3 | 186.9 | 118.5 | 125.3 |
| Chlorophyll | 100 | 74.7 | 43.3 | 87.2 | 72.2 |
| Carotene | 100 | 72.8 | 39.3 | 86.7 | 69.7 |
| Xanthophyll | 100 | 74.1 | 46.9 | 85.8 | 75.2 |

virus on the oxidase system of tobacco leaves. They claim that, 48 to 72 hours after infection, one cyanide-sensitive system is suppressed, a second is unaffected and that a third cyanide-resistant oxidase is stimulated. TAKAHASHI (1947) could not confirm these results and could detect no differences between the effects of cyanide on the respiration of healthy and infected leaves. Increases in diastase have been reported in sandal suffering from spike disease (SREENIVASAYA and SASTRI, 1928) and in tobacco with mosaic (LUDTKE, 1930). WYND (1943) has described increased invertase and changes in oxidising enzymes caused by tobacco mosaic virus. Changes in pectase and protease resulting from systemic infections with mosaic virus have been described by HOLDEN and TRACEY (1948); their results also indicate how different conclusions may be reached, depending on the method whereby the leaves are fractionated and the results expressed, for infection affects the distribution of enzymes and other leaf components and so alters the relative proportions of substances obtained in different fractions. The different methods used by different workers in the past may account for many of the apparent divergent conclusions. The effects analysed by HOLDEN and TRACEY (1948) are summarised in TABLE 30.

Many workers have described changes in carbohydrate content of leaves, and in the carbohydrate/nitrogen ratio, as a result of virus diseases. It is clear that some diseases increase the ratio and others reduce it (TRUE and HAWKINS, 1918; JODIDI *et al.*, 1920; BAILEY, 1924; CAMPBELL, 1925; BREWER, KENDRICK and GARDNER, 1926; ROSA, 1927;

DUNLAP, 1930), an increase being characteristic of potato leaf roll and the yellows type of disease and a decrease of the mosaic diseases. DUNLAP (1930) suggested that viruses divide into two types, one of which increases nitrogen and reduces carbohydrate and the other reduces nitrogen and increases carbohydrate. Differences between the two types, however, may be less than this and changes in carbohydrate only may be adequate to account for the observed effects. Indeed, it is probable that neither leads to an increased production of either carbohydrate or protein. The observed effects on C/N ratios could occur solely because carbohydrate is utilised at different rates in plants with the different kinds

TABLE 30. — *Some effects of tobacco mosaic virus on systemically infected tobacco leaves:—*

| | |
|---|---|
| *Decrease in:* | Wet weight of plants |
| | Wet weight of leaves |
| | Total dry matter |
| | Total phosphorus |
| | Dry matter as % wet weight |
| | Sap sediment N as % non-fibre N |
| | Protease/g. protein N of sap |
| | Pectase/g. N of fibre |
| *Increase in:* | Total N as % dry matter |
| | Fibre N as % dry matter |
| | Sap N as % dry matter |
| | Total P as % dry matter |
| | Fibre P as % dry matter |
| | Sap N/ml. |
| | Protein N/ml. |
| | % total dry matter as fibre |
| | Total protease/g. dry weight |
| | Fibre protease/g. dry weight |
| *No significant change in:* | Total N |
| | % total N on fibre |
| | % total P on fibre |
| | Sap P as % dry matter |
| | Total units protease/g. total protein N |
| | Fibre protease/g. N |
| | Pectase/g. dry matter of fibre |

of disease, carbohydrate moving slowly and accumulating in the yellows type of disease, and so increasing the ratio, but not in the mosaic diseases.

It is well established that a reduced carbohydrate content is characteristic of mosaic diseases, and this would reduce the C/N ratio without any change in nitrogen. Different workers have reported various results for their effect on nitrogen. The changes described by DUNLAP (1930) are small compared with those affecting carbohydrate and are similar to the 10% increase in total nitrogen which CORDINGLEY *et al.* (1934) found to be produced by tobacco mosaic virus. STANLEY (1937), on the other hand, claimed much larger increases, stating that tobacco mosaic and tomato aucuba mosaic viruses stimulate protein production so that, although infected plants are smaller than normal, they may contain twice as much nitrogen and protein. With the mosaic diseases caused by potato virus X, cucumber mosaic, severe etch and tobacco ringspot viruses, he recorded a decrease in protein but gave no information on total nitrogen. MARTIN, BALLS and MCKINNEY (1938) failed to confirm

STANLEY's conclusions with tobacco mosaic virus and pointed out that his results showed only an increased amount of extractable protein and not any effect on total protein. They measured the total protein and total nitrogen of three tobacco varieties when healthy and when infected with three strains of tobacco mosaic virus, and found no appreciable differences. Other workers, too, have found no increase in total nitrogen (HOLDEN and TRACEY, 1948), though the extractable nitrogen and the nitrogen expressed as a proportion of the dry matter is increased (TABLE 30). On an average about one-third of the nitrogen of tobacco mosaic plants is in the form of virus (BAWDEN and PIRIE, 1946), but in differently treated plants the amount ranges from 10 to 60%, being highest in those receiving abundant phosphorus and little nitrogen (HOLDEN and TRACEY, 1948). The occurrence of so large a proportion of the nitrogen in the form of the virus, without any corresponding increase in protein production, suggests that the virus is largely produced at the expense of normal proteins, and there is some evidence to support this idea. MARTIN *et al.* (1938) found that during the first few days of infection the increase in virus is accompanied by a corresponding decrease in normal protein, and TAKAHASHI (1941, 1947) found that virus increased in detached leaves maintained in the dark when the normal proteins were autolyzing. WOODS and DUBUY (1941) also found virus to multiply in nitrogen-starved leaves; they state the multiplication is at the expense of chromoprotein, but their evidence seems inconclusive. By contrast, SPENCER (1941) stated that the virus cannot utilise normal proteins for its synthesis, and that it can multiply only when other supplies of nitrogen are available.

The increased carbon/nitrogen ratio characteristic of plants suffering from yellows type of disease cannot be taken as evidence of any increase in carbon-assimilation. Indeed, BARTON-WRIGHT and MCBAIN (1932) state that potato leaf roll, which causes one of the greatest increases in proportion of carbohydrate, reduces the photosynthesis below that of healthy plants. The change more likely occurs because the translocation of carbohydrate is in some way hindered. QUANJER (1931) found that the phloem becomes necrotic in leaf-roll potatoes and suggested that this caused starch to accumulate in the leaves. MURPHY (1923), however, found that starch accumulated and the leaves often rolled soon after infection, before any phloem necrosis is detectable. He, and others who have made similar observations, suggested that phloem necrosis is a secondary effect, and that the starch accumulates in the leaves because of other changes, perhaps alterations of enzymic activity. THUNG (1928) found no evidence for any enzymic changes; starch in rolled leaves changed readily into sugars when leaves were placed in the dark, and he also found that rolled leaves usually contained more sugar, as well as more starch, than healthy leaves. THUNG recognised that starch accumulation often precedes obvious phloem necrosis, but he considered that some other change occurs in the phloem and reduces translocation, and that this is the primary cause of both leaf rolling and phloem necrosis. BARTON-WRIGHT and MCBAIN (1932) stated that the sugar of translocation in healthy potatoes is sucrose whereas in leaf roll it is hexose, and they suggest that in diseased plants translocation is restricted to diffusion through the ground parenchyma. Although they concluded that the carbohydrate metabolism was greatly altered by leaf roll, they (1933*b*) found only slight effects on nitrogen metabolism. On the other hand, SCHWEIZER (1930) contends that starch accumulation and phloem necrosis both result from derangement of protein metabolism. He states that the nitrogen passes into the young shoots much more rapidly from infected tubers than from uninfected ones; this he suggests stops diastatic activity, so that much starch is unhydrolysed and the tubers remain hard longer than healthy ones.

The destruction of the phloem in leaf-roll potatoes no doubt contributes towards immobilising carbohydrates, but there may also be other factors involved. Infection with many mosaic-producing viruses, which are not known to effect the phloem and which cause no accumulation of carbohydrate, also affects the speed at which starch is hydrolysed and translocated. This fact is amply demonstrated by the production of starch-iodine local lesions (p. 158), for these depend on starch being retained for longer in virus-infected than in virus-free cells. That their effect is on parenchymatous cells and not the conducting vessels, is clearly shown by the different rates at which starch is translocated from nearby cells of the same leaves. The absence of any effect on the main conducting tissues is emphasized by results of BOLAS and BEWLEY (1930) with tomato aucuba mosaic virus; they stated that, although starch is retained in tomato leaves at foci of infection when plants are placed in the dark, other parts of the leaves at first lose starch more rapidly than do healthy leaves. Whether the delay in translocation of starch occurs because of effects of the virus on carbohydrases, or changes in cell permeability, as suggested by WYND (1943), remains undetermined. In Arran Victory potato plants with paracrinkle, BARTON-WRIGHT and MCBAIN (1933*a*) have also reported a reduced translocation rate for carbohydrate, though the phloem shows no abnormalities. Unlike other mosaic diseases, paracrinkle is stated to cause carbohydrate to accumulate, though much less so than does leaf roll. No differences were found between President plants carrying paracrinkle virus and virus-free plants. In Arran Victory plants with crinkle symptoms caused by dual infection with potato viruses A and X BARTON-WRIGHT (1941) found little effect on carbohydrate metabolism, but the diseased plants contained more nitrogen and protein than healthy plants. Spike disease of sandal has been reported to have the unusual effect of increasing the amounts of both nitrogen and carbohydrate (SREENIVASAYA and SASTRI, 1928); calcium deficiency is said to be a constant feature of the disease, and it is suggested that this may account for the reduced movement of carbohydrates. SREENIVASAYA (1930) has found mannitol in all diseased plants but not in healthy sandal, and suggests that this sugar is a specific product of virus activity.

Comparisons of the relative respiration rates of healthy and virus-infected plants have been made by various workers. THUNG (1928) found that potato leaves with leaf roll respired at a higher rate per g. of dry weight than healthy leaves and he suggested this was because of the larger amounts of respirable sugars rather than because of any increase in enzymes. WHITEHEAD (1931) also found a higher rate of respiration characteristic of leaf roll plants, though the differences varied with different tissues and organs at different times. Infected immature tubers at first respired more than healthy ones, but at the end of dormancy they respired slightly less. When sprouting first started, the infected tubers continued to respire at a low rate until leaves were formed, when the rate exceeded that of healthy plants and remained so throughout the growing season. Like THUNG, WHITEHEAD (1934) concluded that the increased respiration rate reflects an increased concentration of respirable substrate in infected plants rather than any other effect, and he states that the respiration rates of healthy and infected plants become similar if the accumulation of carbohydrates is delayed by exposing young infected plants to low light intensities.

Increased respiration rates, however, do not seem to be confined to virus diseases in which carbohydrates accumulate, though the reports of different workers are rather conflicting. DUNLAP (1930) found that the respiration of young leaves was increased by infection with tobacco mosaic virus, but that older infected leaves respired less than healthy ones. CALDWELL (1934b), on the other hand, found that the $CO_2$ output of tomato leaves infected with aucuba mosaic virus was consistently higher than that of healthy leaves. This was so whether the respiration was measured in terms of initial dry weight, residual dry weight or of residual nitrogen, and whether $CO_2$ output was measured in oxygen or in nitrogen. As the carbohydrate content of these diseased leaves is less than that of the controls, CALDWELL attributed the greater respiration to increased activity of the plant oxidases. This is also suggested by WOODS and DUBUY (1942), but TAKAHASHI (1947) found no evidence to support it and described only slight effects of tobacco mosaic virus on respiration. WYND (1942) has described rather different effects. He states that, within four days of inoculating a lower leaf, the whole plant shows a greatly increased respiration rate, the increase occurring in non-inoculated leaves where no virus is yet detectable as well as in inoculated leaves. WYND stated that some days later, when the virus has permeated the upper leaves, these leaves have a lower respiration rate than comparable healthy leaves. WYND (1943*a*) further stated that, during the early days of infection, oxygenase activity fluctuates, being at first higher than normal, dropping below normal on the sixth day and rising to a new maximum about 14 days after infection. Peroxidase was little affected, but catalase was, reaching one maximum about 8 days and a second about 16 days after infection. WYND's claim that the physiology of leaves is affected before they contain any detectable virus is clearly one of great interest, but as yet it remains unconfirmed. Much more work is needed on this problem before any safe conclusions can be made.

It is obvious that viruses which produce necrotic symptoms must have profound effects on the metabolism of their host cells, but these are little understood. Necrosis is often accompanied by dissolution of starch from the neighbouring tissues and by the accumulation of deeply-staining substances; in potato tubers cork cambia often occur around the dying cells. Two kinds of necrotic lesion occur in leaves, one consisting of white spots or rings, the other of black or dark-brown areas. The first consist of collapsed empty cells, while the second contains deposits of gum-like materials, rich in pectin and which absorb infra-red rays strongly (BAWDEN, 1933). Such lesions often also contain material that fluoresces strongly in ultraviolet light (Fig. 56). BEST (1936) found that the substance responsible for this in the lesions caused by tomato spotted wilt virus in tobacco, and tobacco mosaic virus in *N. glutinosa*.

is a water-soluble material which is present in small quantities in normal plants. Later (BEST, 1944) he isolated the substance from spotted wilt plants and identified it as 6-methoxy-7-hydroxy-1:2 benzo-pyrone (scopoletin). He states that its occurrence is directly related to virus concentration and suggests that it accumulates because the physiological activities of the virus breaks some link in a chain of reactions in which scopoletin is normally an intermediate product. Scopoletin is also responsible for the fluorescence in ultraviolet light of potato tubers infected with the leaf roll virus (ANDREAE, 1948).

**The Movement of Viruses:** — The apparent ability of viruses to move freely through plants to cause systemic infections is one of their most striking properties. Because it is the common experience, it is taken for

FIG. 56*a* and *b*. — Photographs of the under surface of a tobacco leaf showing systemic invasion by the virus causing tomato spotted wilt. — *a:* showing the necrosis when photographed by daylight on a panchromatic plate, and *b:* showing fluorescence when photographed under filtered ultra-violet light, with an aesculine filter placed between the leaf and camera to absorb any reflected ultra-violet. (BEST, R. J., 1936, Austral. J. Exp. Biol. and Med. Sci. *14*, 199).

granted, and localised infections often attract more attention, being regarded as exceptional rather than normal, though this behaviour would, in fact, seem more in keeping with what is known of the properties of viruses. They are among the largest of known proteins, do not diffuse through semi-permeable membranes, and yet appear to move much more freely than other cell constituents. We say appear, because the movement of individual virus particles from one cell to another, or from one part of a plant to another, has never been formally demonstrated. Something moves, and as this leads to the production of further virus, it is simplest and most reasonable to assume that the something is virus; like other writers on the subject, we shall make that assumption.

Local infections with no systemic invasion occur with many different virus-host combinations. The examples that have attracted most attention are those in which the host reacts by producing necrotic local lesions, and the phenomenon has been readily explained on the assumption that viruses are obligate parasites which are unable to multiply in or move from dead cells. Although plausible enough, this explanation is not entirely adequate, because it fails to account for the many occasions in which viruses do become systemic in spite of the production of necrotic local lesions. Strains of potato viruses X and Y, for example, are sometimes localised in certain potato varieties and sometimes not, though they produce apparently similar necrotic local lesions in all the plants. So, too, in tobacco plants, strains of virus X that produce ringspot symptoms are sometimes localised and sometimes produce systemic infections, without any obvious differences in the local reaction. Such variable results can be explained by assuming that infected host cells do not all die at the same rate and that some survive long enough to allow movement from them, but the postulate that viruses normally move and are prevented from doing so only by death of host cells is by no means necessarily valid. Indeed, the assumption, which is widely made, that plants not

FIGURE 56*b* (*see* p. 288).

immune from viruses give either necrotic local lesions or become systemically invaded is almost certainly unfounded. It has become accepted because viruses are studied almost exclusively in hosts that react in one of these two ways. There are obvious reasons for this, the main ones being that, unless special tests are made to detect them, viruses remain unnoticed in hosts in which they cause neither local lesions nor a systemic disease, but they are not reasons justifying any conclusions on the general behaviour of viruses when introduced into leaves. As a prerequisite to causing an economic disease, a virus must become systemic, and it is when this occurs that viruses are normally detected, but it does not mean that systemic invasion is the normal behaviour of viruses. In many hosts viruses may be able to multiply but not move with any freedom, and plants which become systemically invaded may even be exceptions rather than the rule.

There is, indeed, much evidence that viruses can be localised at infection sites in the absence of any necrosis of the host. HOLMES (1938, 1946) found that both tobacco mosaic and tobacco etch viruses could infect and multiply locally in leaves of many plants in which they cause neither necrotic local lesions nor a systemic infection. More than half of the 199 species found susceptible to tobacco mosaic virus when it was rubbed over the leaves, failed to become systemically infected, and of the

108 in which the virus remained localised, more than 80 showed no symptoms and most of the others developed only chlorotic local lesions. The tobacco necrosis viruses also seem to move no more freely in *Primula obconica,* in which they cause no symptoms, than in hosts in which they cause necrotic local lesions. Indeed, in French bean, which produces necrotic local lesions, these viruses can invade most of the tissues of inoculated leaves, whereas in the symptomless *P. obconica,* they are more strictly confined to the actual regions initially infected. In both hosts, virus occasionally passes from an inoculated leaf to infect uninoculated ones, but in neither does this movement lead to a fully systemic infection. In beans the movement shows because of necrosis in the stem or by the production of isolated necrotic spots in uninoculated leaves, but in *P. obconica* it is detectable only by specific tests for the presence of virus, which show that the viruses may be present in considerable concentration in small areas, but not in the bulk of the plant (BAWDEN and KASSANIS, 1947). Thus it seems that movement from infection foci is not necessarily an intrinsic property of viruses, interfered with only by death of host tissues, but rather a reflection of some specific virus-host interaction. It may be that viruses only become fully systemic when both parenchymatous cells and vascular tissues can support virus multiplication, and that localisation in non-necrotic hosts occurs when only parenchymatous cells are susceptible, but in the absence of any information it is idle to speculate further.

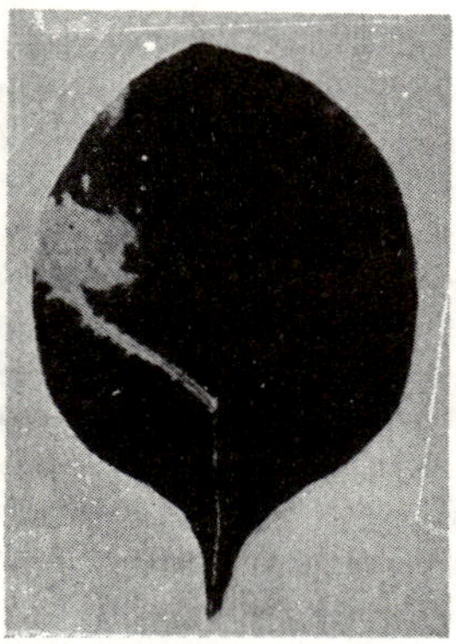

FIG. 57. — Leaf of tobacco plant inoculated with diluted sap from plants infected with yellow tobacco mosaic virus. Treated with iodine eight days after inoculation. The uninfected leaf tissue has stained darkly, and the local lesion and the path taken by the virus down the vein to the petioles show clearly. (SAMUEL, G., 1931, Ann. Appl. Biol. *18*, 494).

In hosts that become systemically infected with sap-inoculable viruses, it seems that there is movement of two kinds, but only a few viruses have been studied and there is some conflict between the conclusions of different workers. CALDWELL (1930) found that tobacco mosaic virus would only pass through living tissues, and not through pieces of stem killed by steaming. When a leaf above the steamed portion was inoculated, all others above this portion ultimately became infected but those below remained uninfected, and *vice versa.* As the plants remained turgid, water movement apparently was unaffected, so CALDWELL concluded that the virus does not normally enter the xylem. GRAINGER (1933) and MATSUMOTO and SOMAZAWA (1933), on the other hand, claimed that all tissues including the xylem become infected. CALDWELL (1931, 1934*a*) claimed that when the virus was experimentally introduced into the xylem, it remained there unless the vessels were mechanically damaged, which, he stated, then caused the leaves to develop symptoms. He concluded that there is movement only between and through living cells, that it is independent of the phloem as well as xylem, is not influenced by the translocation of food-stuffs and occurs at a more or less uniform rate in all directions through any living tissues. GRAINGER (1933) also concluded that tobacco mosaic virus moved at a steady rate and was not influenced by the direction of either transpiration or translocation streams.

Other workers, however, have reached other conclusions; they find, first, a slow movement from infection foci, presumably as the virus diffuses along a concentration gradient set up by virus multiplication, and, secondly, a much more rapid dispersal of virus to distant parts of the plant, apparently occurring when the virus has reached the vascular tissue, and presumed to take place in the phloem. HOLMES (1930, 1931, 1932) found that the virus for the first few days is multiplying at infection foci and moves very slowly and only short distances, but then suddenly and rapidly moves from these foci and occurs at distant points in the stem, roots and apices. He showed that, in the uninoculated leaves, infection first occurred in the vicinity of veins, that cutting the main veins reduced the rate of movement considerably, whereas cutting interveinal areas had little effect, and from the distribution of the virus in partly-shaded plants he concluded that direction of movement was correlated with the translocation of food-stuffs. SAMUEL (1931), like HOLMES, followed the movement of a strain of tobacco mosaic virus in tobacco by iodine staining and obtained similar results. The virus spread slowly from initial infection foci until it reached a vein, when it rapidly passed down into the main stem and shortly afterwards was detectable in the veins of young leaves (Figs. 57 and 58).

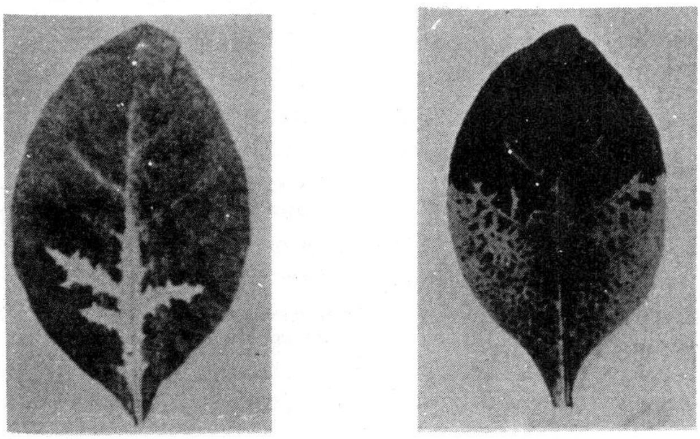

FIG. 58. — Two leaves from the centres of plants becoming systemically invaded with yellow tobacco mosaic virus. Before staining with iodine these showed clearing of the veins. The darkly staining parts at the tips are not infected, the virus being restricted to the areas around the veins which are differentiated by the iodine staining. (SAMUEL, G., 1931, Ann. Appl. Biol. *18*, 494).

From experiments with tobacco mosaic virus in tomato plants, SAMUEL (1934) obtained further evidence for two different rates and kinds of movement. One terminal leaflet was inoculated on each plant, and then at intervals the plants were cut into portions, which were incubated to allow any virus present to multiply and then tested for the presence of virus. No virus passed out of the inoculated leaflets during the first 3 or 4 days, but this period of little movement was followed by one when the virus moved rapidly to distant parts of the plants. Virus was detectable in the roots within 12 hours of first appearing in the petioles of the inoculated leaves, and within a day was also present at the apex. This rapid long-distance movement clearly does not depend on continuous multiplication, as suggested by CALDWELL and GRAINGER, because portions of plants far away from the inoculated leaflet often contained virus while nearer portions did not. In other words, virus particles apparently passed through pieces of stem leaving them virus-free, and when viruses first entered the stem SAMUEL often found pieces of stem several centimetres long to be uninfected though other pieces, both above and below, contained virus. SAMUEL suggested that this rapid movement occurs in the phloem and that it is correlated with transport of food from the leaves. Developing fruit trusses often became infected simultaneously with the roots, but leaves adjacent to the trusses remained virus-free for days or weeks. In small plants, all the leaves became infected soon after the youngest ones, but it took three weeks for a medium-sized plant to be invaded completely and up to two months for plants in the fruiting stage. Fig. 59 is a diagrammatic representation of the invasion of a medium-sized plant; except for limited infection in the main veins, the mature leaves of large plants remained free from virus for as long as three months after the initial inoculation.

Other workers (KUNKEL, 1939; CAPOOR, 1939; MATSUMOTO, 1941, 1941*a*) have obtained results with tobacco mosaic virus essentially similar to those reported by SAMUEL, though KUNKEL did not find that the virus always moved first to the roots. He found that roots and apical points usually became infected simultaneously, and that, when they did not, movement was as likely to be first towards the apex as towards the roots. After first entering the stem, virus sometimes moved more than 17 cm. in an hour, and in these conditions long pieces of virus-free stem often occurred between other pieces containing virus. CAPOOR found that some other sap-transmissible viruses behaved similarly to tobacco mosaic, the only significant differences being in the length of time taken to move from the inoculated leaflet into the stem, this varying from 2-3 days with potato virus Y to 6 days with tomato aucuba mosaic virus. All the viruses moved rapidly when they first entered the stem, and during this period some stem-pieces were apparently virus-free while adjacent ones, both above and below, were infected. These results show that the long-distance movement of virus early in the course of systemic infection is not a continuous chain process depending on multiplication in every cell, but that virus particles (or, at least, something capable of initiating virus particles) move individually over long distances and may pass through many individual cells without becoming established or multiplying there. There is, however, no reason to consider that any large amount of virus moves over long distances in this manner. A few particles distributed to sites in various parts of plants where they become established and multiply would be all that is necessary to lead ultimately to a full systemic invasion, by diffusion from infected to healthy cells or by further occasional movement over long distances. Obviously, if this is so, viruses that multiply readily in the tissues through which they can travel long distances, will be more likely to become systemic than any which need to escape from these tissues to other cells before they can multiply.

Many workers have suggested that the direction in which viruses move over long distances depends on the direction in which products of photosynthesis are moving, but the most detailed observations on this were made by BENNETT (1939, 1940). Using *Nicotiana glauca* plants grafted both at their apices and on basal suckers with tobacco scions, he found that virus normally passed from the top to the bottom scion in 7 days, whereas it took months to pass from the bottom to top scions. By

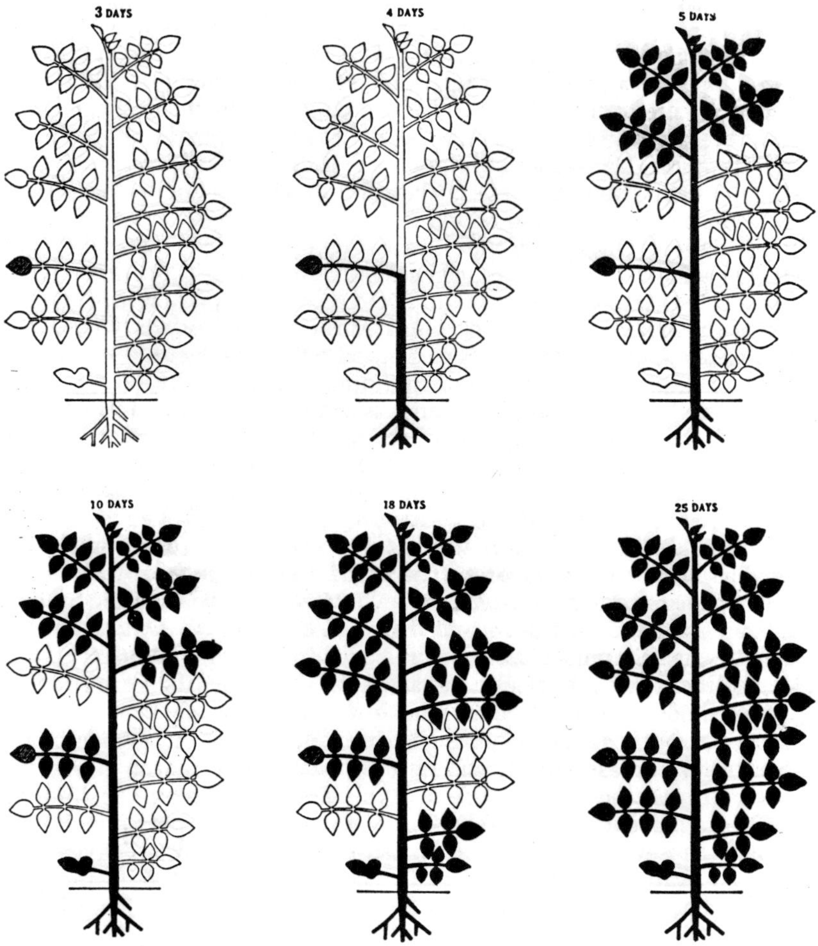


FIG. 59. — Diagram showing the progress of tobacco mosaic virus (in black) through a medium-sized, young tomato plant. Based on tests made with Dwarf Champion plants about 15 in. high, growing in 6 in. pots in an unheated greenhouse. The inoculated leaflet is shown shaded. (SAMUEL, G., 1934, Ann. Appl. Biol. *21*, 90).

contrast, when plants were defoliated or placed in the dark, treatments that would be expected to reverse the normal movement of food-stuffs. virus moved from the bottom to the top scions in a few days. Breaking the phloem continuity in the *N. glauca* plants prevented the passage of tobacco mosaic virus, which does not multiply readily in this species, but not of cucumber mosaic virus, which does. With tobacco plants, in the parenchymatous cells of which tobacco mosaic virus multiplies readily, breaking the phloem continuity delayed the passage of tobacco mosaic virus but did not prevent it. In plants setting seed, when the usual basipetal movement of food-stuffs is partially reversed, tobacco mosaic virus also moved upwards through *N. glauca* more rapidly than usual. In earlier work with raspberry leaf curl and mosaic, BENNETT (1927) also obtained results suggesting a correlation between movement of food and virus. When canes were infected, the viruses soon produced symptoms on apical growth and also passed to the roots, but symptoms occurred only on canes that were individually infected, others on the same plant remained healthy until new growth began in the following spring. By pruning such canes and so stimulating excessive lateral growth, however, they became infected and showed symptoms in the current season. The passage of viruses from infected to healthy plants through dodder (*Cuscuta* sp.) is also affected by treatments that would be expected also to influence movement of plant food. COCHRAN (1946) found that 75% of the healthy plants connected by dodder to plants infected with various viruses became infected if the dodder was heavily pruned and the healthy plants were shaded, whereas there was no transmission with unshaded plants and unpruned dodder.

BENNETT's (1934, 1937) experiments with sugar beet curly-top virus provide still further evidence connecting food movement with long-distance movement of the virus. BENNETT considers that the virus is largely restricted to the phloem, and it does not move across pieces of tobacco stem where the phloem continuity has been broken by ringing. In normal conditions the virus moves towards the roots much more rapidly than towards the apices; the virus passed from apical to basal tobacco scions grafted to *N. glauca* in 30 days, whereas it did not pass from basal to apical scions in 300 days. When the *N. glauca* were defoliated, however, virus passed from basal to apical scions in 18 days. Similar results were obtained in sugar beet. When plants were separated into three shoots united only at the lower end of the tap root, virus introduced into one shoot did not pass into the others in periods of more than 100 days, unless they were defoliated or kept in the dark, when they became infected within a few days. The time taken for this virus to move from infected leaves also seemed to depend on their physiological conditions. In normal conditions, virus moved out of green leaves within 4 hours of being introduced into the distal portions, but no movement occurred from etiolated leaves in the dark until 21 days. When infected etiolated leaves were illuminated the virus moved out of them within 24-48 hours. This fact, and results showing the rapid infection of fruit, is perhaps the best evidence that long-distance movement of viruses is correlated with the translocation stream. The development of symptoms in new shoots forced by pruning, or in shoots kept in the darkness or shade, is less unequivocal, for the conditions created by these treatments also increase the likelihood of tissues becoming infected when directly inoculated with some viruses (BAWDEN and ROBERTS, 1947, 1948). Hence, the results from such treatments may not necessarily mean that virus moves any more readily into tissues treated in these ways, but rather that what does enter has a greater chance of becoming established and producing symptoms. There is, however, much evidence from different sources that many viruses move from one part of a plant to another because of the mechanical carriage of individual particles along with the main food stream.

Cell to cell diffusion following multiplication no doubt occurs, but it is much slower and takes place over only short distances. It is probably responsible for the increasing total of infected cells that develop around entry points, but its limitations are clearly shown by the small part of an inoculated leaf that becomes infected from virus introduced at a single entry point. UPPAL (1934) measured the time taken for tobacco mosaic to pass from the inoculated upper epidermis to the lower one and calculated a rate of 8 $\mu$ per hour, a value compatible with diffusion along a concentration gradient produced as a consequence of multiplication. How the particles pass from one parenchymatous cell to another is unknown, but they may do so along the plasmodesmata. Failure to move long distances by cell to cell diffusion is further suggested by the many reports of the slow movement of virus from roots to tops. With all viruses yet studied, movement from leaves to roots occurs consistently and rapidly, but movement in the opposite direction is delayed and irregular. PRICE (1938) found that when tobacco roots were inoculated with tobacco mosaic virus, the virus entered and multiplied readily, but did not pass into the stems in periods up to three weeks. FULTON (1941), too, found that a number of viruses multiplied when inoculated into roots and rapidly invaded all parts below the inoculation point, but they moved upwards only slowly and did not usually reach the stems. BENNETT (1940) found that cutting off the tops of tobacco plants which had infected roots, increased the rate at which virus passed into the stems. Similarly, we have found at Rothamsted that the chances of root infection leading to infection of shoots are increased if the shoots are largely defoliated or kept in the dark. In good growing conditions and bright light, shoots have remained virus-free for periods of months while the roots support a high virus concentration. All these phenomena indicate that movement by diffusion is a slow and limited process, and that long-distance carriage of virus is a separate process, consisting of the dispersal of occasional particles in a direction largely determined by the transport of food-stuffs.

The transport of most virus over large distances probably occurs in the phloem, and the delay of some days before mechanically-transmitted viruses move out of inoculated leaves is probably the time taken for virus to multiply and diffuse from initial entry points into the phloem. Some insect-transmitted viruses move over long distances much more promptly. SEVERIN (1924) detected sugar beet curly-top virus 18 cm. from the vector's feeding place within half an hour, and BENNETT (1934) records movement through beet leaves at the rate of 2.5 cm. per minute. In tobacco, where less carbohydrate is translocated than in beet, movement was less rapid, approximately 1.5 cm. per hour. STOREY (1928) obtained similar results with maize streak virus, detecting it 40 cm. from the infection point in 2 hours. These values contrast strikingly with the periods of days required for mechanically-transmitted viruses to move out of an inoculated leaf, but they resemble the rates found for such viruses once they have passed into the stem and they are of the same order as those recorded for movement of metabolites through the phloem. There is good evidence that, to cause infection, curly-top and maize streak viruses must be injected directly into the phloem; if this is so, they will immediately find themselves in tissues through which metabolites are constantly flowing from leaves to stem, a position only reached by mechanically-transmitted viruses after several days.

The regular occurrence of systemic infection and spread of viruses through large plants suggests at first sight that viruses have a peculiar power of movement that distinguishes them sharply from other plant proteins. Fruit-proteins do not occur in leaves, nor leaf-proteins in roots, yet viruses occur in all tissues. This difference may be apparent rather than real, for viruses may move no more freely than other cell constituents of comparable size. The difference may lie simply in the fact that viruses multiply in all these different parts of plants, whereas other proteins do not. If a particle, or indeed many particles, of a leaf-protein moves into, say, a fruit, there is no method of knowing that it has done so. If virus moves, however, it multiplies and makes its presence obvious by producing symptoms, or its presence can be ascertained by infectivity tests. There is no reason to consider that any large amount of virus ever moves from one part of a plant to another. On the contrary, there is evidence that it does not. In potato plants setting tubers, there is a large and continuous flow of metabolites from haulm to tubers, conditions that all the evidence suggests should direct virus into the tubers. Yet, in plants infected with virus X, the virus content of immature tubers is usually less than one-thousandth that of the leaves. Only when the tubers start to sprout and conditions are favourable for virus multiplication does the content increase, and it is clear that the amount that moves into tubers from the leaves, although it may be all-important in leading to infection, is a negligible amount relative to that produced later by multiplication locally (BAWDEN, KASSANIS and ROBERTS, 1948). Hence, the apparent ease of movement may be largely illusory, a misleading consequence of the ability of the relatively few particles that happen to be carried long distances, to multiply at the sites they reach.

### *References:-*


ANDREAE, W. A. (1948): Can. J. Research, C. *26*, 31.
BAILEY, E. M. (1924): Connecticut Agr. Exp. Sta. Bull. *258*.
BARTON-WRIGHT, E. (1941): Ann. Appl. Biol. *28*, 299.
—— and MCBAIN, A. M. (1932): Trans. Roy. Soc. Edin. *57*, 309.
—— and —— (1933*a*): Ann. Appl. Biol. *20*, 525.
—— and —— (1933*b*): *ibid. 20*, 549.
BAWDEN, F. C. (1933): Nature *123*, 168.
——, KASSANIS, B. and ROBERTS, F. M. (1948): Ann. Appl. Biol. *35*, 250.
—— and KASSANIS, B. (1947): Ann. Appl. Biol. *34*, 127.
—— and —— (1948): Unpublished.
—— and PIRIE, N. W. (1945): Brit. J. exp. Path. *26*, 294.
—— and —— (1946): *ibid. 27*, 81.
—— and ROBERTS, F. M. (1947): Ann. Appl. Biol. *34*, 286.
—— and ——(1948): *ibid. 35*, 418.
BENNETT, C. W. (1927): Mich. Agr. Exp. Sta. Tech. Bul. 80.
—— (1934): J. Agric. Res. *48*, 665.
—— (1937): *ibid. 54*, 479.
—— (1939): Phytopath. *29*, 1.
—— (1940): J. Agr. Res. *60*, 361.
BEST, R. J. (1936): Austral. J. Exp. Biol. and Med. Sci. *14*, 199.
—— (1937): *ibid. 15*, 191.
—— (1944): *ibid. 22*, 251.
BOLAS, B. D. and BEWLEY, W. F. (1930): Nature *126*, 471.
BREWER, P. H., KENDRICK, J. B. and GARDNER, M. W. (1926): Phytopath. *16*, 843.
BUNZEL, H. H. (1913): U. S. Dept. Agric. Bull. *277*.
CALDWELL, J. (1930): Ann. Appl. Biol. *17*, 429.
—— (1931): *ibid. 18*, 279.

—— (1934a): *ibid.* *21*, 191.
—— (1934b): *ibid.* *21*, 206.
CAMPBELL, E. G. (1925): Phytopath. *15*, 427.
CAPOOR, S. P. (1939): Thesis, London University.
COCHRAN, G. W. (1946): Phytopath. *36*, 396.
CORDINGLEY, H., GRAINGER, J., PEARSALL, W. H. and WRIGHT, A. (1934): Ann Appl. Biol. *21*, 78.
DUNLAP, A. A. (1930): Amer. Journ. Bot. *17*, 348.
ELMER, O. H. (1925): Iowa Agr. Exp. Sta. Res. Bull. *82*.
FULTON, R. W. (1941): Phytopath. *31*, 575.
GRAINGER, J. (1933): Ann. Appl. Biol. *20*, 236.
HOLDEN, M. and TRACEY, M. V. (1948): Biochem. J. *43*, 151.
HOLMES, F. O. (1930): Amer. Journ. Bot. *17*, 789.
—— (1931): Contrib. Boyce Thompson Inst. *3*, 163.
—— (1932): *ibid.* *4*, 297.
—— (1938): Phytopath. *28*, 58.
—— (1946): *ibid.* *36*, 643.
JODIDI, S. L., MOULTON, S. C., and MASKLEY, K. S. (1920): J. Amer. Chem. Soc. *42*, 1061.
KALMUS, H. and KASSANIS, B. (1944): Nature *154*, 641.
KRAYBILL, H. R., BREWER, P. H., SAMSON, R. W. and GARDNER, M. W. (1932): Phytopath. *22*, 629.
KUNKEL, L. O. (1939): Phytopath. *29*, 684.
LUDTKE, M. (1930): Phytopath. Zeitschr. *2*, 341.
MARTIN, L. F., BALLS, A. K. and MCKINNEY, H. H. (1938): Science *87*, 329.
MATSUMOTO, T. (1941): Trans. Nat. Hist. Soc. Formosa *31*, 201.
—— (1941a): *ibid.* *31*, 275.
—— and SOMAZAWA, K. (1933): J. Soc. Trop. Agric. *5*, 37.
MURPHY, P. A. (1923): Sci. Proc. Roy. Dublin Soc. *17*, 163.
PETERSON, P. D. and MCKINNEY, H. H. (1938): Phytopath. *28*, 329.
PRICE, W. C. (1938): Amer. J. Bot. *25*, 603.
QUANJER, H. M. (1913): Meded. Landbouwhoogeschool, Wageningen, *6*, 41.
RISCHKOV, V. L. and GROMIKO, E. P. (1941): Microbiologia, *10*, 898.
——, SMIRNOVA, V. A. and GORODSKAJA, O. S. (1946): Biochimia *11*, 197.
—— and SMIRNOVA, V. A. (1939): Compt. Rend. Acad. Sci. U. R. S. S. *23*, 95.
ROSA, J. T. (1927): Plant Physiol. *2*, 163.
ROUZINOFF, P. G. (1930): Morbi Plantarum, Leningrad, *19*, 148.
SAMUEL, G. (1931): Ann. Appl. Biol. *18*, 494.
—— (1934): *ibid.* *21*, 90.
SCHWEIZER, G. (1930): Phytopath. Zeitschr. *6*, 557.
SEVERIN, H. H. P. (1924): Phytopath. *14*, 80.
SPENCER, E. L. (1935a): Phytopath. *25*, 178.
—— (1935b): *ibid.* *25*, 493.
—— (1937): Plant Physiol. *12*, 825.
—— (1939): *ibid.* *14*, 769.
—— (1941): *ibid.* *16*, 227.
—— (1941a): *ibid.* *16*, 663.
—— (1941b): Science, *94*, 96.
—— (1942): Plant Physiol. *17*, 210.
SREENIVASAYA, M. (1930): Nature *126*, 438.
—— and SASTRI, B. N. (1928): J. Indian Inst. Sci. *11*, 23.
STANLEY, W. M. (1937): Phytopath. *27*, 1152.
STOREY, H. H. (1928): Ann. Appl. Biol. *15*, 1.
TAKAHASHI, W. N. (1941): Phytopath. *31*, 1117.
—— (1947): Amer. J. Bot. *34*, 496.
—— (1948): Science *107*, 226.
THORNTON, M. H. and KRAYBILL, H. R. (1934): Phytopath. *24*, 19.
THUNG, T. H. (1928): Tijdschr. over Plantenziekten *33*, 1.
TRUE, R. H. and HAWKINS, L. A. (1918): J. Agric. Res. *15*, 381.
UPPAL, B. N. (1934): Indian J. Agric. Sci. *4*, 865.
WHITEHEAD, T. (1931): Nature *128*, 967.
—— (1934): Ann. Appl. Biol. *21*, 48.
WOODS, A. F. (1899): Central. Bakt. Par. Krankheit. *2*, 745.
—— (1900): Science *11*, 17.
WOODS, M. W. (1940): Science *91*, 295.
—— (1943): Phytopath. *33*, 77.
—— and DUBUY, H. G. (1941): Phytopath. *31*, 978.
—— and —— (1942): *ibid.* *32*, 288.
WYND, F. L. (1942): J. gen. Physiol. *25*, 649.
—— (1943): Bot. Rev. *9*, 395.
—— (1943a): Plant Physiol. *18*, 90.

*Chapter 16*

# CONTROL MEASURES AGAINST VIRUS DISEASES

The economic importance of viruses derives largely from their ability to cause systemic diseases and to persist in the vegetative parts of plants for as long as these remain alive. Once infected, perennial plants continue to be so indefinitely, and progeny of vegetatively-propagated plants are also usually infected in perpetuity. Hence, losses are not confined to the season in which infection occurs, but continue as long as the infected line is in culture. Symptoms may fluctuate with the age of the host or the environment, and in some diseases an acute initial reaction regularly gives place to a less severe condition, in the latter stages of which infected plants may be barely distinguishable from healthy ones. Along with this apparent recovery, the virus content of the plants may fall, but such plants still contain virus and, though they may not themselves be greatly affected, supply sources of infection for other plants. In the absence of true recovery, the control of plant virus diseases lies almost wholly in preventive measures to protect healthy plants, and to this end the most generally applicable precaution is the destruction of infected plants, whether or not they themselves show symptoms.

As with most general statements about virus diseases, there are exceptions to the non-recovery rule, probably because infection is not always completely systemic. A proportion of suckers taken from certain varieties of sugar cane with mosaic, for example, regularly give rise to symptomless plants which also seem to be virus-free (SUMMERS, BRANDES and RANDS, 1948). Similarly, by propagating cuttings from emerging shoots, HOLMES (1948) has obtained healthy dahlia plants from tubers infected with tomato spotted wilt virus. All shoots are at first symptomless; they become diseased if left attached to the tubers, but many can be obtained virus-free if detached early enough. Some shoots from potato tubers infected with potato virus Y are also symptomless and virus-free at first, but they become infected if left attached to the parent tuber. Such incompletely systemic infections probably occur with other viruses and propagating from suitably chosen pieces of tissues may enable virus-free lines to be regenerated from infected parents.

**Curative Treatments:** — Virus-free material can also be obtained from some diseased plants by subjecting them to treatments that destroy the viruses without imparing the viability of the host cells. This possibility was first indicated by work with two suspected virus diseases of sugar cane, the sereh disease (HOUTMAN, 1925) and the chlorotic-streak disease (WILBRINK, 1923; MARTIN, 1930; BELL, 1933). Sugar cane cuttings survived immersion in water at 52°C. for periods up to one hour, and cuttings from diseased parents usually grew into healthy plants if immersed for 20-30 minutes. The first established virus disease to be cured by heat was peach yellows, from which KUNKEL (1935) reported trees recovered if kept for a fortnight or more at 35°C. Similar treatments also freed peach trees from little peach, red suture and rosette, but were unavailing against peach mosaic (KUNKEL, 1936*a*, 1936*b*). Large trees needed to be kept at 35°C. for longer than small ones and the treatment freed the tops of the trees from virus more rapidly than it did the roots. KUNKEL showed that the recovery was not masking of symptoms by high temperatures and that it did not occur because avirulent strains were segregated; scions from treated trees produced no symptoms when grafted to healthy ones, and the cured trees developed usual symptoms when again grafted with scions from diseased trees. He also showed that infected peach trees could be cured by hot-water treatments similar to those effective against the sugar cane diseases. Dormant trees were freed from the viruses by immersion in water at 50°C. for about 10 minutes, and bud-sticks were freed at lower temperatures; after 5 days at 34°C., 40 minutes at 42°C. or 4 minutes at 50°C., bud-wood from diseased trees developed into healthy shoots. Heat treatments also free peach trees from phony disease (HUTCHINS and LA RUE, 1939) and yellow-red disease (HILDEBRAND, 1941).

Heat therapy is unlikely to be generally applicable, but by seeking suitable conditions of heating it could probably be extended to many virus-host combinations. It is obvious that its use depends on normal host constituents resisting the effects of heat better than the viruses, and its likelihood of success increases with the ability of a host to withstand high temperatures. Obviously, too, success is more likely with viruses that have low than high thermal inactivation points, but it is by no means necessarily restricted to the former. Some viruses have high thermal inactivation points but small temperature coefficients of thermal inactivation. With these, although short exposures to high temperatures would probably be unavailing, long exposures at temperatures well below the thermal inactivation point might produce cures. Against viruses that have thermal inactivation points above about 55°C. and that also have high temperature coefficients of inactivation, the method will almost certainly be inapplicable. KUNKEL (1941, 1943, 1945) has extended heat therapy to hosts other than peach. *Vinca rosea* and *Nicotiana rustica* infected with aster yellows virus were freed from it after growing for two weeks at 40°C., and *V. rosea* was also cured by immersion in water at 45°C. for a few hours. *V. rosea* was also freed from potato witches' broom and cranberry false blossom viruses by growing for 14 days or more at 42°C. Potato tubers with diameters up to 2 cm., infected with witches' broom virus, were cured by storing at 36°C. for 6 days, and cranberry plants were cured of false blossom when maintained for 8 days at 43°C., though this treatment injured the plants and some died. KASSANIS (1949) has also freed potato tubers from the leaf-roll virus by keeping them at 37°C. in a moist atmosphere for periods of 10 to 20 days.

The possibility of chemotherapeutic treatments has been raised by STODDARD (1942), who states that bud-wood from peach trees affected with X-disease can be cured by soaking in water solutions of quinhydrone, urea and sodium thiosulphate. This claim still awaits confirmation, but with increasing knowledge of substances that render viruses non-infective, it can be expected that some will be found to be capable of inactivating viruses *in vivo* without killing the host cells. Various kinds of radiations inactivate viruses *in vitro*, and though there is no report of curing infected plants by irradiation, it would seem worth investigation.

Although they are unlikely to replace other control measures, effective therapeutic treatments would have several practical applications. They would make possible the regeneration of desirable varieties of vegetatively-propagated plants which have fallen into disuse because of virus infection. They would also provide nuclear stocks of healthy propagating material and ensure a constant supply of such material. Routine treatment for freeing bud-wood from viruses would also do much to prevent the unwitting use of infected scions for plants propagated by grafts, which has probably spread many viruses widely in the past.

**General Measures:**—One general precaution, the destruction of infected plants, has already been mentioned, and other precautionary measures include isolation from sources of infection and the use of virus-free planting material. Could these hygienic precautions be operated thoroughly, other control measures would not be needed, but complete eradication of sources of viruses is practically impossible. Their efficacy will vary from disease to disease, depending on how rapidly spread occurs, over what distance it can happen and on how easily infected plants can be detected. Removing diseased plants (roguing) as they are noticed within a crop is all that is necessary to control some diseases that spread slowly, rarely come into the crop from outside sources and that are readily detectable soon after plants become infected. On the other hand, it may be valueless and no more than wasted effort, if spread is rapid or virus is frequently reintroduced into a crop from outside. With one and the same disease, roguing may be valuable in one set of conditions, but valueless in others. The health of potato stocks in the "seed-growing" parts of Britain, for example, can not only be maintained, but often improved, by roguing, whereas in other parts, where spread occurs more frequently and earlier, roguing as soon as plants show symptoms of leaf roll and rugose mosaic is largely ineffective (DONCASTER and GREGORY, 1948). This is so despite the fact that spread is mostly local, restricted to plants relatively near to infected ones within the crop, and few infections come in from outside sources. Roguing would be still less effective in controlling sugar beet yellows, which spreads much more rapidly than the potato viruses and readily comes into the crop from outside. Isolation from sources of infection, such as the over-wintering mangold and beet crops grown for seed, spinach beet that has stood through the winter, or stored mangolds, is likely to be a more useful measure than any attempt at roguing in the standing crop (HULL and WATSON, 1945).

Most virus diseases are not seed-transmitted, so that most annual crops start healthy and contract viruses from outside sources, usually infected perennial or, as with sugar beet yellows, biennial hosts. To control such diseases by hygienic measures calls for eradication of these alternative hosts rather than roguing in the annual crop itself. Celery mosaic, caused by cucumber mosaic virus, for example, is reported to be wholly controlled in Florida by eradicating *Commelina nudiflora*, a perennial weed-host, from the neighbourhood of celery crops (TOWNSEND, 1948). In Britain this virus is common in tomato and marrow crops where these are growing near to infected herbaceous perennials, but is negligible elsewhere. So, too, infected dahlias, or other vegetatively-propagated or perennial hosts, provide the source of tomato spotted wilt virus for tomato and tobacco crops. Unfortunately, cucumber mosaic and tomato spotted wilt viruses have wide host ranges and do not always produce clear symptoms in their perennial hosts, but much might be done to reduce their incidence by adequate weed-control and the destruction of all suspicious-looking or unthrifty perennial plants. No general rule can be laid down for the practicability of either roguing within a crop or eradication of alternative hosts; conditions will vary from disease to disease and, often, from district to district. Local experience on the rate and distance of spread, and knowledge of the sources of infection, are essential before it can be assumed that measures which have worked with other diseases, or in other places, can be usefully adopted.

Information on the sources of infection can often be gained by studying the distribution of infected plants within a crop, which may in turn suggest whether roguing within the crop is likely to be of any value as a control measure. To be so, it is obvious that most of the spread must be from plant to plant within the crop, with relatively few plants becoming infected from outside sources. Sometimes diseased plants occur in such clearly defined groups that it is obvious at a glance that spread to neighbouring plants has occurred, but this is not always so. On the assumption that plants becoming infected from outside sources will be distributed at random, whereas those becoming infected by spread from diseased plants within the crop will be grouped, VAN DER PLANK (1946) has developed a simple test for distinguishing between the two kinds of spread. The frequency with which infected plants occur in pairs is determined, pairs being any two adjacent plants so that a run of three infected plants is two pairs. Of a sequence of $n$ plants examined, in which $x$ are infected, the number of pairs to be expected at random is given by the expression $p = \frac{x(x-1)}{n}$, and for high values of $n$, $p$ will have a standard error of $\sqrt{p}$. If the observed value for $p$ differs significantly from the calculated, spread to neighbour plants within the crop can be assumed. The converse may not always hold, for spread within a crop may not always be predominantly to neighbour plants, but the method applied to spotted wilt in South Africa and Australia (BALD, 1937) supported other suggestions that plants become infected individually from sources external to the tobacco or tomato crops. Measuring the incidence of infected plants in different parts of a crop will often help to identify these sources if they are not too remote, for disease gradients may be found that will show the main direction in which the virus is entering the crop.

The use of healthy planting material is an obvious first precaution, for sources of virus scattered through a crop from the start of the growing season are ideally placed to ensure the maximum amount of spread. Most care is obviously needed with vegetatively-propagated crops, particularly those meant to stand for many years, and to safeguard growers many countries now operate health certification schemes for potatoes and certain other crops. Such schemes might well be extended to all important crops that are vegetatively propagated and to those known to suffer from seed-transmitted viruses. Most severe attacks of lettuce mosaic, for example, are attributable to infected seed, as also are many outbreaks of virus diseases in leguminous crops. As yet the certification schemes have been mainly operated by field inspections, so that they select healthy-looking rather than necessarily virus-free stocks. Supplementary tests for virus-freedom are now being introduced into some schemes, and these are highly desirable, for vigorous plants of one variety are often carrying viruses that cause serious diseases in other varieties. Hence, with such crops as strawberries, raspberries and hops, it is a wise precaution not to grow different varieties adjacent to one another, unless it is known that they are all virus-free and react similarly to the same viruses. With hops, too, in which male plants are scattered through the crop to act as pollinators, it is important that both types of plant should react similarly towards viruses, for outbreaks of mosaic often result from the use of males that are infected carriers (KEYWORTH, 1947).

Practices implied by the phrase "good farming" will help to reduce the incidence of some virus diseases. One of these is weed destruction, for this may eradicate alternative hosts of both viruses and their insect vectors. Another is crop rotation. The probability that a long rotation will reduce soil-borne virus diseases, which now appear to be far more numerous and important than was thought some years ago, will be too obvious to need stressing, but its value is not restricted to such diseases. Many crops can survive from year to year on the same land, and by doing so also perpetuate the viruses of that crop. The lifting of potato tubers, for example, is rarely complete and large numbers often remain to produce plants in subsequent seasons. When these grow in other crops they are of little importance, but in another potato crop they are dangerous sources of infection and may largely vitiate the value of planting certified seed (DONCASTER and GREGORY, 1948). When a rotation adequate to eliminate self-set plants is not feasible, other eradication methods should be used. The pulling of all living roots of cotton plants, for example, greatly reduces the incidence of leaf curl in the Sudan, by reducing the carry-over of virus from one season to the next (BOUGHEY, 1947).

The control of virus diseases by variations of normal cultural practices has not yet been widely tried, though it would seem to offer considerable scope. The reduction in yield caused by infection often depends greatly on the stage of development of plants when they contract the disease, losses being greater the younger the plants are. Changes in the physiological state of plants with age also sometimes increase their resistance to infection. Hence varying the sowing date may affect the incidence of certain diseases, or reduce the losses suffered even if the incidence is unaffected. Early-sown sugar beet suffer less from both curly top (WALLACE and MURPHY, 1938) and yellows (WATSON, 1942) than late-sown crops, because the latter are at a more susceptible stage when vectors become active. Increased rates of sowing, and singling later than normal, are other practices that might help to control virus diseases that enter crops from outside sources, and might also reduce losses produced by secondary spread from plant to plant within the crops. Too few infective insects, or other sources of viruses, usually come into crops to infect all the plants, and the number entering will presumably be independent of the plant density. Thus doubling the number of plants per unit area could be expected approximately to halve the proportion of the crop that becomes infected. VAN DER PLANK and ANDERSSEN (1945) have shown that this expectation is realised with tobacco Kromnek (tomato spotted wilt virus) in South Africa, where the disease enters the crop early in the season and there is little subsequent spread. By planting at half the usual spacing, or by setting two plants per hill instead of one, and then thinning to the normal density after the thrips invasion has occurred, the diseased plants can be removed leaving a full stand of healthy plants.

When spread occurs within a crop, the position is rather more complicated, but again increased density of planting would be expected to reduce losses. Unless a denser stand in some way increases the susceptibility of the plants or increases rate of spread, no more plants will be infected from each infection focus than in a thinner stand. Thus, with a constant number of sources of virus per area, a smaller proportion of a dense crop will become infected than of a thin one. As VAN DER PLANK (1947, 1948) has emphasized, denser planting, by producing smaller plants, would be expected *a priori* to reduce rate of spread, for a connection between size of plant and rate of spread is implicit in the concept of a systemic disease. Only one initial infection is needed to infect a whole plant, regardless of its size. Other things being equal, merely because of its larger surface area, a large plant is more likely than a small one to become infected; further, as this also leads to a larger total volume of infected tissue, spread is more likely to occur from a large plant to neighbours than from a small one. Hence, unless plants become increasingly resistant as they get larger, two factors, increased likelihood of becoming infected and the provision of larger sources of potential infection, will both tend to increase the vulnerability of plants as their size increases. No doubt many factors operate against this simple relationship holding generally in the field, but several workers on potatoes and sugar beet have noted that large plants are more liable to infection than small ones (WHITEHEAD, 1927; EDMUNDSON, 1940; WATSON, WATSON and HULL, 1946; DONCASTER and GREGORY, 1948). It is also known that close spacing of groundnut plants controls rosette (STOREY, 1935); this has been attributed to some effect on the activities of the aphis vector, but in the absence of any evidence to support such an explanation, it is probable that at least some of the benefit derives from reducing the proportion of the plants that become infected, because of changes in the number per area and in their size. Clearly there are practical limits to the extent to which reducing plant size, by such methods as increasing density or withholding fertilisers, can be applied, and more experiments will be needed with a range of diseases before their general usefulness can be gauged, but the possibility of controlling virus diseases by such variations of cultural practices warrants further study.

The simplest method of avoiding virus diseases is, of course, to grow disease-resistant varieties, and a number of diseases have now been combated by raising such varieties. Mosaic of sugar cane, for example, once threatened the existence of the crop, but the development of the P.O.J. varieties, whose resistance was first noticed in Java, allowed it to survive (SUMMERS, BRANDES and RANDS, 1948). Similarly, in parts of the U. S. A. where curly top made sugar beet growing impossible, new varieties now give profitable crops. In conditions when European varieties gave no measurable yield, MURPHY (1946) found that U.S. lines Nos. 1, 33, 12, 22 and 22 (second release) cropped 6.3, 8.4, 11.2, 14.3 and 16.6 tons per acre respectively. The term resistance is usually applied to varieties that withstand losses in the field, and from published descriptions it is often difficult to ascertain to what type of behaviour they owe this property. The results of yield trials do not show whether the varieties resist infection, or whether they become infected but tolerate the viruses. Tolerance rather than resistance was probably the character of beet varieties first used to combat curly top, but varieties produced later seem to combine resistance to infection with tolerance.

The ideal at which to aim is immunity, but unfortunately with few of our crops is there any evidence for the requisite genes. The only success of this type that yet seems to have been achieved is the production in America of a potato (U.S.D.A. seedling 41956) immune from potato virus X, which is universally present in most commercial varieties. Short of immunity, however, there are other properties that confer important measures of resistance. Different varieties of plants not only differ in the ease with which they become infected, but also differ widely in the ways in which they react when infected; with the same virus, one variety may localise infection, a second may die rapidly from a systemic disease, a third may suffer a chronic, crippling disease, a fourth may show symptoms but suffer relatively little loss of crop, and a fifth may be a carrier and show no symptoms. Most of the "resistant" varieties that have been brought into commerce so far probably fall into the fourth and fifth types, that is, they are tolerant.

The practical value of ability to carry or tolerate viruses is shown by the wide scale on which carrier-varieties of some plants are now grown, and by the fact that they produce good yields where intolerant varieties do not. For example, in districts where the strawberry Royal Sovereign soon loses its cropping power, the carrier Huxley will remain vigorous for long periods. Nevertheless, the use of tolerant varieties is far from an ideal control measure. Unless such varieties also resist infection, viruses spread in them rapidly and soon achieve populations unattainable in plants whose vigour is reduced. Tolerant varieties of perennial or vegetatively-propagated plants often provide the main sources whereby viruses are perpetuated, and their potential danger to other varieties, or to other crops, often passes unrecognised. With the increased plantings of Huxley strawberries in Britain, it has become increasingly difficult to grow Royal Sovereign and in some districts the cultivation of Royal Sovereign, in spite of its many superior qualities, has now almost ceased. In crops raised annually from seed, the use of tolerant varieties may be less dangerous, but even so it may have unfortunate consequences by allowing large virus populations to develop and so encouraging the opportunities for mutations. Since tolerant sugar beet have been introduced in America, it is noticeable that curly top has been recorded as a serious disease of many different crops. The reasons for this widening of host range cannot be given with any certainty, but it is worth reflecting on the increased opportunities for mutation that were supplied to the virus by providing it with large numbers of hosts in which it could multiply and maintain high populations for much longer than in the older intolerant varieties. It is well to realise that the introduction of tolerant varieties may solve a problem in one crop, only to raise similar problems in other crops.

The cultivation of tolerant hosts makes roguing impossible and so provides conditions that favour the perpetuation and dissemination of viruses, which in turn increases the opportunities for new strains to evolve. Because of this, tolerant varieties may provide only temporary remedies, whose use is inadvisable unless no other control measure is available and then only if there is good reason to consider that the virus concerned is unlikely to pass into other crops. Less dangerous features to seek in breeding new varieties are resistance to infection and extreme sensitivity, for these will reduce the rate of spread and the numbers of infected plants. Genes determining these two characters probably exist in many species, though only the potato, tomato and *Nicotiana* sp., have yet been studied in any detail and these with a few viruses only. For example, HOLMES (1943) has found that the Chilean tomato, *Lycopersicum chilense,* is difficult to infect with tobacco mosaic virus and that this property occurs in certain derivatives from its hybrid with the cultivated tomato. Differences, amply large enough to be economically important in determining the rate at which degeneration occurs, have been found between the susceptibility of different potato varieties to a number of common viruses (SCHULTZ *et al.*, 1937; JONES and VINCENT, 1937; STEVENSON, FOLSOM and DYKSTRA, 1943; COCKERHAM, 1943; LARSON, 1945; BAWDEN and KASSANIS, 1946; BALD, NORRIS and HELSON, 1946; SCHULTZ, STEVENSON and AKELEY, 1947). Resistance to one virus occurs independently of resistance to another, and differences between varieties are not because they are colonised to different degrees by insect vectors. The viruses that cause leaf roll and rugose mosaic are both transmitted by *Myzus persicae,* yet one variety may be more susceptible than another to leaf roll but less susceptible to rugose mosaic.

The relative effects of tolerant and intolerant varieties in encouraging the spread of viruses is shown by the distribution of certain viruses in different potato varieties. Potato viruses A and X, for example, occur almost universally in commercial stocks of tolerant varieties, whereas they are rare in intolerant varieties. The best type of intolerance is the rapid death of the first infected cells, when the only result of infection is a necrotic local lesion, but failing this a systemic necrosis provides a useful method of checking spread. Varieties which die when infected in effect do their own roguing, and sources of virus are automatically eliminated from the stock instead of remaining as infection foci as they do in tolerant varieties. The loss of the individual is complete, but the effect on the stock is transient and negligible. Many potato varieties and seedlings are known which are hypersensitive, reacting with local lesions only or with a systemic lethal disease, to viruses A, B, C, X, and Y (SCHULTZ *et al.*, 1940; COCKERHAM, 1943*a*; HUTTON and BALD, 1945; HUTTON, 1945), and one variety, Craig's Defiance, has been raised which is hypersensitive to the first four (BLACK, 1939). Local hypersensitivity to tobacco mosaic virus occurs in *Nicotiana glutinosa,* and HOLMES (1938) has found this to be due to a single pair of dominant genes, which he transferred to *N. Tabacum* through *N. digluta.* By repeated backcrosses, he has produced tobacco-like derivatives with the hypersensitivity of *N. glutinosa.* Unfortunately the character is linked with genes that reduce yield (GERSTEL, 1948), but VALLEAU (1946) states that after 9 backcrosses with Kentucky 16, hypersensitive varieties have been produced that are somewhat earlier and lower yielding than Ky. 16, but otherwise satisfactory. Resistant varieties of this type, he states, are now being grown extensively in Kentucky and suffering no loss from mosaic.

When temperatures are high, or when concentrated inocula are used, the hypersensitive tobacco varieties develop systemic necrotic diseases. Because of this, CLAYTON and MCKINNEY (1941) suggested that the hypersensitive reaction had little practical value and that another type of resistance possessed by the variety Ambalena (NOLLA and ROQUE, 1933) and T. I. 448 (CLAYTON, SMITH and FOSTER, 1938; MCKINNEY, 1939) offered more promise. However, as already discussed with potatoes, a lethal systemic reaction is itself a type of behaviour that minimises spread by eliminating infection foci within the crop, and the fact that artificially-inoculated plants die is no reason to assume that there will be a high death rate in field crops. The Ambalena type of resistance is determined by two recessive pairs of genes, and seems to combine some degree of tolerance with the ability to reduce multiplication and movement of the virus below that usual in other varieties. VALLEAU (1946) reports that he has been unable to incorporate this type of resistance into commercial varieties, as it is linked with undesirable genes which give plants whose leaves are unduly susceptible to wilting and scalding.

There are clearly many opportunities for the plant breeder to produce new varieties that will suffer less from virus diseases than those now currently in cultivation. It is, nevertheless, a vain hope to expect that, within any foreseeable future, virus diseases will be conquered by the plant breeder alone, or to expect that any one variety, however great its apparent virtues, will provide a permanent control to any single disease. Viruses give every sign of being extremely mutable things, and can be expected to produce new strains that differ in virulence and host range from those that now cause losses. A succession of new varieties may well be needed to combat these new strains; similarly, different varieties may be needed in different districts, for the prevalent strains may differ from district to district. There is already evidence for such variability from "breakdowns" of resistance. The sugar beet varieties first introduced as resisting curly top now suffer losses (GIDDINGS, 1946); the tomato, Pearl Harbour, found in Hawaii to resist tomato spotted wilt, is not resistant in the U. S. A. (HOLMES, 1948*a*); and potato seedlings that localise infection with potato virus Y in Australia, are not hypersensitive to all the strains that occur in Britain (HUTTON, 1948). With such variable pathogens as viruses, it is always uncertain for how long, and in what conditions, new varieties bred for resistance, hypersensitivity, tolerance, or some other feature that reduces loss, will retain this property after passing into cultivation. Mutations of viruses, with consequent reduction in the useful life of such new varieties, are likely to be correlated with the extent to which viruses are permitted to multiply. They will be reduced by varieties that resist infection, are hypersensitive, or show clear symptoms so that diseased plants can be readily identified and destroyed, whereas they will be increased by varieties that are tolerant but easily infected. The work of the plant breeder will be made easier and more lasting if it proceeds together with measures for destroying infected plants, and these cannot be operated if there are large numbers of tolerant hosts.

The discovery that infection with one strain of a virus can protect plants against the effects of other strains, and that strains of some viruses, for example, tobacco mosaic (HOLMES, 1934) and potato X (SALAMAN, 1938), cause few or no symptoms in their common hosts, raises the possibility of combating virulent strains by deliberately infecting plants with avirulent ones. The method still awaits practical tests, but small plots of tomatoes infected with masked tobacco mosaic virus were found to crop better than plots not protected against symptom-causing strains; SALAMAN, too, points out that the process has in effect been going on in potatoes unwittingly, for stocks of many old varieties still in existence are infected with, to them, avirulent strains of virus X. The problems of infecting plants on a large scale with suitable protecting strains could probably be overcome without great difficulty, but the method should be proceeded with cautiously and thoroughly tested before put into practice. It superficially resembles the vaccination of animals, but the two phenomena differ fundamentally and the protection of plants has intrinsic dangers which are absent from true vaccination. First, plants are protected only so long as they are fully infected with one strain, so that this strain would be distributed widely, and the treated plants would be comparable with tolerant varieties and provide potential breeding places for new strains. Secondly, virulence, as often as not, is far from an intrinsic property of a virus strain, but is a reflection of specific host-parasite reactions, so that a strain may be avirulent towards one host but virulent towards another. Hence the use of "protected" plants might prevent a disease in one crop only to lead to outbreaks in another. Thirdly, the protection operates against strains of only one virus, and the plants can still become infected with other viruses. Should this occur, the plants will probably suffer more severely than would normal plants, for many pairs of viruses together cause much more serious diseases than either causes alone. Such interactions also render it impossible to "protect" plants simultaneously against more than one virus, for infection with two viruses, each on its own sufficiently innocuous to be used as a "vaccine", may lead to severe diseases.

**Insect-Borne Viruses:** — Most common virus diseases are spread by insects, and the manner in which their incidence varies from season to season, or from region to region, largely reflects variations in the populations and activities of their insect vectors. The different rates at which different diseases spread also largely reflect the behaviour of the particular insect involved. The rates will be determined by the numbers of vectors present, the ease with which they become infective by feeding on infected plants, the frequency with which they move from plant to plant, and the length of time for which they remain infective. All these diseases would presumably cease to spread in the absence of the vectors, and an attack on the vectors seems an obvious method of control. Indeed, such diseases as spotted wilt are readily prevented under glass when freedom from insects is ensured by regular fumigation, but controlling insect vectors in the field is a more complicated problem. There are, of course, established methods for dealing with insect pests, such as spraying or dusting with suitable insecticides, and fumigation under portable tents or towed covers. The control of an insect that is causing damage as a direct pest, however, is not the same problem as the control of an insect vector. The first calls only for reducing the infestation to a level at which the pest causes no economic loss, but much more than this may be needed to prevent the spread of a virus, and as yet there have been few claims for the successful use of insecticides in controlling virus diseases. WATSON (1937) found that weekly spraying with nicotine reduced the incidence of henbane mosaic sufficiently to produce a significant increase in yield of *Hyoscyamus niger* and to justify the cost, and in South Africa MOORE (1941) reduced the incidence of spotted wilt in tomatoes from 100% to 5% by spraying with tartar emetic. On the other hand, DONCASTER and GREGORY (1948) found that fumigating potatoes with nicotine successfully destroyed the aphid infestation but was without effect on the spread of leaf roll and rugose mosaic, and other workers have reported apparently successful insecticidal treatments which did not stop the spread of virus diseases. Indeed, BRONSON *et al.* (1946) state that potato plots treated with DDT (dichloro-diphenyl-trichloroethane), and with various other substances that effectively checked aphid infestations, contracted more leaf roll than unsprayed plots.

Such seeming anomalies are not entirely unexpected from what is known about the behaviour of vectors, though much more information will be needed before they can be fully explained and the chances of controlling virus diseases by customary insecticidal treatments can be judged. DONCASTER and GREGORY (1948) considered that most of the spread of leaf roll and rugose mosaic in potato crops in south-east England was brought about by the winged migrants and that the wingless individuals forming the mass of the infestation were relatively unimportant; fumigation with nicotine would have no effect on the winged individuals entering the crop, and could only prevent their progeny from developing into a large population. BRONSON *et al.* (1946) attribute their result to the fact that the plots treated with DDT remained green for longer and attracted more winged migrants than the unsprayed plots. Insecticides that do not kill rapidly enough to prevent insects from feeding, clearly cannot be expected to prevent diseases caused by infective vectors which enter the crops from external sources. What would seem necessary to control these, is a persistent substance that acts as a deterrent to the vectors and prevents them from feeding or that kills them before they have time to cause infection. Current insecticides are more likely to produce results when applied to the alternative hosts from which vectors enter the crop. SUKHOV and PETLYUK (1940), for example, report that winter-spraying of hedge-rows, in which the virus and vector *(Delphax striatella)* over-winter, reduces the incidence of pupation disease in oats. Applying insecticides, even persistent ones such as DDT, to a standing crop can be expected to control virus diseases only if these are spreading from sources within the crop and by insects that develop within the crop. Even then, their efficacy is far from assured, because a treatment that kills most of the insects, but disturbs the survivors so that they move excessively, may increase the rate of spread. It is important to remember that, although stationary insects may be serious pests, they are not vectors of viruses; any treatment that stimulates movement, even perhaps the disturbances inseparable from roguing, spraying or other seemingly beneficial practices, may result in changing potential into actual vectors and increase the amount of spread.

The most reliable control measure is to prevent the vectors, whether or not they arrive infective, from entering the crops. In the U.S.A. muslin screens have been found effective in protecting asters from yellows transmitted by hoppers and potatoes from aphid-transmitted viruses, but such precautions are unlikely to be practical on a large scale. Potato virus diseases are mainly controlled by separating crops intended to supply "seed" from ordinary crops and raising them in districts where aphids are least common, the ware-grower regularly obtaining new "seed" from these districts. At high latitudes suitable regions are usually found in the cooler, wetter and more windy districts; in Britain for example, seed-growing is practised in the north and west, and, long before the existence of virus diseases was recognised, it was appreciated that, in general, seed potatoes from those parts yielded better than those from the south-east. There are probably many reasons for the relative scarcity of aphids in these regions, but the one that has been stressed is that spring and summer weather are not conducive to aphid flight. DAVIES (1935, 1936) reported that *Myzus persicae,* the main vector of leaf roll and rugose mosaic, does not fly readily when the temperature is below 18°C., the wind-speed is more than 8 m.p.h., the relative humidity is above 70, and in the absence of sunshine. Conditions favouring migration certainly occur less frequently in the north-west than in the south-east of Britain, but an important factor contributing to the health of potato stocks in the north-west is the scarcity of overwintering hosts for the aphid. The normal winter host is the peach, on which *M. persicae* survives as eggs, but it can also survive as viviparous individuals on many different herbaceous plants, garden species as well as brassica and sugar beet seed crops. Its ability to survive in this manner will depend on the severity of the winter, and a mild winter is often followed by a year in which virus diseases of the potato, and of other hosts of *M. persicae,* are more prevalent than usual. After a cold winter, few viviparous individuals survive except in glasshouses, sheltered gardens, or on stored mangolds (BROADBENT and HULL, 1947), and these sources, together with those that develop from eggs on peach trees, will supply the spring migrants. There are more of all these sources in the south-east, where spring also comes earlier than in the north-west, so that the potato crops are invaded by more winged aphids early in the season, which probably accounts for the failure of roguing and early-lifting to safeguard the stocks as it does in the seed-growing districts; in the ware-growing districts there are also more infected plants (self-sets and others) from which the migrating aphids can become infective.

Except in years unusually conducive to aphids, field inspections to determine the incidence of leaf roll and rugose mosaic suffice, in the seed-growing districts of Britain, to indicate the suitability of a stock for seed. When conditions are more favourable for aphids, field inspections may be inadequate, for symptoms, especially of leaf roll, often do not show in the current season and many infected plants may pass undetected. In such conditions some check on the health of the stock is needed to ensure its suitability for seed. In some American certification schemes, the stock is tested by the tuber-index method (DYKSTRA, 1941), whereby a sample of the stock is grown during the winter either under glass or in the Southern States. So far the production of seed potatoes has been largely confined to cool regions, but hot, dry weather also inhibits the development of *M. persicae* (PORTER, 1935; NORRIS and BALD, 1943), and VAN DER PLANK (1944) has reported that satisfactory seed can be produced in South Africa when the mean daily temperature is about 32°C.

A possible method of controlling insect-borne viruses that has received little attention is the breeding of varieties resistant, or unattractive, to the vector. ADAMS (1946) found that different potato varieties differ greatly in their susceptibility to aphis attack, but it is noticeable that susceptibility to aphids was in no way correlated with susceptibility to viruses. On the contrary, Katahdin, which is among the most resistant varieties to rugose mosaic and leaf roll, was more heavily infested than any other variety. The combination of the virus-resistance of Katahdin with the aphid-resistance of other varieties would seem well worth seeking.

**Viruses Transmitted Mechanically:** — There are, no doubt, many different viruses that come into this category, but two exceedingly common ones, potato X and tobacco mosaic, will serve adequately as examples. There is no evidence that insects play any part in spreading either of these in the field, and spread seems to occur solely because they are so readily transmitted by sap-inoculation that healthy plants can become infected if they are brought into contact with infected ones or with fragments or residues from infected plants. The two viruses raise rather different problems, for virus X is mainly important in vegetatively-propagated crops and tobacco mosaic in annual crops raised from seed, and they have different properties that affect their survival.

Virus X is less stable than tobacco mosaic virus and probably survives from year to year only in viable infected tissues, of which by far the most important are potato tubers. It occurs to some extent in annual crops like the tomato, in which it spreads more rapidly than in the potato (ROBERTS, 1948), but rarely causes any serious losses. Its economic importance depends on its great prevalence in potatoes, for though the common strains cause slight reductions in yield compared with leaf roll and rugose mosaic, commercial stocks of most varieties are 100% infected. This has come about because the field certification schemes that have successfully controlled leaf roll and rugose mosaic are largely ineffective in freeing stocks from virus X. Such schemes are based on roguing and depend on the ease with which infected plants can be recognised and removed; they have been effective in reducing the numbers of plants infected with virulent strains of virus X, which has reduced the losses that otherwise might be caused, but have allowed strains avirulent towards a variety to spread unchecked and to permeate the whole seed stocks. Strains that produce no obvious symptoms in field crops may yet reduce yield by 10% or more (BALD and NORRIS, 1940; SCOTT, 1941; BALD, 1943, 1944; SMITH and MARKHAM, 1945; BAWDEN, KASSANIS and ROBERTS, 1948), so that the use of such stocks annually cause a great aggregate loss. Their use also complicates the operation of field inspection schemes, for virulent strains often segregate and cause symptoms in some plants; also, strains that cause mottles in some conditions do not in others, so that a stock may seem healthy at one inspection but not at another.

To prevent the losses caused by the widespread occurrence of virus X and to increase the reliability of field inspections, the use of new seed stocks is essential. These could be either new varieties preferably immune from or hypersensitive to virus X, or existing ones propagated from specially tested plants known to be, not only healthy in appearance, but virus-free. Both these methods are being pursued and virus-free lines of many varieties have now been found and are being multiplied (BALD, 1944; SMITH and MARKHAM, 1945). The value of the latter will depend on how readily they can be maintained virus-free under field conditions. Virus X can spread from infected to healthy plants by leaf contact (LOUGHNANE and MURPHY, 1938) or by contact below ground (ROBERTS, 1946), but spread is slow, less than 10% of the plants adjacent to infected ones usually becoming infected in one season. Provided such stocks are kept from contact with other potatoes, they should remain uncontaminated long enough to have a useful commercial life, for CLINCH, LOUGHNANE and MURPHY (1938) found that such stocks in isolation remained virus-free for six years. A prime necessity will be planting on ground where potatoes have not been taken for several years, for although there is no evidence that virus X is soil-borne, groundkeepers can survive for many years and these probably constitute the greatest danger as sources of infection. The stocks will need separating from other potatoes, not only during the growing season, but also during storage, for virus X is transmitted by contact between sprouts of infected and uninfected tubers (MCINTOSH, 1944). When cut seed-pieces are used, care will also be needed that the virus is not introduced on the cutting knife. The ease with which this happens may depend on the potato variety and virus strain. BAWDEN, KASSANIS and ROBERTS (1948) found that infection occurred only when a contaminated knife passed through a sprout, whereas MAI (1947) found that 10% or more of the seed-pieces cut with a contaminated knife produced infected plants even when the sprouts were avoided.

Tobacco mosaic viruses reach higher concentrations in infected tobacco and tomato plants than virus X reaches in the potato, and spread between plants in contact, or by inoculation with contaminated hands and tools, occurs correspondingly more easily. Only a few plants need to be infected initially from external sources because plant-to-plant transfer can soon infect a large proportion of the crop. These external sources are largely products from infected plants, either residues from previous crops or processed tobacco, though there are always additional sources in infected viable plants. Control depends on reducing these sources of infection and so reducing the chances of introducing the virus, together with modifications of cultural practices to reduce the chances of spread.

One common source of virus is from the hands of workers who smoke or chew tobacco. VALLEAU and JOHNSON (1937) and BERKELEY (1942) have shown that many plants become infected, especially during transplanting or weeding of the seedbeds, unless precautions are taken to ensure that workers' hands are not contaminated. Barn-cured tobacco is particularly likely to be contaminated, but virus is also often present in commercially prepared tobacco. While handling tobacco or tomato plants, workers should not use tobacco, and before starting work they should wash their hands thoroughly with soap and water.

Another important source of tobacco mosaic virus is contaminated plant debris, roots or haulm, remaining in the soil from previous crops (JOHNSON, 1937). The virus can remain active in such debris for years and seems to infect plants less through the roots than through abrasions in the stems and lower leaves produced during cultural operations. Few plants become infected in contaminated soil if left undisturbed, but such operations as weeding lead to many infections. BERKELEY (1942) has emphasized the need for efficient soil sterilisation when a permanent seed-bed is used and of taking every other precaution to prevent the introduction of virus into the seed-bed; he found that, because of the handling entailed during transplanting, only a few infected plants in the seed-bed produced a field crop with 75% plants infected. On a field scale, the elimination of soil-borne virus presents greater difficulties, but the residues from infected plants should not be ploughed in or used as manures for susceptible crops. JOHNSON (1937) found that growing non-susceptible crops, and cultural operations that expose the debris to weathering, freed the soil of the virus.

The loss caused by tobacco mosaic virus depends greatly on the time at which plants become infected, and every precaution should be taken to safeguard seedlings. When the disease occurs in a standing crop, it may be worth roguing when only a few individuals are affected, but not otherwise. Cultural operations should, if possible, always be carried out when plants are dry, for BERKELEY (1942) found that more than four times as many plants became infected during cultivations when crops were wet with rain or dew as when they were dry. When tobacco plants are being topped, or tomato plants tied or dis-budded, plants known to be infected should be treated last, their neighbours immediately before them, and healthy plants furthest from them treated first. VAN DER WEIJ (1940) states that dipping hands and implements periodically in solutions of $Na_3PO_4$ reduces the amount of spread, but such a strong alkali is likely to be harmful to the skin. VAN SCHREVEN (1941) states that 0.5% solutions of commercial tanning substances are as effective as the phosphate, and some of the many substances that are known to act as virus-inhibitors, such as whey or extracts of *Phytolacca* sp., might also have similar effects while being otherwise innocuous.

*References:-*

ADAMS, J. H. (1946): Amer. Pot. J. *23*, 1.
BALD, J. G. (1937): Council Sci. Ind. Res. (Australia), Bull. 106.
—— (1943): *ibid.* Bull. 165.
—— (1944): J. Counc. Sci. Indus. Res. Austral. *17*, 258.
—— and NORRIS, D. O. (1940): *ibid. 13*, 252.
——, —— and HELSON, G. A. H. (1946): Coun. Sci. Ind. Res. (Australia), Bull. 196.
BAWDEN, F. C. and KASSANIS, B. (1946): Ann appl. Biol. *33*, 46.
——, —— and ROBERTS, F. M. (1948): *ibid. 35*, 250.
BELL, A. F. (1933): Queensland Agr. J. *40*, 460.
BERKELEY, G. H. (1942): Sci. Agric. *22*, 465.
BLACK, W. (1939): Ann. Rep. Scot. Soc. Plant Breeding *5*.
BOUGHEY, A. S. (1947): Imp. Mycol. Institute Mycol. Paper No. 22.
BROADBENT, L. and HULL, R. (1947): Agriculture *54*, 319.
BRONSON, T. E., SMITH, F. F., SHANDS, W. A. and SIMPSON, G. W. (1946): Maine Agr. Exp. Sta. Bull. 442.
CLAYTON, E. E. and MCKINNEY, H. H. (1941): Phytopath. *31*, 1140.

——, SMITH, H. H. and FOSTER, H. H. (1938) : *ibid.* **28**, 286.
CLINCH, P., LOUGHNANE, J. B. and MURPHY, P. A. (1938) : Sci. Proc. R. Dublin Soc. **22**, 17.
COCKERHAM, G. (1943) : Ann. appl. Biol. **30**, 80.
—— (1943*a*) : *ibid.* **30**, 338.
DAVIES, W. M. (1935) : *ibid.* **22**, 106.
—— (1936) : *ibid.* **23**, 401.
DONCASTER, J. P. and GREGORY, P. H. (1948) : Agric. Res. Council Report Series No. 7.
DYKSTRA, T. P. (1941) : U.S.D.A. Farmer's Bull. 1881.
EDMUNDSON, W. C. (1940) : U.S. Dept. Agric. Bull. 1843.
GERSTEL, D. U. (1948) : J. agric. Res. **76**, 219.
GIDDINGS, N. J. (1946) : Proc. IV General Meeting Amer. Soc. Sugarbeet Technologists, 405.
HILDEBRAND, E. M. (1941) : Contrib. Boyce Thompson Inst. **11**, 485.
HOLMES, F. O. (1934) : Phytopath. **24**, 845.
—— (1938) : *ibid.* **28**, 553.
—— (1943) : *ibid.* **33**, 691.
—— (1948) : *ibid.* **38**, 314.
—— (1948*a*) : *ibid.* **38**, 467.
HOUTMAN, P. W. (1925) : Arch. Suikerind. Nederlandsch-Indië **33**, 631.
HULL, R. and WATSON, M. A. (1945) : Agriculture **52**, 66.
HUTCHINS, L. M. and LA RUE, J. L. (1939) : Phytopath. **29**, 12.
HUTTON, E. M. (1945) : J. Council Sci. Ind. Res. (Austral.) **18**, 219.
—— (1948) : Aust. J. Sci. Res. B, **1**, 416.
—— and BALD, J. G. (1945) : J. Council Sci. Ind. Res. (Austral.) **18**, 48.
JOHNSON, J. (1937) : J. agr. Res. **54**, 239.
JONES, L. K. and VINCENT, C. L. (1937) : *ibid.* **55**, 69.
KASSANIS, B. (1949) : Nature **164**, 881.
KEYWORTH, W. G. (1947) : Annual Rep. East Malling Research Station, 1946.
KUNKEL, L. O. (1935) : Phytopath. **25**, 24.
—— (1936*a*) : *ibid.* **26**, 809.
—— (1936*b*) : Amer. J. Bot. **23**, 683.
—— (1941) : *ibid.* **28**, 761.
—— (1943) : Proc. Amer. Philos. Soc. **86**, 470.
—— (1945) : Phytopath. **36**, 805.
LARSON, R. H. (1945) : J. agric. Res. **71**, 441.
LOUGHNANE, J. B. and MURPHY, P. A. (1938) : Sci. Proc. R. Dublin Soc. **22**, 1.
MAI, W. F. (1947) : Amer. Pot. J. **24**, 341.
MARTIN, J. P. (1930) : Hawaii Planter's Record **34**, 375.
MCINTOSH, T. P. (1944) : Gardners' Chron. **116**, 87.
MCKINNEY, H. H. (1939) : Phytopath. **29**, 16.
MOORE, E. S. (1941) : Nature **147**, 480.
MURPHY, A. M. (1946) : Proc. IV General Meeting Amer. Soc. Sugarbeet Technologists, 408.
NOLLA, J. A. B. and ROQUE, A. (1933) : J. Dept. Agr. P. R. **19**, 301.
NORRIS, D. O. and BALD, J. G. (1943) : Austral. Coun. Sci. Ind. Res. Bull. 163.
PLANK, J. E. VAN DER (1944) : Nature **153**, 589.
—— (1946) : Trans. Roy. Soc. S. Africa **31**, 269.
—— (1947) : Ann. appl. Biol. **34**, 376.
—— (1948) : *ibid.* **35**, 45.
PLANK, J. E. VAN DER and ANDERSSEN, E. E. (1945) : Sci. Bull. Dep. Agric. S. Africa No. 240.
PORTER, D. R. (1935) : Univ. Calif. Agr. Exp. Sta. Bull. 587.
ROBERTS, F. M. (1946) : Nature **158**, 663.
—— (1948) : Ann. appl. Biol. **35**, 266.
SALAMAN, R. N. (1938) : Phil. Trans. Soc. B. **229**, 137.
SCHREVEN, D. A. VAN (1941) : Natuurwet. Tijdschr. Ned. Ind. **101**, 113.
SCHULTZ, E. S., CLARK, C. F., RALEIGH, W. P., STEVENSON, F. J., BONDE, R. and BEAUMONT, J. H. (1937) : Phytopath. **27**, 109.
——, CLARK, C. F. and STEVENSON, F. J. (1940) : *ibid.* **30**, 944.
——, STEVENSON, F. J. and AKELEY, R. V. (1947) : Amer. Pot. J. **24**, 413.
SCOTT, R. J. (1941) : Scottish J. Agr. **23**, 98.
SMITH, K. M. and MARKHAM, R. (1945) : Nature **155**, 38.
STEVENSON, F. J., FOLSOM, D. and DYKSTRA, T. P. (1943) : Amer. Pot. J. **20**, 1.
STODDARD, E. M. (1942) : Phytopath. **32**, 17.
STOREY, H. H. (1935) : East Afr. Agric. J. **1**, 206.
SUKHOV, K. S. and PETLYUK, P. T. (1940) : C. R. Acad. Sci. U.S.S.R. **26**, 483.
SUMMERS, E. M., BRANDES, E. W. and RANDS, R. D. (1948) : U. S. Dept. Agric. Tech. Bull. No. 955.
TOWNSEND, G. R. (1948) : Plant Disease Reporter **31**, 118.
VALLEAU, W. D. (1946) : Phytopath. **36**, 412.

—— and JOHNSON, E. M. (1937): Kentucky Agr. Exp. Sta. Bull. 46.
WALLACE, J. M. and MURPHY, A. M. (1938): U. S. Dept. Agric. Tech. Bull. No. 624
WATSON, M. A. (1937): Ann. appl. Biol. *24*, 557.
—— (1942): *ibid.* *29*, 358.
——, WATSON, D. J. and HULL, R. (1946): J. Agric. Sci. *36*, 151.
WEIJ, H. G. VAN DER (1940): Meded. Deli Proefstation Medan *3*, 6.
WHITEHEAD, T. (1927): Welsh J. Agric. *3*, 169.
WILBRINK, G. (1923): Arch. Suikerind. Nederlandsch-Indië *31*, 1.

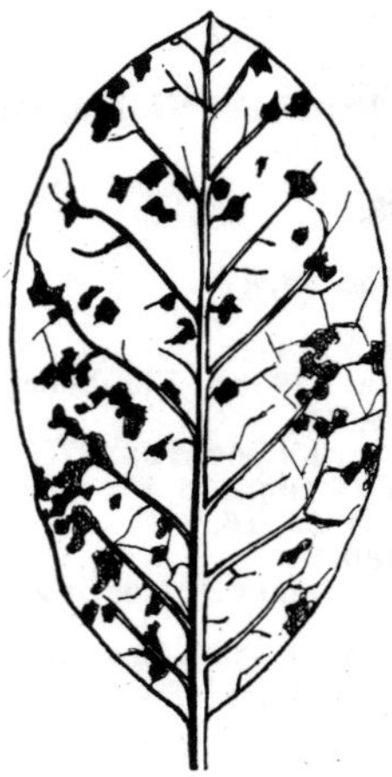

*Chapter 17*

# SPECULATIONS ON THE ORIGINS OF VIRUSES

Two questions, *Are viruses living?* and *Where do they come from?*, are probably put to virus workers more often than any others, and they have continually raised controversies since the existence of viruses was first recognised. That they are unanswerable is implicit in the fact that they continue to be asked and is wholly unaffected by the fact that answers have often been provided in no uncertain manner. Although such answers may satisfy their propounders, they do little else, because, as often as not, those provided by different workers are mutually contradictory. These disagreements themselves show that the questions are not susceptible to direct experimentation and that answers depend on personal opinions rather than observable facts. The questions provide endless opportunities for speculation, some valuable because it stimulates investigation, but much that can be only sterile and inconclusive. Particularly is this true of the first, for the second may become answerable as future work reveals biological units that resemble viruses in some way or other. Answering the second will not necessarily also answer the first, for the difficulties with it arise less from ignorance about viruses than from uncertainties intrinsic to the question itself.

PIRIE (1937) has pointed out that the words "life" and "living" have been introduced into science from every-day use without being rigidly defined. Consequently, they lack the precision needed to decide unequivocally whether or not certain systems are "alive". Everyone appreciates the broad manifestations of life, and when applied to things as different as dogs and kennels, or mosses and stones, the words create no problems, but there is no single criterion whereby systems less obviously living or lifeless can be identified as one or the other. A dog that breathes and moves will be accepted unquestionably as alive, but it is less easy to decide how many of the numerous components that make up the whole dog should individually be considered as living. This is precisely the difficulty with viruses, which appear manifestly living only when functioning as units in other living systems.

Most of the phenomena customarily associated with life, such as multiplication, growth, variability, respiration and movement, can be simulated by systems that would not generally be considered as alive, and there are many intracellular units on whose status no decision can be reached. A rigid definition of life, which is as likely to emerge from work on viruses as from else-where, will enable such units, and viruses, to be called living or non-living, but may achieve little else, because it will be based on arbitrary criteria largely selected for this purpose. The use of such arbitrary criteria is already evident from the fact that phenomena accepted as conclusive with bacteria would be quite unacceptable with higher organisms. HERRIOT and PRICE (1948), for example, found that bacteria subjected to mustard gas would not multiply but were still susceptible to bacteriophage. They advance this as evidence that viruses can multiply in dead cells, but no-one would make a similar claim because a castrated dog developed distemper, though, by the same criterion, the dog is as dead as the bacteria.

Whether workers conclude that viruses are living depends, not only on their personal opinions of the most important manifestations of life, but also on the viruses they have studied and the type of work they have done. The last is particularly important, because the predominant characters of viruses vary greatly depending on their environment. When studied exclusively as pathogens, there is little in their behaviour to suggest that they differ in any essential way from pathogenic organisms; they raise the same epidemiological problems and in susceptible hosts they multiply and occasionally give rise to new variants. On the other hand, *in vitro* some at least behave much more like substances than as organisms and they are more amenable to study by the techniques of the protein chemist than by those of the bacteriologist; their small size, chemical simplicity, regular structure and apparent lack of any independent metabolism, all seem to separate them sharply from even the smallest bacteria. This dual set of characters is vividly illustrated by the contrasting views of those who call viruses organisms and those who call them molecules.

A prime purpose of any speculations on the origins of viruses must be to reconcile these contrasting properties. The difficulty is not avoided by assuming that the agents now grouped together as viruses are a mixed collection, some being organisms and others not, because this raises more problems than it solves. Such an assumption ignores the many characters shared by all viruses and the fact that, even in size, the character that seems most variable, there is a steady graduation from small to large viruses; it therefore only poses the question at which stage one type passes into another and raises the need to postulate widely different evolutionary processes for different viruses. It is simpler to assume some degree of uniformity through the group and to attempt the reconciliation along other lines. This is less difficult than it may seem, for although the gap between organisms and molecules may be wide and unbridgeable, the contrast as introduced into virus research is largely false. It is based not on facts, but on conclusions arrived at prematurely and largely on irrelevant data. Phenomena such as apparent ability to reproduce and vary, no more prove that viruses are organisms than the apparent homogeneity and inertness of purified preparations prove that they are molecules. To place viruses into either category is to pre-judge what is still an open question, and to do so against the bulk of the evidence. The word organism is best restricted to systems which obviously have independent biological activities separable from those of other organisms. The word molecule is best restricted to simple substances whose structure can be determined, because it implies a precise reproducibility of structure that cannot be demonstrated in large particles. As Pirie (1948) has stated, the word is not only inappropriate for viruses, but for all biological materials with particles so large that slight chemical changes are undetectable because they involve too small a portion of the whole particle. Between unequivocal molecules on the one hand and organisms on the other, there occur in living cells many materials with a range of particle sizes that overlaps viruses and which may also resemble them in other ways. To equate viruses with organisms or with molecules is to ignore both the existence of these particles and the meagreness of our knowledge on their constitution and properties.

As intracellular pathogens, viruses are, in fact, cellular constituents and their individuality becomes merged into that of the infected cell. Their biological activities show only through changes in the customary activities of other organisms, and there is no method of differentiating between effects brought about specifically and solely by the virus particles themselves and those that result from the combined activities of all the constituents of the infected cell. From studying viruses *in vivo,* their biological properties are easily over-estimated and activities may well be attributed to them which should more properly be regarded as those of the whole cell, in which the virus is only one determining component. If studies *in vivo* do this, it is equally certain that studies of purified viruses *in vitro* tend to over-simplify the picture of a virus. Preparations of viruses can be obtained that differ in no obvious manner from preparations of other proteins, and from behaviour *in vitro* it is impossible to tell whether a protein is capable of acting as a virus or not. This becomes evident only when a susceptible host is infected, and to make the distinction entails the use of the complex mechanism of living cells. The infective nucleoproteins of purified virus preparations are probably the bare minima necessary to divert the metabolism of susceptible cells into new channels, and they need bear no closer resemblance to the final dynamic system they stimulate than a fertilised ovum does to the organism into which it ultimately develops. To attribute any independent existence or activities to viruses may be a misconception; outside their normal habitat of living cells they are simply substances with the potentialities of becoming functional only when returned to this habitat. The purified virus may be only a part of a larger active system, and in studying the part it is important not to overlook the whole. To treat the purified viruses as complete and self-sufficient systems, or to differentiate between their biological activities and those of the total components of an infected cell, is to go further than present knowledge allows.

It is worth remembering that the change from a potential to an actual virus may to a large extent undo what has been laboriously achieved in the laboratory during the purification procedures. A first and essential step for the purified particles may be combination with other cell constituents; resistance or susceptibility in a host may depend on the presence of suitable components and the various effects produced by one virus in different hosts may result from combination with different ones. If this is so, and cells certainly contain a variety of substances with which viruses could combine, the chemical simplicity of viruses suggested by studies on purified preparations is largely illusory, a temporary and artificial status which is lost before the particles start to function as viruses. Such combinations may also explain the fact that purified preparations of some animal viruses appear to be chemically more complex than those of the purified plant viruses. The differences may mean merely that, with the plant viruses, other constituents of the infected cells are more readily separable from the minimal reproductive parts than with the animal viruses. The only activity of viruses that is customarily measured is infectivity, and the removal of material inessential for this is regarded as purification, whereas material that cannot be removed without reducing infectivity is regarded as an intrinsic part of the virus particle.

The limitations of current techniques make this approach a logical necessity, but it is by no means necessarily valid. Although it may be essential to the functioning of a virus, any material that can be obtained from a host cell might be removable without necessarily affecting infectivity, for it could be re-acquired when the particle entered another cell. On the other hand, material completely irrelevant to the functioning of a virus would seem essential were it so tightly bound to the reproductive parts that its removal damaged these parts. The most that studies of purified virus preparations can do is to set certain chemical minima below which particles cannot go and still be infective. The uncertainties in interpreting results and the difficulties entailed in distinguishing between what is "virus" and what is host is illustrated by work with vaccinia and influenza. The soluble antigens of vaccinia (HUGHES, PARKER and RIVERS, 1935; SALAMAN, 1937) are specific to infected cells but are inessential for infectivity, whereas purified preparations of influenza virus contain substances which also occur in uninfected egg or mouse lung, depending on whether the virus is propagated in one or the other, and which have not yet been removed from the particles without causing inactivation (KNIGHT, 1946). Thus the apparent differences in complexity between different viruses may to a large extent reflect the fact that their basic particles adhere with varying degrees of tenacity to other constituents of host cells, rather than that their basic particles differ widely. Apparently these other constituents may be peculiar to infected cells or common to both infected and uninfected cells, a fact emphasising the intimate manner in which viruses merge into the general metabolism of the host cells and the impossibility of differentiating clearly between what is a product of virus activity and what is a product of the host.

To account for variations in size and chemical complexity of different viruses, the idea is often expressed that viruses represent a transition stage between substances and organisms, some writers considering that they represent an upward stage in evolution, being descendants of primitive pre-cellular forms of "life", and others considering that they have evolved by a process of retrograde evolution from larger organisms. Until saprophytic agents resembling viruses have been shown to exist, the first view is barely worth considering, for obligate parasitism, which appears to be one of the chief characters of viruses, is hardly a property that would have favoured the continuation of the first forms of "life". The second is more plausible and has been ably propounded by GREEN (1935), LAIDLAW (1938) and BURNET (1945). It is known that saprophytic organisms sometimes lose the ability to synthesise certain essential growth requirements and so cannot multiply unless these are supplied pre-formed. Extrapolating from this fact, it is argued that viruses have evolved from larger organisms as a result of losing synthetic powers that confer no survival value in an intracellular environment which provides all necessary requirements ready-made. Existing viruses thus would represent descendants of pathogenic organisms at various stages of degeneration, the larger ones having lost fewer synthetic powers than the smaller. The strongest argument advanced in support of this theory is the existence of what appears to be an almost continuous range of forms between unequivocal bacteria and the viruses that have been isolated as nucleoproteins. However, an alternative explanation for this, namely, combination with different degrees of firmness of certain basic particles with various host constituents, has already been advanced in the preceding paragraph, so that it is by no means conclusive. Also, the assumption that inability to synthesise all essential food necessarily indicates a retrograde evolution and is accompanied by reduction in size raises wide implications, for it would be necessary to consider that all organisms except autotrophic bacteria and green plants, have evolved in this manner from larger organisms.

Another objection to this theory of the evolution of viruses is that it is unnecessarily restrictive, for there would seem no reason to assume that all viruses need have evolved from pathogenic organisms. The concept of the infected host cell as the functioning entity, in which the virus particles are simply one component, not only reconciles many of the seemingly contrasting properties of viruses, but also provides a wide choice for the development of viruses from cell components, without the need to postulate a long and complicated evolution from micro-organisms by degenerative processes involving a progressive loss of form and function. When the differences between viruses and other cell constituents are examined, they are seen to be slight, merely that viruses sometimes cause conditions that we recognise as diseases and that viruses multiply in cells that are normally free from them. Because viruses appear to act as self-reproducing units, they are often regarded as differing fundamentally from most other cell components. But chromosomes and determinants of cytoplasmic inheritance equally appear to be self-reproducing, and we do not know that viruses differ from normal proteins or other cell components in their method of multiplication. Indeed, there is no method of telling whether any normal cell component reproduces itself or is produced by something else, and attempts to make the distinction take us from science to philosophy. There is little new in the idea of viruses being merely determinant parts of host cells. BORDET (1931) and others have suggested that viruses are metabolic products of their hosts that do not reproduce directly but modify the behaviour of the cells so that these produce virus instead of, or in addition to, their usual metabolic products. And the concept of viruses as cell constituents is implicit in the "free gene" theory advanced by MULLER (1922) and DUGGAR and ARMSTRONG (1923), who pointed out several analogies between genes and viruses and suggested that viruses might be genes that had escaped from the nucleus and were acting independently of it. Like other theories for the origin of viruses, this cannot be disproved, but there is some evidence against it. There is much to suggest that viruses may contain several characters that are inherited independently (HOLMES, 1936; NORVAL, 1938; SALAMAN, 1938), so that it would seem more appropriate to compare viruses with groups of genes or fragments of chromosomes rather than with single genes, which may be chemically much simpler than the chromatin base on which they are carried. It is not unlikely that viruses and chromatin reproduce in a similar manner and both may produce their effects through similar active groups, but the chemical evidence does not support the idea that viruses are nuclear fragments, for the nucleic acid characteristic of nuclei is desoxypentose, whereas all the plant viruses yet isolated contain ribose nucleic acid. This theory, too, seems unnecessarily restrictive, as also do those which postulate viruses as originating from specified cytoplasmic units such as plasmagenes (DARLINGTON, 1944) or by loss of form and function from plastids and mitochondria (DUBUY and WOODS, 1943). Too little is known about any of these bodies for one rather than another to be chosen as a likely source, and there seems no reason why different viruses should not have originated from different cell constituents.

In previous editions of this book it was pointed out that, as viruses in their simplest forms can be obtained as nucleoproteins, a theory of their origin as probable as any other is that they arise during the protein metabolism of organisms, and it was suggested that viruses which cause diseases in some hosts may be natural constituents of others. There was no positive evidence to support this, but it was suggested that the most plausible explanation of paracrinkle virus (SALAMAN and LE PELLEY, 1930), which has been found in every plant of the potato variety King Edward examined but not elsewhere, was that it is a normal product of the metabolism of King Edward with the characters of a virus when transferred to other potato varieties. This virus seems to have no natural method of spread and has been transmitted to other plants only by the artificial procedure of grafting them with scions of King Edward.* It does not occur in the seedlings that have been raised from King Edward (CARSON *et al.*, 1944) and its existence at the present time would appear to depend solely on the fact that its original host is propagated vegetatively. There is no reason to assume that paracrinkle virus and King Edward is unique. The circumstance that makes it exceptional is that the host has been cultivated on a large scale apparently without the virus spreading to other plants, which provides circumstantial evidence that the virus did not infect the original King Edward plant from outside sources, but originated within it as an consequence of the genetical constitution of the host. Such occurrences are probably infrequent, but may be less rare than they appear because most would be untraceable. A virus (or potential virus) arising in a host in which it causes no symptoms would normally remain undetected and its existence would become apparent only if spread occurred to a host that shows symptoms. A virus originating in a host in which it causes symptoms is also unlikely to be noticed unless it has spread to other plants, by which time the original source becomes obscured. Viruses that cause lethal diseases in their originating hosts would have little chance of survival unless they also have exceptional means of spread. Many viruses or potential viruses have probably arisen only to disappear with the death of their host. Only those that find some natural method of spread, such as a vector, so that they invade fresh hosts before the death of their original ones, or which arise either in perennial plants, or in those propagated vegetatively, will be able to survive and become sufficiently widespread for their existence to attract attention.

---

*Paracrinkle virus has now been transmitted mechanically. Tomato plants readily become infected when inoculated with sap from King Edward potato plants, but they show no symptoms (BAWDEN, F. C., KASSANIS, B., and NIXON, H. L. 1950: J. gen. Microbiol., in the press). This does not disprove the hypothesis that the virus is intrinsic to King Edward, but does make it less plausible. If there are many hosts that behave like tomato, natural spread could occur but would pass unnoticed, and the virus may have been originally introduced into King Edward from such a host.

Once it is allowed that viruses resemble products of cellular metabolism more than they do organisms, their origin ceases to raise any great evolutionary problems and their synthesis by organisms can be suggested without raising the discredited bogey of spontaneous generation of life. Viruses may be merely changed forms of normal cell constituents, or, what is equally possible, normal, perhaps even essential, components of one organism may act as viruses in others. Some may have arisen in species in which they now cause diseases, whereas others may have come from different ones or even quite different types of organism. New virus diseases of plants are being described with an almost frightening frequency; some of these have no doubt occurred for long but have gone unrecognised, but there is reasonably good evidence that some have developed only recently. Nevertheless, the causative viruses may not be new; they may have existed as constituents of other species in which they are not pathogenic, but they may have been transmitted only recently to plants in which their pathogenicity allows their recognition. When one reflects on the fact that whenever cellular constituents from one organism are introduced into another, there is the possibility that something may be introduced that can behave as a virus, the occurrence of new virus diseases ceases to be surprising. The great movement of plants to and from all parts of the world in modern times has afforded abundant opportunities for such potential viruses to spread to susceptible hosts. Insects have been provided with many new plants on which to feed, and the practice of grafting has also favoured the intermingling of constituents from different plants. The extension of plant breeding also may play a part in producing viruses, for new combinations of genes may lead to the synthesis of particles that act as viruses in the parent plant, in other species, or in both.

The example of paracrinkle virus from King Edward potatoes shows that one plant may always contain something that acts as a pathogen in another. That this may not be exceptional is suggested by BENNETT's (1944) discovery of dodder latent mosaic virus, contained in apparently normal *Cuscuta californica* and capable of causing serious diseases in sugar beet and other plants parasitised by it. These diseases are unknown in field crops, and dodder latent mosaic virus, like paracrinkle, seems as yet a harmless constituent of its host, which will remain so until it finds an effective method of transmission when it may become an economically-important virus. Indeed the ability to spread, or to be actively infective, is the one character that clearly distinguishes viruses from other cell constituents. Many viruses affecting flowering plants have probably originated in them, but there is no reason to assume that all have. Pathogenic fungi and bacteria enter into such intimate contact with their hosts that there must be many opportunities for interchange of cellular components, from which virus diseases may have developed. No conclusive examples can be given, but the production of bacteria-free galls, on parts of plants remote from those actually inoculated with *Bacterium tumefaciens* (WHITE and BRAUN, 1942), is as plausibly explained on the theory that the bacterium contains something that acts like a virus in the higher plant, as on any other.

Before an agent is recognisable as a virus it must be transmitted to an organism in which it causes symptoms, but the causing of disease may be incidental to other activities and with little permanent significance in the evolution of the agent. By extending their habitats, spread to further organisms may temporarily increase the number of viruses, but they are ill-equipped to survive except in living cells, and virulence is a character as fatal to a virus as to the host it kills. Viruses can persist in intolerant hosts only by having some efficient method of transmission that continually transfers them to fresh individuals, and their continued survival is more assured in hosts in which they are harmless or beneficial. Excessive virulence towards a host probably means a newly acquired habitat, or that the virus itself has changed to a form that is unlikely to survive. The serious disease-outbreaks by which viruses attract attention may be passing events with a more lasting effect on the development of the organisms they attack than on their own evolution. Some degree of tolerance in a host is essential if a virus is to be long associated with it, and it is probably to tolerant organisms that we should look for the origins of our current pathogens. One possible source for plant viruses would be the insect pests that act as vectors, for these seem to be completely tolerant, but it is not an origin worth serious consideration until better evidence is forthcoming that the vectors can support multiplication of the viruses they transmit. Responsibility for initiating new diseases may be theirs, even though they do not supply the causes, for they must provide one of the commonest methods whereby the substance of one plant gets introduced into another.

One of the most cogent reasons advanced in favour of identifying the specific nucleoproteins isolated from infected plants with the viruses themselves was that nothing similar could be isolated by similar methods from uninfected plants of the same species. At first sight this fact might seem to exclude the possibility that viruses may have arisen from normal plant constituents, but it really has little value in this context. First, only a few species of plants have yet been examined at all critically and the viruses that have been isolated from them may have originated in different species or even quite different kinds of organisms. Secondly, even with the few species that have been studied, the examination has been far from exhaustive; only the sap that comes from macerated leaves has been examined and the results merely show that proteins similar to the purified viruses do not occur in sap from healthy plants to anything like the same extent as they do in sap from infected plants. They may occur, but not be released into the sap, like much of the virus, or they may occur in sap in much smaller quantities, for our techniques are not well adapted for isolating small quantities and would probably fail at concentrations less than 1 mg/l. Also, attempts using methods similar to those successful with tobacco mosaic and tomato bushy stunt virus have failed to isolate anything similar from plants infected with many other viruses. This may be because these other viruses are chemically different, but as the sap containing them has much smaller dilution end-points, they more probably occur at too great a dilution for the methods to be applicable. The possibility of isolating tomato bushy stunt or tobacco necrosis viruses depends on an adequate virus content and methods that work readily with highly infective lots of sap fail with others, though these contain enough virus to cause infection when inoculated to healthy plants. When our techniques are so limited that they fail to isolate anything from extracts known to contain viruses, there is obviously no justification for assuming that their failure to isolate anything from healthy plants proves that these contain nothing similar to viruses. Indeed, it is by no means impossible that some virus diseases result from quantitative changes in some cell constituent rather than from any fundamental qualitative change, for increasing concentration would increase the chance of the constituent moving to other cells or of being transmitted to other plants.

Infectivity is the only character that distinguishes viruses from other cell components and it is the only one for which tests are normally made. It is a character that is readily lost, for there are many treatments that destroy it without producing any gross changes in the serological or physico-chemical properties of the particles, and there is no reason to assume it is the only biological activity of viruses. The loss of infectivity produced by ultra-violet irradiation, for example, does not prevent bacterial viruses from entering bacteria, nor from having some effects within the bacteria. Such inactivated particles of one virus may interfere with the establishment of another (LURIA and DELBRÜCK, 1942) and the entry into one bacterium of two or more inactivated particles of one virus can lead to the reappearance of infective particles (LURIA, 1947). To produce the whole sequence of events by which we recognise infection, it seems that particles need powers to carry out a number of activities which proceed separately, and if a power is lacking from one particle it may be supplied by another. Failure to acquire all the essential activities may account for the presence, in preparations of the Rothamsted culture of tobacco necrosis virus and of tobacco mosaic virus (BAWDEN and PIRIE, 1945, 1945*a*), of non-infective particles with serological and physico-chemical properties similar to those of infective particles. Though non-infective, such particles may not have been devoid of all activities in the cells where they occurred. They fall short of the full stature of virus particles and are distinguishable from other cell constituents and referable to the viruses only from knowledge of *in vitro* properties of the viruses. The structural differences between such infective and non-infective particles are too slight to be ascertained by current techniques, but during the synthesis of large particles there must be opportunities for many structural variations to occur. Such variations in viruses presumably account for the production of new strains and for non-infective particles; by analogy, if variations in viruses can render them non-infective, there would seem no reason why variations in normally non-infective proteins should not render them infective.

As all the viruses that have yet been purified seem to be nucleoproteins or to be largely so, in seeking their origins it is natural to look to normal cell constituents which are also rich in nucleoprotein. Too little is known about this subject for any one source to be selected rather than another, and the viruses that are known to be nucleoproteins differ from one another so widely that there is no reason to assume that they will all have derived from similar sources. Nor may all viruses be nucleoproteins, for anything with the power to stimulate cells so that more of itself is produced could come into the category of viruses. In the definition proposed in Chapter 1, disease production is demanded of a virus, but this is included largely as a matter of convenience. Were changes in type of pneumococci regarded as "disease" symptoms, the factor responsible for the transformation from one type to another would be regarded as a virus, and it is highly probable that this seemingly self-reproducing agent is a nucleic acid and not a nucleoprotein (AVERY, MACLEOD and MCCARTY, 1944). There is no reason to assume that other chemical types may not also "multiply", and it may be that other phenomena now not considered to be connected with viruses, because the effects produced are not "diseases", will be found to have such causes. The behaviour of "florigens", agents of unknown structure that stimulate flowering and are produced in the leaves of some plants when exposed to day-light of an appropriate duration (MURNEEK and WHYTE, 1948), is of interest in this connection. Like viruses, they move from the cells in which they occur initially and they can be transmitted between plants by grafting. Their production shows how sensitive is the leaf metabolism to external influences and that transmissible agents are regularly synthesised by normal cells under suitable conditions. In animals, too, stimuli may call forth unusual cell constituents, for hormone-induced mammary tumours in mice have been shown by electron microscopy to contain particles of about 100 m$\mu$ in diameter that are not present in normal cells (PORTER and THOMPSON, 1948). The connection between such phenomena, and many others that could be quoted, and virus diseases will not become clear without much further work, but potential sources of viruses would seem to exist widely in different kinds of organisms, and no long evolutionary processes need to be invoked to account for the outbreaks of new virus diseases.

By analogy with certain proteolytic enzymes, STANLEY (1936) and NORTHROP (1938) have suggested that viruses are autocatalysts, which reproduce by changing inactive precursors present in susceptible cells into infective particles. For example, the pancreas does not produce trypsin, but an inactive protein, trypsinogen, which changes to trypsin on contact with active trypsin. Similarly, the inert protein, pepsinogen, produced by the gastric mucosa is transformed by pepsin into more pepsin. Such phenomena may be fairly general with inter-cellular enzymes, because some mechanism for their removal from the parent cell in an inactive form would seem necessary if the cell proteins are to escape hydrolysis. This theory obviously fits well with the concept of viruses as changed constituents of normal cells, but it is highly doubtful that viruses are simple autocatalysts of this type. It is difficult to see what benefit organisms could derive from manufacturing virus precursors, but there are other than teleological grounds for thinking the analogy with proteolytic enzymes inappropriate.

Pepsinogen and trypsinogen resemble pepsin and trypsin in many respects and pass into them readily. The transformation is quantitative and there is no reason to assume that the enzymes play any part in producing their precursors. The position is obviously different with viruses; tobacco mosaic virus may account for one-third of the protein of infected tobacco plants, yet nothing resembling the virus has been isolated from healthy plants. If a precursor exists, it must occur in much smaller quantities than the virus attains and as that initially present is transformed into virus more must be produced. In other words, the virus must stimulate the production of its precursor. The inability of viruses to multiply in extracts of susceptible plants, too, contrasts strikingly with the manner in which pepsin and trypsin can increase *in vitro* at the expense of their precursors. Susceptible cells clearly contain within themselves the seeds of their destruction by providing all that is necessary for the synthesis of virus particles, but there is no reason to suppose they keep these particles already largely manufactured. Were this so, it would be necessary to assume that all hosts susceptible to a given virus contain complex particles of an identical type, and the host range of some viruses cover such widely-different species that this is improbable. Also, some plants would need to contain vast numbers of different precursors to account for their susceptibility to so many different viruses. The tobacco plant, for example, is susceptible to so many viruses and so many strains of some viruses, that the imagination rather boggles at the thought of cells containing pre-formed precursors to all of them, and by analogy with the proteolytic enzymes this would be needed. The enzyme produced by a precursor is determined solely by the precursor and not by the enzyme that activates it. Thus trypsin turns chymotrypsinogen into chymotrypsin and not into trypsin, and swine pepsin turns chicken pepsinogen into chicken pepsin and not into swine pepsin. A very different state of affairs holds with viruses, for with these the inoculum alone determines the end result. Many strains of tobacco mosaic virus have been isolated and maintained with reasonably constant properties over many years. Cultures of tobacco mosaic and aucuba mosaic viruses retain their characteristic differences when transferred repeatedly and such constancy seems incompatible with the precursor theory. If this theory were true, each plant would contain precursors to all the strains to which it is susceptible, and any strain should transform them, not into itself, but into their appropriate strains. Thus, whether plants were inoculated with tobacco mosaic or aucuba mosaic virus, the final result should be the same, a mixture of all the strains to which the inoculated plant is susceptible, with each occurring in the proportions in which its precursor existed. That this does not happen is, perhaps, the most cogent reason for rejecting the autocatalytic theory until some evidence for virus precursors is forthcoming. It seems probable that viruses multiply at the expense of normal plant proteins, for despite the large amount of virus produced in tobacco mosaic plants, there is no increase in total protein. However, this does not mean that the virus merely changes some previously-formed large particle into a virus, but rather that normal proteins and viruses can be synthesised from similar small units.

Viruses are so different structurally from organisms that they probably also multiply in a manner distinct from binary fission. The uniformity in size of viruses such as tomato bushy stunt in itself suggests that there is no increase in size followed by fission into smaller units. Binary fission, however, is only the last stage whereby one organism, or one cell, becomes two. Both before and after this stage, there must be increases in the quantities of cell constituents, and the multiplication of viruses is probably more appropriately compared with their increase rather than with the fission of whole cells. Saying this does not carry us far, because little is known about how proteins and other large particles are produced in cells, but it does suggest that virus particles need not be produced as linear descendants of one another, increasing geometrically and each generation partaking of some substance from the previous one. One particle may stimulate a cell to produce many particles simultaneously, all of which may resemble the original but owe none of their substance to it directly.

In normal cells protein synthesis proceeds along specific cells, each type of cell producing certain definite proteins and no others, except perhaps for occasional "accidents" such as those we have suggested may sometimes originate viruses. BERGMANN and NIEMANN (1937) attribute this maintenance of characteristic proteins to the specificity of the intracellular proteinases. They suggest that the enzymes have at their disposal many protein fragments of different sizes and structures, which are subjected to a series of transformations by synthesis, hydrolysis and replacement, so building one structure after another until a protein results that is stable in the presence of the enzymes. As intracellular proteases are probably mostly proteins, this postulates the synthesis of one set of proteins by another, and BERGMANN and NIEMANN point out that a proteinase with the ability to synthesise a replicate of itself would behave in the same way as a virus. There is no evidence that any purified virus possesses protease activity *in vitro*, but this picture of viruses influencing the general direction of cell metabolism so that more virus is produced seems at present the most useful.

Except for the production of "disease" symptoms, little is known for certain about the metabolic effects of plant viruses on their hosts, but much information is being accumulated with bacterial viruses. COHEN (1948) states that when *Escherichia coli* are infected with $T_2$ bacteriophage, the normal synthesis of nucleic acid is modified, so that instead of producing three times as much ribonucleic acid as desoxyribonucleic acid, only the latter, which is the one characteristic of the bacteriophage, is produced. Working with the same bacterium and virus, LATARJET (1948) found that during the first 7 minutes after infection $T_2$ is insusceptible to X-rays and the bacteria synthesise ultra-violet-absorbing material, possibly destined to produce future virus particles. Multiplication of $T_2$ begins 9 minutes after infection and is complete in 4 minutes, during which time about 130 particles are produced. From the thirteenth to the twenty-first minute, when the bacterium bursts, no further multiplication occurs but the particles seem to undergo some changes because their resistance to X-rays alters. These results are not necessarily incompatible with the idea of multiplication by a doubling of existing particles, though they call for a fission every 30 seconds during the period of increase. It is at least equally probable that, after a period during which the components of virus particles are synthesised and accumulate to critical concentrations, these components then come together to form new particles without ever entering into the structure of the originally-infecting one. The infecting particle after diverting the cell metabolism into new channels may later act as a template determining the form in which certain products of this metabolism ultimately associate. An analogy is provided by crystal growth, the beginning and form of which in solutions is often determined by the type of "seed" crystal that is introduced.

Although crystallisation provides a convenient picture for the final formation of a virus particle from its constituent units, it is almost certainly much over-simplified. Symptoms do not result simply from a kind of metabolic "fatigue" of cells because necessary products are diverted into viruses, for virulence does not seem in any way associated with the ability of viruses to reach high concentrations. Tobacco plants are much more severely affected by severe etch virus than by most strains of tobacco mosaic virus, yet the latter occurs in quantities a thousand times as great as the former. Also, with different strains of one virus, those virulent towards a host often reach lower concentrations than others that are relatively innocuous. It is clear that viruses have many activities *in vivo* in addition to self-duplication, some of which appear not only irrelevant but definitely inimical to this fundamental character. Much more work is needed comparing the metabolic products of normal and infected cells before any definite conclusions will be possible on the causes of virulence, how viruses originate and how they multiply. The relationships between viruses and other invisible agents that influence the metabolism of cells affords at the present time one of the most fascinating biological problems. Because of their infectivity, viruses are particularly favourable units for study and it can confidently be expected that advances in knowledge of their behaviour will have repercussions in subjects now considered far remote from pathology.

## *References:-*


AVERY, O. T., MACLEOD, M. C. and MCCARTY, M. (1944) : J. exp. Med. *79*, 137.
BAWDEN, F. C. and PIRIE, N. W. (1945) : Brit. J. exp. Path. *26*, 277.
—— and —— (1945*a*) : *ibid. 26*, 294.
BENNETT, C. W. (1944) : Phytopath. *34*, 77.
BERGMANN, N. and NIEMANN, C. (1937) : Science *86*, 187.
BORDET, J. (1931) : Proc. Roy. Soc. B. *107*, 398.
BURNET, F. M. (1945) : *Virus as Organism.* Harvard University Press, Cambridge, Mass.
CARSON, G. P., HOWARD, H. W., MARKHAM, R. and SMITH, K. M. (1944) : Nature, *154*, 334.
COHEN, S. S. (1948) : J. Biol. Chem. *174*, 281.
DARLINGTON, C. D. (1944) : Nature, *154*, 164.
DUBUY, H. G. and WOODS, M. W. (1943) : Phytopath. *33*, 766.
DUGGAR, B. M. and ARMSTRONG, J. K. (1923) : Ann. Missouri Bot. Gard. *10*, 191.
GREEN, R. G. (1935) : Science *82*, 443.
HERRIOT, R. M. and PRICE, W. H. (1948) : J. gen. Physiol. *32*, 63.
HOLMES, F. O. (1936) : Phytopath. *26*, 896.
HUGHES, T. P., PARKER, R. F. and RIVERS, T. M. (1935) : J. Exp. Med. *62*, 349.
KNIGHT, C. A. (1946) : J. Exp. Med. *83*, 281.
LAIDLAW, P. P. (1938) : *Virus diseases and viruses.* Cambridge.
LATARJET, R. (1948) : J. gen. Physiol. *31*, 529.
LURIA, S. E. (1947) : Proc. Nat. Acad. Sci. *33*, 253.
—— and DELBRÜCK, M. (1942) : Arch Biochem. *1*, 207.
MULLER, H. J. (1922) : Amer. Naturalist *56*, 32.
MURNEEK, A. E. and WHYTE, R. O. (1948) : *Vernalization and Photoperiodism.* Chronica Botanica Co., Waltham, Mass.
NORTHROP, J. H. (1938) : J. Gen. Physiol. *21*, 335.
NORVAL, I. P. (1938) : Phytopath. *28*, 685.
PIRIE, N. W. (1937) : *Perspectives in Biochemistry*, p. 11. Cambridge.
—— (1948) : Medical Bulletin *5*, 329.
PORTER, K. R. and THOMPSON, H. P. (1948) : J. Exp. Med. *88*, 15.
SALAMAN, M. H. (1937) : Brit. J. Exp. Path. *131*, 468.
SALAMAN, R. N. (1938) : Phil. Trans. Roy. Soc. B. *229*, 137.
—— and LE PELLEY, R. H. (1930) : Proc. Roy. Soc. B. *106*, 140.
STANLEY, W. M. (1936) : Phytopath. *26*, 305.
WHITE, P. R. and BRAUN, A. C. (1942) : Cancer Research *2*, 597.

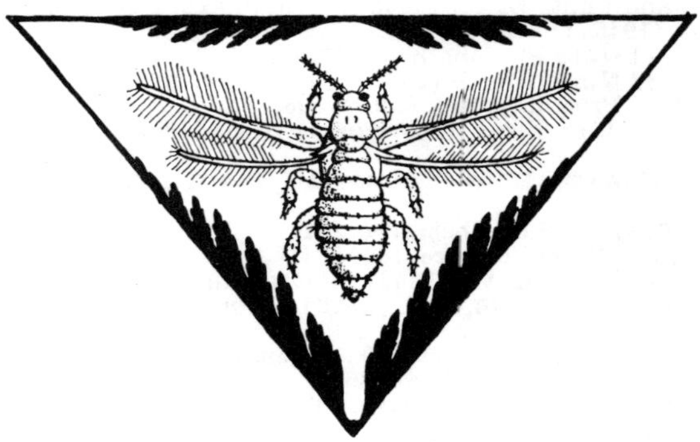

# GENERAL  INDEX

# AUTHOR INDEX

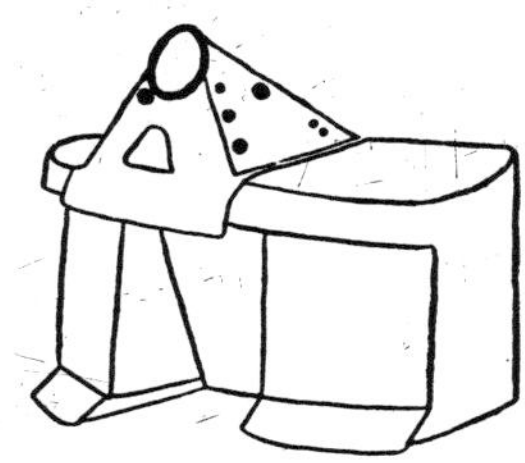